Sharon Pastor Simson, PhD
Martha C. Straus, BA, BS, HTM
Editors

Horticulture as Therapy
Principles and Practice

"*Horticulture as Therapy: Principles and Practice* is the most complete book on horticultural therapy I have seen. This book should be incorporated into the library of anyone with an interest in horticulture and how it relates to people. It would also be an excellent reference book for a course in horticultural therapy.

A wide audience, from the beginner to the more advanced professional, will find this book extremely valuable. It discusses the development of horticultural therapy from ancient times through today with a look to the future.

People with little gardening experience or individuals with limited knowledge of specialized populations will find this book very informative. This book presents up-to-date statistics and descriptions of the many populations that could benefit from horticultural therapy—including individuals with physical or developmental disabilities or mental illnesses; children, youth, and the elderly; substance abusers; and criminal offenders—by outstanding contributors. Horticultural activities and several interesting case studies are discussed for each population.

There are informative chapters on design, techniques, and tools for outdoor and indoor gardening and botanical and community gardening.

Information on applied research, program evaluation, and assessment, as well as an extensive list of resources at the end of each chapter, will be particularly beneficial to professional health care specialists and horticultural therapists."

Richard J. Shaw, PhD, HTT
Associate Professor,
Department of Plant Sciences,
University of Rhode Island

The Food Products Press
An Imprint of The Haworth Press, Inc.

Horticulture as Therapy
Principles and Practice

New, Recent, and Forthcoming Titles
of Related Interest

Horticulture as Therapy
Principles and Practice

Sharon Pastor Simson, PhD
Martha C. Straus, BA, BS, HTM
Editors

The Food Products Press
An Imprint of The Haworth Press, Inc.
New York • London

Published by

The Food Products Press, an imprint of The Haworth Press, Inc., 10 Alice Street, Binghamton, NY 13904-1580

Cover design by Marylouise E. Doyle

Library of Congress Cataloging-in-Publication Data

Horticulture as therapy : principles and practice / Sharon Simson, Martha C. Straus, editors.
    p. cm.
  Includes bibliographical references and index.
  ISBN 1-56022-859-8 (alk. paper)
  1. Gardening—Therapeutic use. I. Simson, Sharon. II. Straus, Martha C.
RM735.7.G37H68 1997
615.8′515—dc21

                              97-7516
                               CIP

To Michael B. Simson, MD, Michael D. Simson, MD '99,
and Laura B. Wilson, PhD.
And to the Magyar gardeners:
Elizabeth Horvath, Ella Horvath Pastor, Michael Pastor,
Olga Horvath, Elizabeth Pastor Sharkady, Nan Pastor Beebe,
Michael C. Pastor, and Brooke Elizabeth Beebe Bartels.

Sharon Pastor Simson

To William Schulze III, my husband,
and to my mother, Katherine C. Straus,
who first told me about horticultural therapy
(and of course, I did not listen).

Martha C. Straus

CONTENTS

Chapter 7. Mental Illness and Horticultural Therapy Practice   157

Barbara A. Shapiro
Maxine Jewel Kaplan

Chapter 8. Children and Youth and Horticultural Therapy Practice   199

Thom Pentz
Martha C. Straus

ABOUT THE EDITORS

Sharon Pastor Simson, PhD, is a lecturer in the Department of Health Education at the University of Maryland, College Park, and an adjunct professor of gerontology at its University College. The former associate director of geriatrics and associate professor of mental health sciences at the Hahnemann University Medical Center, she is the co-editor of the *Handbook of Geriatric Emergency Care; Prevention and Aging; Minority Health and Aging;* and *Therapeutic Landscapes: Health and Healing.* She was awarded her PhD in health services research by the University of Pennsylvania and graduated from the Barnes Foundation Arboretum School. She is the recipient of fellowships and grants from the National Institute of Mental Health, the Administration on Aging of the Department of Health and Human Services, and the Gerontological Society of America. Dr. Simson is a member of the board of directors of the American Horticultural Therapy Association and editor of the *Journal of Therapeutic Horticulture.*

Martha C. Straus, BA, BS, HTM, is a horticultural therapist and teacher at the Forbush School, which provides special education and therapeutic services for children and adolescents of the Sheppard Pratt Health System in Baltimore. She has been working in the field of horticultural therapy for twenty years and is the former supervising director of horticultural therapy at Friends Hospital in Philadelphia. She has given more than forty national workshops and presentations and has granted interviews to national magazines and journals including *National Geographic, Longevity, Good Housekeeping, Vegetarian Times,* and *Prevention.* Ms. Straus has published articles in *Green Scene, American Nurseryman,* the *Journal of Home and Consumer Horticulture, Therapeutic Horticulture,* and *People/Plant Connection.* She is the president of the board of directors of the American Horticultural Therapy Association.

CONTRIBUTORS

Douglas L. Airhart, PhD, HTM, is a professor of horticulture at Tennessee Technological University. He received his BS from the University of California at Davis, MS from the University of Wisconsin at Madison, and PhD from the University of Georgia at Athens. He has been a member of the American Horticultural Therapy Association (AHTA) since 1977 and has served as president and on the board of directors and executive committee. He has published numerous articles about horticultural therapy and compiled the proceedings for the 1995 American Horticultural Therapy Association Conference. Dr. Airhart currently teaches a horticultural therapy course and offers horticultural therapy program-development workshops.

Kathleen M. Airhart, EdS, HTR, teaches students with special needs at Monterey High School in Tennessee. She received a BS in agricultural natural resource management, an MA in special education, and most recently an EdS in special education from Tennessee Technological University. Her work experience over the past fifteen years has been devoted to horticultural therapy and persons with disabilities. She has been actively involved in AHTA, serving on the executive committee, board of directors, and numerous work teams since 1987. She continues to promote horticultural therapy through presentations, publications, and grant development.

Pamela Catlin, HTR, has been working in the field of horticultural therapy since receiving a bachelor's degree from Washington State University in 1976. She has helped to establish over thirty horticultural therapy programs in Washington, Oregon, Arizona, and in Illinois where she was employed with Chicago Botanic Garden as supervisor of horticultural therapy services. Ms. Catlin is currently a private contractor and part-time community college instructor in Prescott, AZ. She also served as the continuing education coordinator for Forward Challenge, Inc., a networking organization for people with developmental disabilities and their families. An active member of the American Horticultural Therapy Association, she served two terms on the board of directors, three as a division director, chaired several committees, and has presented at five AHTA annual conferences as well as many area workshops.

Nancy K. Chambers, HTR, is director of the Enid A. Haupt Glass Garden at The Howard A. Rusk Institute of Rehabilitation Medicine, New York University Medical Center. A practicing horticultural therapist for over twenty years, Ms. Chambers consults with programs and facilities, both here and abroad, on program development and accessible gardening for special populations. Ms. Chambers has received distinguished honors from the American Horticultural Therapy Association, the Community Service Award from the Parks Council of New York City, and the Howard A. Rusk, MD Memorial Award from the Auxiliary of the Rusk Institute. Ms. Chambers received at BA from New York University. She has participated in AHTA board activities for twenty years and is actively involved with Metro Hort Group (an association of professional horticulturists in New York City) and with the American Association of Botanical Gardens and Arboreta.

Linda M. Ciccantelli, HTR, is the horticultural therapy coordinator at Magee Rehabilitation Hospital in Philadelphia and instructor at Temple University in Ambler, Pennsylvania. She is also a consultant to numerous specialized programs for older adults, those with physical disabilities, and children with developmental disabilities. Ms. Ciccantelli was awarded a BA in psychology from Muhlenberg College and attended the Arboretum School of the Barnes Foundation in Merion, Pennsylvania. She has served on the board of directors of the American Horticultural Therapy Association for ten years and was the organization's vice president. She also served as the president of the Delaware Valley Chapter of AHTA.

Steven Davis, MS, CAE, is the executive director of the American Horticultural Therapy Association. Previously, he was director of horticulture for the American Horticultural Society and botanical editor of the society's Plant Sciences Data Center. He received his MS degree in botany from Old Dominion University. He has contributed to many publications, including serving as primary author of the *Time-Life Gardening and Landscaping* series' volume on wildflowers; serving as the author of the "Propagation," "Ferns," and "Wildflowers" chapters of *Rodale's Illustrated Encyclopedia of Landscaping and Gardening Techniques;* and serving as founding editor for the Water Lily Society, an international plant society. Mr. Davis is also a frequent lecturer, instructor, and consultant in horticulture, and has served in the capacity of board member or advisor to a number of organizations.

Matthew Frazel, HTR, supervises the Chicago Botanic Garden's Horticultural Therapy Services. Previously, he worked at the Menninger Foundation as horticultural supervisor for British diplomatic residences in Berlin, and with the Peace Corps establishing fruit tree nurseries in Senegal, West Africa. He graduated from Kansas State University.

Maria Gabaldo, MEd, OTR, HTR, has a bachelor's degree in ornamental horticulture from the University of Illinois, a master's degree in education from Illinois State University, and a master's degree in occupational therapy from Western Michigan University. She had been a registered horticultural therapist since 1982. She has provided horticultural therapy to survivors of brain injury since 1984, developing programs for two post-acute transitional rehabilitation facilities in Michigan. A registered occupational therapist since February 1996, she continues to work with survivors and advocates the use of horticultural therapy through her private practice in southwestern Michigan.

Karen Haas, HTR, is the horticultural therapist at The Holden Arboretum. Previously, she was assistant horticulturist at the Garden Center of Greater Cleveland, now known as the Cleveland Botanical Garden. Ms. Haas received the Rhea McCandliss Professional Service Award from AHTA. She serves on the AHTA board of directors as secretary and was previously a strategic area manager. She has served on the AHTA Board of Directors, Ohio Chapter; now serves on the advisory councils for both "Cultivating Our Community," a program of Ohio State University Cooperative Extension's Urban Gardening Program in Cuyahoga County, and the federally funded program "Horticulture Hiring People with Disabilities"; and chairs Holden Arboretum's committee on the Americans with Disabilities Act. Ms. Haas has made numerous presentations to professional groups, including AHTA, the Canadian Horticultural Therapy Association, the Nursing Home Area Training Center, and Resident Activity Area personnel. With a concentration in horticultural therapy, she graduated with a BS in agriculture from Kansas State University.

Rebecca Haller, HTM, provides education, consultation, and information to health care and human service providers, horticultural therapy students, and the public at Denver Botanic Gardens. She worked with adults with developmental disabilities for over ten years, establishing a vocational horticultural therapy program in Glenwood Springs, Colorado, which generated significant sales revenues. An active leader of the American Horticultural Therapy Association since 1986, she has served as president, secretary, and a division director. She initiated the foundation of the Central Rocky Mountain Chapter of AHTA and served two terms as president of that organization. She has taught continuing education courses, provided numerous trainings and workshops on horticultural therapy, and has lectured to allied professionals and the general public. She earned an MS degree in horticultural therapy and a BS in sociology and psychology from Kansas State University.

Maxine Jewel Kaplan, HTM, received a BA degree in elementary education with a concentration in psychology from Hunter College and an MA degree in teaching in special education from Manhattanville College.

She is a graduate of the New York Botanical Garden Horticultural Therapy Certificate Program and is a Cornell Cooperative Extension master gardener. Ms. Kaplan has been a horticultural therapist at the Westchester County Medical Center, Valhalla, New York, for the past fifteen years, providing services to children, adolescent, and adult inpatient programs; an adolescent substance abuse day treatment program; and a geriatric program. Ms. Kaplan has been an active board member of many community service organizations. She serves on the AHTA board of directors and received the AHTA 1995 Rhea McCandliss Professional Service Award. Ms. Kaplan has lectured and written extensively and is on the teaching staff of the New York Botanical Garden.

Jean Kavanagh, ASLA, is a licensed landscape architect and member of the graduate faculty in landscape architecture at Texas Tech University, and holds a BSLA and an MLA from Cornell University. Professor Kavanagh's work in therapeutic landscapes, universal design, and the design of horticultural therapy landscapes has foreshadowed the growing collaboration between landscape architects and horticultural therapists across the country. She serves as a member of the AHTA board of directors and was vice president of the association.

Thom Pentz, PhD, is a senior psychologist at the Sheppard Pratt Health System in Towson, Maryland. He received his PhD in developmental psychology from the Johns Hopkins University and obtained clinical training at Sheppard Pratt. He is currently a psychotherapist and supervisor in the Forbush School, a special education program serving emotionally disturbed children and adolescents. Dr. Pentz has written and lectured on providing treatment in the context of a developmental psychology model.

Paula Diane Relf, PhD, HTM, is professor and extension specialist in environmental horticulture at Virginia Polytechnic Institute and Virginia State University. She was awarded her PhD and MS in horticulture by the University of Maryland and a BS in horticulture by Texas Tech University. Recipient of many grants, Dr. Relf is known nationally and internationally as a leading researcher, educator, and horticultural therapist. She has published over thirty articles, edited *The Role of Horticulture in Human Well-Being and Social Development* (Timber Press), and presented at conferences in New Zealand, Australia, Brazil, Japan, England, Germany, and Finland. Among Dr. Relf's awards are the American Society for Horticultural Science Outstanding Extension Educator for career accomplishments, and the Extension Division Distinguished Achievement, the John Walker Award for Professional Service from the AHTA (formerly the National Council for Therapy and Rehabilitation), National Science

Foundation Fellowship, and Phi Kappa Phi National Honor Society. She is chair of the People-Plant Council.

Linda L. Remy, PhD, is the senior statistician for the Family Health Outcomes Project at the Institute for Health Policy Studies, University of California at San Francisco. She has been the principal investigator, project director, senior methodologist, or senior statistician on dozens of studies influenced by social ecology. Dr. Remy has taught undergraduate and graduate research methods, and supervises students for their theses and dissertations.

Jay Stone Rice, PhD, was awarded his doctorate in clinical psychology from the San Francisco School of Psychology. He was the principal investigator for an exploratory study of the effectiveness of the San Francisco Sheriff's Departments' innovative horticultural therapy program. Dr. Rice co-edited *The Healing Dimensions of People-Plant Relations,* published by the Center for Design Research, University of California at Davis, and has written about the social ecology of inner-city family trauma, trauma's relationship to substance abuse and crime, and gardening as a treatment intervention. Dr. Rice has consulted with the Center for Mental Health Services (CMHS), the National Institute for Corrections (NIC), and San Francisco Sheriff's Department on the development of ecologically sensitive treatment programs. Dr. Rice teaches at the Institute for Imaginal Studies in Petaluma, California and is a clinical supervisor and family therapist in private practice.

Vera Roth, MA, CRC, NCC, CPC, is a senior-level vocational counselor and vocational program coordinator at the Forbush School, which provides special education to emotionally disturbed children and adolescents. The school is a division of the Sheppard Pratt Health System. Previously, she was the vocational services-section chief for the hospital's inpatient programs. Ms. Roth has a MA from Towson State University and a BA from the University of Maryland. She is a certified rehabilitation counselor, a nationally certified counselor, and a certified professional counselor in the State of Maryland. She has been employed in the mental health field for twenty years and as a vocational counselor for twelve of those years. She has experience in the areas of vocational assessment, testing, career awareness and exploration, job hunting, work readiness, transition planning, supervision, and program development.

Patricia Schrieber manages the newly formed outreach department of the Pennsylvania Horticultural Society (PHS). This department offers environmental programs for youth, specialized training in horticulture, community organizing, and technical assistance on community greening issues to a city-wide, regional, and national audience. Previously, she held a number of field positions with PHS's Philadelphia Green and organized and subsequently managed its education department. Ms. Schrieber is an associate member of the American Horticultural Therapy Association. She has spoken at professional conferences of AHTA and the American Community Gardening Association. Her articles have appeared in PHS's Green Scene, Green Prints, and the Friends of Horticultural Therapy newsletter.

Barbara A. Shapiro, HTM, received her horticulture therapy degree from Kansas State University. For the past ten years her work at the Menninger Foundation has focused on specializing crisis admissions for young adults and special evaluations for geriatric patients. Young children are also involved in the extensive horticultural therapy facility. Ms. Shapiro is working toward earning a master's degree in education administration. She is active with various horticulture clubs and organizations and has a particular interest in aromatherapy. Ms Shapiro has incorporated aromatherapy into the Horticulture Therapy Program at the Menninger Clinic.

Nancy C. Stevenson, HTR, served as horticultural therapy coordinator at the Cleveland Botanical Garden from 1981 to 1994. Recently retired, she continues to consult with the garden and local agencies on horticultural therapy program development. Ms. Stevenson is a past president of both AHTA and its Ohio Chapter and is past chairperson of the Friends of Horticultural Therapy. She was awarded the Medal of Merit from the Garden Club of America and the Garden Club of America Horticultural Award in Zone X for work in horticultural therapy. She was the recipient of AHTA's Alice Burlingame Humanitarian Service Award and the American Horticultural Society's Horticultural Therapy Award. Ms. Stevenson has a BA from Smith College and an MA in human services from John Carroll University. She is a graduate of the Cuyahoga County Cooperative Extension Service Master Gardener Program.

David Strauss, PhD, is vice president of clinical services for ReMed (Conshohocken, Pennsylvania), a continuum of post-acute neurologic rehabilitation programs in community-based settings. Dr. Strauss's primary responsibilities include the development and operation of ReMed's autism, community, reentry, outpatient, homecare, behavioral, long-term, and vocational programs. He provides staff supervision and training; client evaluation and education; as well as individual, group, and family therapy specifically designed for people coping with brain injuries. Dr. Strauss serves as chairperson for the National Head Injury Foundation's (NHIF) Task Force on Substance Abuse and Brain Injury. Dr. Strauss is a nationally known speaker and is frequently asked to present his practical ideas on post-acute rehabilitation, outcomes, substance abuse, sexuality, and community reentry.

Melanie Trelaine is the activity director for a leading retirement community in Ohio, as well as an active consultant in the development of horticultural therapy programs. She is the former executive director of Hands, Heart, and Health, where she pioneered the development of contracted services with numerous Arizona health care providers as well as city, county, state, and federal funding authorities. She is a member of the AHTA Board of Directors. Her monograph appears in *Provider*, the national publication of the American Health Care Association. She has published pamphlets, brochures, and other material for distribution at regional and national horticultural therapy and related conferences and seminars.

Lisa Ann Whittlesey, MA, HTM, is an extension associate in horticulture with the Texas Agricultural Extension Service at a Federal Prison Camp in Bryan, Texas. She has provided leadership to the development, implementation, management, and overall evaluation of the extension service's Master Gardener/Horticulture Program. Ms. Whittlesey received an MA in agriculture in horticulture from Texas A&M University and a BS in Horticulture from Texas A&M University. She has presented numerous talks related to horticulture and master gardening and her work has been featured in *Progressive Farmer* and *Cosmopolitan Magazine,* and by the Associated Press.

Matthew Wichrowski, HTR, is senior horticultural therapist at the Enid A. Haupt Glass Garden, New York University Medical Center, supervising the clinical program. He received a bachelor's degree in psychology and philosophy at the State University of New York at Stony Brook. Mr. Wichrowski is an instructor at the New York Botanical Garden and a lecturer and consultant on integrated pest management.

Patricia Myroniuk Williams, LMS, lives in Bryan, Texas with her husband Pat and their three cats. She received her MSW from the University of Connecticut as well as a BS in environmental horticulture. She worked as a horticultural therapist for ten years at Greenbrier, an Easter Seals Goodwill Industries Rehabilitation Center program in New Haven, Connecticut. She later went on to a career in clinical social work and human service administration, while volunteering for the AIDS Project, coordinating regional conferences, and consulting as a horticultural therapist. Ms. Williams has been an active member of AHTA for fifteen years and the New England chapter since its inception in 1980. She has served as a board member for both associations.

Patrick Neal Williams, PhD, HTM, completed his doctorate at Texas A&M University studying humanistic horticulture. He received his MS from Kansas State University and a BS from California Polytechnic State University at San Luis Obispo. He has taught at the Bancroft School and the New York Botanical Gardens. He worked for five years as a senior horticultural therapist at the Rusk Institute of Rehabilitation Medicine at the New York University Medical Center. Mr. Williams has been a member of AHTA since 1986 and spent five of those years serving on the board of directors. He has published several articles and made numerous presentations regarding horticultural therapy.

Preface

People have been dependent on plants since the beginning of time. Plants provide food, clothing, shelter, and medicine essential for human survival. This relationship between people and plants has been taken a step further by the discipline of horticultural therapy. Horticultural therapy is a treatment modality that uses plants and plant products to improve the social, cognitive, physical, psychological, and general health and well-being of its participants. While treatment and rehabilitation typically have been offered in health care facilities, many have found that a garden offers a complementary health care setting that helps to restore physical and mental health to those who work the soil and watch seeds grow.

Horticultural therapy is a relatively new discipline that has been developing rapidly during the last twenty-five years. The significance of this discipline and the contribution it can make to advancing health and well-being has just begun to be understood. This book seeks to help realize the potential of this discipline by providing a definitive, state-of-the-art textbook on the principles and practice of horticulture as therapy.

The text contains seventeen chapters organized into four sections:

- Part One: The Practice of Horticultural Therapy
- Part Two: Special Populations for Horticultural Therapy Practice
- Part Three: Settings for Horticultural Therapy Practice
- Part Four: Skills for Horticultural Therapy Practice

Chapters use common formats to provide an overall structure to the text, standardize content, and ensure readability. The chapters present current information, concepts, and skills that will enable readers to study, learn, and practice the discipline of horticulture as therapy.

In addition to its value to horticultural therapists, this textbook holds interest for many other audiences, including occupational, physical, and recreational therapists; health personnel such as nurses, physicians, social workers, and technicians; students who are studying horticulture and related areas; horticulturists; educators working with school and community horticultural programs; garden club members; and volunteers involved with activities related to horticultural therapy.

Contributors to this textbook are recognized leaders in horticultural therapy and related fields. They hold professional positions in established and well-known organizations throughout the United States. They are affiliated with universities, health institutions, arboreta and botanical gardens, horticultural societies, social agencies, government, community programs, and the corporate sector. They have been officers of national organizations such as the American Horticultural Therapy Association, have published literature advancing the discipline, and have conducted notable scientific and practice projects.

It is our hope that this textbook will provide the reader with the principles and practice needed so that horticulture as therapy can continue to make a difference in the health and well-being of so many people, today and tomorrow.

Acknowledgments

This textbook was supported in part by the financial contributions of Michael B. Simson, MD; William Schulze III; the Forbush School at the Sheppard Pratt Health System in Baltimore, Maryland; the Friends of Horticultural Therapy and its past president, Mrs. Nancy C. Stevenson, HTR; and Mrs. Barbara Brainerd's gift in memory of her husband, Dr. John Whitney Brainerd.

We are grateful to all the contributors, who were the gardeners of this textbook. We are especially appreciative of the staff at The Haworth Press for their encouragement, enthusiasm, outstanding editorial assistance, and their support of horticultural therapy and this textbook: Bill Palmer, managing editor; Donna Biesecker, typesetter; Patricia Brown, editorial production manager; Susan Trzeciak Gibson, administrative editor; Dawn Krisko, production editor; Amy Lamitie, copyeditor; Peg Marr, senior production editor; and Marylouise Doyle, cover artist.

Sharon P. Simson
Martha C. Straus

For their inspiration and friendship, I thank the members of the Villanova Garden Club, the Barnes Foundation Arboretum School, the American Horticultural Therapy Association, and Martha Straus, premier horticultural therapist.

Sharon P. Simson

REFLECTIONS

In 1993, I met Sharon Simson and my life has never been the same. She was interested in learning more about horticultural therapy and wanted to volunteer in my program. After several major life changes, Sharon and I still remained friends and wished to work together on a project. I had the good fortune to tell her my fantasy of wanting a textbook for horticultural therapy, and she said immediately, "We can do that."

The first two years were spent planning, dreaming, exploring content and format, and doing lots and lots of retyping. We finally developed the prospectus from which a contract with The Haworth Press developed. The rest is history. This book has been a labor of love, of babies (several of our authors have had children during this process), of challenges, of moves, and hopefully of success.

I would like to thank all the chapter authors who may not have realized what they were getting into when they signed on. Horticultural therapists are great at working with people and plants. Getting us inside long enough to write a chapter has been a stimulating experience.

I would also like to thank my husband, who is my biggest support, and finally, I would liked to thank all the patients I have seen over the last twenty years. They have been my teachers and my inspiration.

Martha C. Straus

PART ONE:
THE PRACTICE
OF HORTICULTURAL THERAPY

Chapter 1

Development of the Profession of Horticultural Therapy

Steven Davis

INTRODUCTION

Although horticultural therapy is a comparatively young profession, the concepts upon which the profession is built are as ancient as the pyramids. These concepts were evolved many centuries ago simply because they make good sense. Each of us who has marveled at the perfect flower, taken pride in growing the perfect plant, or felt excitingly renewed upon discovering the first blooms of spring has experienced these founding principles which gave rise to the profession of horticultural therapy. There is a special connection between people and plants, between people and the gardening pursuit, and between people and natural surroundings. The therapist who is trained in horticulture simply uses these realities as non-threatening ways in which to introduce and facilitate therapy and rehabilitation. The profession of horticultural therapy is the result of two disciplines merging complementarily: horticulture in support of therapy and rehabilitation.

> Horticultural therapy is a process through which plants, gardening activities, and the innate closeness we all feel toward nature are used as vehicles in professionally conducted programs of therapy and rehabilitation. (Davis, 1994)

This opening chapter provides an overview of the significant occurrences and individuals who shaped the profession of horticultural therapy. Without a historical perspective, it is difficult to realize the direction in which the profession is moving and the place within that movement where

3

the prospective horticultural therapist will find her or his professional niche. Understand the momentum of the past, which shaped the profession of today, and you will be well on the road to becoming a part of horticultural therapy's influence on the future.

The learning objectives of this chapter are to

1. define and explain the profession of horticultural therapy;
2. review the significant occurrences in the history of horticultural therapy and its professional organization;
3. describe horticultural therapy's position within the health care community; and
4. provide an understanding of the existing and potential growth within the profession of horticultural therapy.

HISTORY

Although we can guess that people have found solace in nature from the beginnings of time, the first recorded use of horticulture in a treatment context occurred in ancient Egypt, when court physicians prescribed walks in palace gardens for royalty who were mentally disturbed (Lewis, 1976). Even at the time of Christ, it was understood by these court physicians that the peaceful, nonthreatening environment of the garden had a quieting effect on people. This realization did not progress to a greater level of use, however, for a number of centuries. It was not until the late 1700s and the early 1800s, in clinical settings in the United States, England, and Spain, that this understanding of a people-plant connection began to evolve into something greater—an accepted approach to treatment.

The 1800s: Horticulture's Use in Treating Mental Illness Evolves

Dr. Benjamin Rush, a professor at the Institute of Medicine and Clinical Practice in Philadelphia, Pennsylvania, distinguished as one of the signers of the Declaration of Independence and considered the first psychiatrist, opened the door to the active use of horticulture in the treatment of mental illness (Tereshkovich, 1975). In 1798, Dr. Rush announced that he had found field labor in a farm setting to have curative effects on people who were mentally ill. His findings were sufficiently received by colleagues elsewhere in the United States and in Europe to initiate a rush of further testing in the early 1800s. In 1806, hospital staffs in Spain began emphasizing the use of agricultural and horticultural activities in their programming for patients with mental disabilities. The favorable results from these replicated studies soon evolved into the common practice of the time of building mental institutions in rural settings and of actively involving patients in the growing and harvesting of field crops. As more and more institutions followed this lead, more and more evidence accumulated in support of the benefits of working the soil for people with mental disabilities.

Another important step in the evolution of horticultural therapy was initiated in 1817, when the first private psychiatric institution in the United States opened in Philadelphia. Then named the "Asylum for Persons Deprived of their Reason," Friends Hospital went well beyond the concept of therapy through field labor. Instead of a farm setting, Friends Hospital created a parklike setting, carefully designing into the landscape shaded walks, quiet forest paths, and open grassy meadows. Although patients still were involved in vegetable and fruit growing, this direct pursuit of the calming effects of the natural environment, as a passive form of therapy, was a new and innovative use of horticulture as a treatment tool (Straus, 1987). Through horticultural pleasures, patients' senses were awakened and their feelings redirected, and through the peaceful and safe setting of the hospital's landscape, an environment conducive to recovery was provided. This concept was taken to another level in 1879, when Friends Hospital installed the first greenhouse solely for therapeutic purposes (Lewis, 1976).

In 1845, in the *American Journal of Insanity,* Daniel Trezvant wrote that exercise and diversion were important to the successful treatment of people with mental disabilities, helping to prevent them from thinking about their own situations. One year later, in the same journal, Trezvant's findings were related specifically to the act of gardening by Isaac Ray (Lewis, 1976).

The founder of the American Psychiatric Association, Dr. Thomas Kirkbride, in an 1880 publication titled *Hospital for the Insane,* described labor as "one of the best remedies; it is as useful in improving the health of the insane, as in maintaining that of the sane." He went on to describe gardening and farming as "admirable means" for the pursuit of outdoor labor (Lewis, 1976).

In an 1896 book titled *Darkness and Daylight or Lights and Shadows of New York Life,* the authors described horticultural activities of the Children's Aid Society, through which tenement children were helped to experience the joys of growing flowering plants. This is one of the earliest mentions of using plants and gardening as uplifting activities for disadvantaged young people. This book also described the horticultural good works of New York City's "flower missions," which collected flowers from different sources and delivered them to hospitals and "homes for the aged and infirm." In 1895 alone, it was purported that more than 100,000 bouquets and bunches of flowers had been distributed through these missions (Campbell, Knox, and Byrnes, 1896).

E. R. Johnston, in the 1899 *Journal of Psycho-Aesthenics,* described sensory stimulation through working with plants and gardens as an important way in which to "enhance learning (in) mentally handicapped children" (Johnston, 1899). These findings were supported one year later by G. M. Lawrence in an article he authored in the same journal, titled "Principles of Education for the Feeble Minded" (Lawrence, 1900).

The Early 1900s: The Beginning Use of Horticulture in Physical Disability Programming

Throughout the 1800s it had been well-established that horticulture was an important addition to treatment programs for persons with psychological or mental disabilities, but it was not until the world wars of the twentieth century that its value in programming for persons with physical disabilities was thoroughly tested and validated. With tremendous numbers of returning wounded entering hospitals, these world wars provided an important opportunity for horticulture to be integrated into programming for persons with physical disabilities. Even so, the use of horticulture in clinical settings during World War I was largely relegated to providing diversions for long-term patients (McDonald, 1995). Plants and gardening activities were introduced into hospitals primarily for occupational and recreational purposes. It was during World War II that horticulture moved beyond "diversionary" status and became an important part of therapy and rehabilitation programming (Lewis, 1976).

Near the end of World War I, in 1917, the women's occupational therapy department of Bloomingdale Hospital in White Plains, New York, offered an educational opportunity in horticulture (Tereshkovich, 1975), which stands out as significant. This offering represents the first instance in which actual training in horticulture for health care professionals was made available and was actively pursued. At this point, horticulture became not just an addition to the treatment program but a valid addition.

Two years later, in 1919, therapy through horticulture received another important level of validation. In that year, the acclaimed Dr. C. F. Menninger and his son Karl established the Menninger Foundation in Topeka, Kansas. From the first day the foundation opened its doors, plants, gardening, and nature study were made integral parts of patients' daily activities (Lewis, 1976).

Each such use of horticulture within treatment programs and the results they generated further validated the effectiveness of the horticultural vehicle. As these validations accumulated, so too did they find their way into the occupational therapy books and textbooks of the 1920s. In 1936, the acceptance of horticulture as a treatment tool by the occupational therapy community moved an important step forward when the just-founded Association of Occupational Therapists in England formally acknowledged the use of horticulture as a specific treatment for physical and psychiatric disorders (McDonald, 1995). This acceptance reached another height in 1942 when Milwaukee Downer College, the first college to award a degree in occupational therapy, became the first institution of higher learning to offer a course in horticulture within an occupational therapy program (Tereshkovich, 1975).

The Influence of a World War and Garden Club Volunteers

The second World War saw extensive use of horticulture in hospital programming, where is was much more accepted as a treatment vehicle than it was at the time of the previous World War. Even more important than its general acceptance was the active participation of throngs of garden club volunteers in hospital wards across the nation, which enabled horticultural programming to prosper both in its application and in its degree of use (Lewis, 1976). Armed with this significant corps of volunteers, occupational therapists actively employed plants and gardening activities in their therapy and rehabilitation programming. This extensive opportunity to test the validity of using horticulture as a treatment vehicle, relative to both physical and mental disabilities, provided resounding testimony that therapy through horticulture was effective and resulted in reduced hospital stays (McDonald, 1995). The involvement of garden clubs in horticultural therapy programming has served and continues to serve as an important support element. In 1951, the National Council of State Garden Clubs named horticultural therapy as one of the major objectives of its member clubs, which it remains today. In 1968, a total of 4,609 of the council's clubs reported involvement in this area.

The 1950s and 1960s: Growth in Programming and in Educational Opportunities

In 1951, Alice Burlingame, a trained psychiatric social worker at the Pontiac, Michigan State Hospital, started a horticulture program in the hospital's geriatric ward (Lewis, 1976). This program is noteworthy for two important reasons: (1) it provided the opportunity to validate the use of horticulture with another population of persons with disabilities—older adults—and (2) its successes convinced Alice Burlingame that horticultural therapy warranted a separate and distinct profession. One year later, Alice Burlingame and Dr. Donald Watson jointly convened the first week-long workshop in horticultural therapy at Michigan State University. The success realized in this workshop and those that followed it led to the 1955 award by Michigan State University of the first master of science degree in horticultural therapy. The recipient, Genevieve Jones, was an occupational therapist at the Hines Veterans Administration Hospital in Chicago, who pursued her degree through a scholarship provided by the Federated Garden Clubs of Michigan, which also published her *Handbook on Horticultural Therapy* (Lewis, 1976).

In 1953, horticultural therapy programming was initiated through a public garden for the first time, when Louis Lipp, a propagator for the Arnold Arboretum of Harvard University, developed a program at a nearby veterans hospital. Three years later, in 1956, he pursued a similar outreach program through the Holden Arboretum in Kirtland, Ohio, at the Golden Age Center in Cleveland.

In 1959, New York University Medical Center's renowned Rusk Institute for Rehabilitative Medicine started a horticultural therapy program within an attached greenhouse, further influencing the growth and acceptance of horticultural therapy. This program quickly and effectively moved into new frontiers, making the horticultural therapist a part of a treatment team along with doctors and psychologists and using horticulture both diagnostically and rehabilitatively (Lewis, 1976). Since its inception, the Rusk Institute's Glass Garden program has served as an important proving ground and example of the effective use of horticultural programming in the treatment of people with physical disabilities.

The first textbook in horticultural therapy, *Therapy Through Horticulture* by Dr. Donald Watson and Alice Burlingame, was published in 1960. This initial academic publication was followed two years later with a publication designed to support volunteers in horticultural therapy programming, specifically garden club members. *The Handbook of Horticultural Therapy* was published by the National Council of State Garden Clubs for its member clubs.

Also in the early 1960s, new activity in support of horticultural therapy programming for people with physical disabilities occurred in England. Although horticultural therapy pursuits in England and in the United States were not closed pursuits, they did not always proceed in parallel directions. Horticultural therapy activity in each nation ultimately brought benefit to all such programs in the form of enhanced acceptance and validation. Cultural influences relative to gardening helped define different paths through which to pursue the common goal of benefiting people with disabilities through horticulture. In England, the emphasis was not toward a separate profession but toward (1) the provision of horticultural assistance to people who were disabled, and (2) programming and educational assistance to the individual programs in horticultural therapy (McDonald, 1995). In support of those intentions, several new organizations were created that would contribute to all of horticultural therapy.

The Disabled Living Foundation was one of these organizations. Devoted to the development and application of horticultural therapy, the Foundation commissioned research on the design of gardens and gardening tools for people who had physical disabilities. The foundation at the Mary Marlborough Lodge of the Nuffield Orthopedic Center in Oxford initiated the first well-documented horticultural training program for persons with disabilities (McDonald, 1995). Both of these initiatives proved very supportive of the entire horticultural therapy movement, helping to define the horticultural path toward independence for persons with disabilities.

Although in both England and the United States the 1960s represented a period of growth in new programming and in knowledge, each new program tended to exist on its own energies and innovations, and there was limited contact and learning between programs. In 1968, Rhea McCandliss, a horticultural therapist at the Menninger Clinic, took an important next step by documenting the existence of and interest in horticultural therapy programming in the United States. She surveyed 500 hospitals and facilities across the nation and found the considerable presence of horticultural therapy programming, interest in program development, and a deficiency in the number of trained, qualified persons available to meet the existing demand (Lewis, 1976). These findings pointed to a profession in the making.

THE RESPONSE: EDUCATIONAL OPPORTUNITIES AND A PROFESSIONAL ASSOCIATION

The greatest response to this need came from the Menninger Foundation in 1972 when it developed a cooperative educational agreement between its activity therapy department and the horticulture department of Kansas State University. Through this agreement, the first horticultural therapy curriculum in support of the mental health field was created and put into effect, providing students with formal training in psychology and horticulture at Kansas State University and a seven-month clinical internship at the Menninger Foundation (Lewis, 1976). Dr. Richard Mattson, HTM, who developed this award-winning program, has expanded and enhanced it through the years. The success of this program soon led to other academic options, including Clemson University's 1973 establishment of a graduate degree program in horticultural therapy.

Also in 1973, Howard Brooks and Dr. Charles Oppenheim provided important medical endorsement of horticultural therapy when their monograph, *Horticulture as a Therapeutic Aid,* was published by the Institute of Rehabilitation Medicine of the New York University Medical Center.

Within this same time frame, a number of individuals involved in horticultural therapy began to gravitate toward the need to create a national forum, through which the profession could be formally established. In 1973, this need evolved into the gathering of twenty individuals at the Melwood Horticultural Training Center (for persons who are developmentally disabled) in Upper Marlboro, Maryland. Their self-imposed charge was straightforward—to evaluate the need to institute a formal organization in the United States that might unite all interests in horticultural therapy. They chose unanimously to develop the first professional organization for horticultural therapists, the National Council for Therapy and Rehabilitation Through Horticulture (NCTRH) (Lewis, 1976). For its first few years of operation, the council was guided by its president, Earl Copus, and was housed on the campus of the Melwood Horticultural Training Center. The NCTRH held its first annual conference in November, 1973, at the USDA National Agricultural Library in Beltsville, Maryland. At the conclusion of that meeting, the council created a membership opportunity that resulted in an initial membership of 85 individuals and programs. In 1975, the council relocated to the River Farm address of the American Horticultural Society in Mt. Vernon, Virginia, where Diane Relf, PhD, one of the Council founders and then its executive secretary, helped this organization to define its mission and purposes. Also, and in the capacity of the council's next president, Dr. Relf brought into support of the NCTRH for the first time a number of the trade associations within the horticulture industry, most notably the Society of American Florists and the American Association of Nurserymen.

Through the influence of the NCTRH, horticultural therapy programming in the United States rallied beneath this new organization's banner. Through the consolidated voice and energies that resulted, the pursuit of the profession became readily achievable. Over the next several years, the NCTRH arrived at important documents that continue to drive the organization, the profession, and the professional.

NEW ORGANIZATIONS IN ENGLAND

While the professional association in the United States was unfolding its registration program, a new organization was created in England in 1978 that would impact the continued development of horticultural therapy in the United Kingdom. This organization was the Society for Horticultural Therapy and Rural Training. The society later simplified its name to Horticultural Therapy and provided a national forum for interdisciplinary communication (McDonald, 1995). The Horticultural Therapy organization also facilitated the coordination of support to programs in horticultural therapy under its charitable mission:

> To relieve persons who are physically or mentally ill, disabled or handicapped, or who are in necessitous circumstances, by the advancement of education in the use of land through horticulture, agriculture, farming and gardening in all their forms.

Although this umbrella organization proved an important and effective unifying factor, its lack of emphasis on the establishment of a profession and on the professional horticultural therapist set it apart from its U.S. counterpart, the NCTRH. One year later, in 1979, this absence was partially addressed through the establishment in England of another charity, the Federation to Promote Horticulture for Disabled People (McDonald, 1995). The initial (and continuing) purpose of this organization was to: "provide . . . a focus for people professionally involved in the practice, promotion and study of open-air activities involving land users from all categories of disabled people."

Although the federation stimulated research and the exchange of information, findings, and approaches between practitioners, its lack of focus on the establishment of an actual profession helped to keep the horticultural therapy movement in England moving in a separate (but equally important) direction from the much more profession-oriented movement in the United States. An academic program in horticultural therapy did not materialize in an institution of higher learning until 1993, when an affiliation between Coventry University and the Horticultural Therapy organization resulted in a diploma program (McDonald, 1995).

THE SIGNIFICANCE OF A PROFESSIONAL ASSOCIATION

The establishment of a professional organization in the United States in 1973 was a significant force in the creation of the profession of horticultural therapy in the United States. As the NCTRH grew in size, so too did its impact on this growing profession. The elected, appointed, and hired leaders of the NCTRH propelled forward the profession and the professional. The NCTRH developed publications, educational programming, national and international networking with health care and horticultural professional and trade associations, and strategic planning and the structured pursuit of goals. As a result of NCTRH's efforts, the profession became qualified, better understood, and more supported. As it began its active pursuit of image clarification and name recognition, NCTRH realized that its cumbersome name inhibited that process. Just as the Society for Horticultural Therapy and Rural Training in England found it appropriate to simplify its name, so too did NCTRH. The organization formally adopted the replacement name of the "American Horticultural Therapy Association" (AHTA) in 1988.

The AHTA, like its predecessor organization, is a member-driven organization and is responsive to the needs of the professional and the profession. The AHTA membership annually elects one-third of its fifteen-member board of directors (for three-year terms) and every three years elects four of the five officers comprising the organization's executive committee (the immediate past-president is the fifth officer). Together with board-appointed volunteers, a board-hired executive director, and a hired staff, the AHTA elected leaders work on behalf of the association's membership toward an enhanced profession, enhanced professional opportunities and status, and a better-educated and trained professional.

MISSION OF THE ORGANIZATION

In the capacity of a professional organization, the AHTA exists to support and to strengthen the profession and the professional. In horticultural terms, the association was created and exists today to "grow" these two entities.

The American Horticultural Therapy Association, Inc. is dedicated to promoting and encouraging national and international interest in developing horticulture and related activities as a therapeutic and rehabilitation medium. AHTA strives to improve the performance of programs utilizing horticulture activities in human development through communication, coordination, knowledge dissemination, and promoting education/training. Furthermore, AHTA is dedicated to enhancing the professionalism of horticultural therapists and to improving the performance of programs providing horticultural therapy services.

The association's code of professional ethics follows the resources section of this chapter.

What Does the Future Hold?

Horticultural therapy is a comparatively young profession. The path traveled by AHTA through its first several decades is not unlike the early experiences of other therapies. The profession of horticultural therapy has made important leaps forward and is positioned upon a sturdy foundation ready for the next surge in growth. This next level of evolution will elevate the profession out of its youth and into a recognized position within the health care community. Validation and growth of horticultural therapy will occur through (1) clinical practice, (2) education, and (3) research.

(1) Validation and Growth Through Clinical Practice

Horticultural therapy offers flexibility in regard to how it can be carried out and in what setting. This enhances its cost-effectiveness potential. Horticulture is a diverse medium that affords therapists the opportunity to employ considerable creativity and diversity into their programming. This flexibility makes it possible for effective programming to occur in many settings: within a hospital's rooftop greenhouse, from a plant cart that is wheeled from ward to ward, or within an outdoor garden.

Horticultural therapy also offers a range of application. This enhances its cost-effectiveness at the facility level. Horticultural therapy can be used just as effectively with

- people who have physical disabilities, as well as with people who have mental, psychological, or developmental disabilities;
- the youngest of children as with the oldest of adults;
- victims of abuse, as well as their abusers;
- people who are recovering from illness as with those who will not recover and who are seeking quality of life in their closing days; and
- accomplished gardeners as with novice gardeners—and even with people who have never gardened.

Although horticultural therapy is not an answer-all therapy, the fact that it is appropriate with so many disability populations will make it a very attractive pursuit for facility administrators of the future.

Horticultural therapy offers a living medium with which the vast majority of Americans are very familiar. Gardening represents a primary hobby for adult Americans. As the U.S. health care system continues its movement toward consumer choice, it is logical to assume that these consumers will show a preference for nonthreatening, familiar modes of therapy and rehabilitation. Horticultural therapy will be viewed favorably by those many thousands of consumers who also are gardeners.

(2) Validation and Growth Through Higher Education

Just as initial interest in therapy through horticulture spawned the first educational opportunities in horticultural therapy, so too has growth within the profession and the professional organization given rise to new educational opportunities. Although these offerings vary considerably relative to content, the association-approved core curriculum (Table 1.1) represents an important influence.

Horticultural therapy coursework, one-year programs, and options within horticulture also are available through institutions of higher learning and community colleges (Table 1.2). Educational opportunities also are offered by botanic gardens and arboreta (Table 1.3).

(3) Validation and Growth Through Research

Documentation of existing research has been achieved by an independent organization known as the People-Plant Council, which was formed in 1990 as a direct result of its national interdisciplinary symposium, "The Role of Horticulture in Human Well-Being and Social Development." The council has focused on increasing research in the people-plant area and encouraging documentation of this research and its dissemination and use. Through the council, a base body of knowledge has been identified and the receptivity of researchers toward this area of research has been kindled. The future holds opportunity to add substantially to that body of research-based knowledge. This expanded body of research will substantively validate the anecdotal findings that have for so long served as the primary proving point of horticultural therapy.

As a consequence of research in horticultural therapy, scholarly and informational publications are being produced by the association. Publications have important impacts: encouraging the pursuit of new research validating horticultural therapy to the medical and health care communities, and presenting horticultural therapy to the general public, the media, potential funders, and to prospective horticultural therapists.

TABLE 1.1. Core Curriculum (Minimum Requirements: 78 Credits + 1,000-Hour Internship)

Horticultural Therapy (HT)	Horticultural Science	Therapy and Human Science	Management
(8 credits + 1000-hour internship)	(40 credits)	(24 credits)	(6 credits)
Horticultural Therapy HT techniques HT programming Special topics in HT Internship	Horticulture Plant propagation Plant materials Greenhouse or nursery production/management Landscape design/construction Botany Soil science Entomology Plant pathology Plant physiology Fruit and vegetable crops gardening Basic floral design Specialization plants	Introduction to psychology Abnormal psychology Sociology Special courses Physical disabilities Developmental disabilities Emotional disabilities Geriatrics Corrections Psychiatry Community-based programs Multiple disabilities Group/process Counseling Vocational rehabilitation Special education Recreation Therapy skills and services Psychology Anatomy/physiology Sign language First aid/CPR Crisis intervention	Communication and public speaking Research methods/statistics Computers Business management/economics

TABLE 1.2. Horticultural Therapy Coursework

Horticultural Therapy Degree
(BS and MS Curricula)
Kansas State University, Manhattan, KS
Virginia Polytechnic Institute and State University, Blacksburg, VA
Options with Horticulture
(BS, MS, and PhD)
Texas A&M University, College Station, TX
University of Rhode Island, Kingston, RI
One-Year Program
Edmonds Community College, Lynnwood, WA
Coursework
College of DuPage, Glen Ellyn, IL
Rockland Community College, Suffern, NY
SUNY Cobleskill, Cobleskill, NY
Temple University, Ambler, PA
Tennessee Technological University, Cookeville, TN
Tulsa Junior College, Tulsa, OK
University of Massachusetts, Amherst, MA
Certification Program
The New York Botanical Garden, Bronx, NY
Additional Opportunities are Developing at
Western Piedmont Community College, Morgantown, NC
University of Florida, Gainesville, FL
Northern Virginia Community College, Loudon, VA

TABLE 1.3. Training Available Through Botanical Gardens and Arboreta

Brooklyn Botanic Garden, Brooklyn, NY
Chicago Botanic Garden, Glencoe, IL
Cleveland Botanical Garden, Cleveland, OH
Denver Botanic Gardens, Denver, CO
Enid A. Haupt Glass Garden, New York, NY
Fairchild Tropical Garden, Miami, FL
Fernwood, Niles, MI
Minnesota Landscape Arboretum, Chanhassen, MN
Norfolk Botanical Garden, Norfolk, VA
North Carolina Botanical Garden, Chapel Hill, NC
Red Butte Gardens and Arboretum, Salt Lake City, UT
Royal Botanical Gardens, Hamilton, Ontario, Canada
Sherman Library and Gardens, Corona Del Mar, CA
The Frelinghuysen Arboretum, Morristown, NJ
The Holden Arboretum, Kirtland, OH
The New York Botanical Garden, Bronx, NY
The Pittsburgh Civic Garden Center, Pittsburgh, PA
The University of British Columbia, Vancouver, British Columbia, Canada

OVERVIEW

Centuries in the making, horticultural therapy has matured significantly. Because it is unique in its composition, nonthreatening in its appearance, flexible in its application, cost effective, and effective in its use—it occupies a necessary position within the health care realm. This position will strengthen in the future as the professional association expands the ranks of horticultural therapists through clinical practice, education, and research initiatives. Horticultural therapy's future could not be brighter!

KEY ORGANIZATIONS

National Council for Therapy and Rehabilitation Through Horticulture: Incorporated in the state of Maryland in 1973 as the professional association for horticultural therapists, it changed its name to the American Horticultural Therapy Association in 1988.

American Horticultural Therapy Association: Formed in 1973 under the original name of the National Council for Therapy and Rehabilitation Through Horticulture, it serves as the professional association for horticultural therapists. Address: 362A Christopher Avenue, Gaithersburg, Maryland 20879.

People-Plant Council: Formed in 1990 for the purpose of encouraging and documenting research in the area of people-plant connections. Address: Department of Horticulture, Virginia Tech, Blacksburg, Virginia 24061-0327.

Horticultural Therapy: Formed in 1978 under the name of The Society for Horticultural Therapy and Rural Training. Address: Goulds Grounds, Vallis Way, Frome, Somerset BA11 3DW England.

BIBLIOGRAPHY

Campbell, H., Knox, T.W., and T. Byrnes. 1896. *Darkness and daylight, or lights and shadows of New York life*. Hartford, CT: Hartford Publishing Co. pp. 307-310.

Davis, S. 1994. Ninth annual congressional initiatives award ceremonies. April 19, Senate Russell Office Building, Washington, DC.

Lawrence, G.M. 1900. Principles of education for the feeble minded. *Journal of Psycho-Aesthenics*. 4(3):100-108.

Lewis, C. 1976. Fourth annual meeting of the national council for therapy and rehabilitation through horticulture. September 6, Philadelphia, PA.

McDonald, J. 1995. "A Comparative Study of the Horticultural Therapy Professions in the United Kingdom and the United States of America." Master's thesis, University of Reading, England.

Straus, M. 1987. *Horticulture as therapy at Friends Hospital,* Pamphlet. Philadelphia, PA.

Tereshkovich, G. 1975. *Horticultural therapy: A review,* National Council for Therapy and Rehabilitation Through Horticulture lecture and publication series. February 1(1):1-4.

ADDITIONAL RESOURCES

Brooks, H.D. and C.J. Oppenheim. 1973. *Horticulture as a therapeutic aid.* Rehabilitation Monograph 4a, Institute of Rehabilitation Medicine. New York: New York University Medical Center.

Burlingame, Alice W. 1974. *Hoe for health.* Alice Burlingame. Birmingham, Michigan.

Johnston, E.R. 1899. The value of sense training in nature study. *Journal of Psycho-aesthenics.* 4(3):231-217.

Watson, D.P. and A.W. Burlingame. 1960. *Therapy through horticulture.* Macmillan, New York.

American Horticultural Therapy Association

CODE OF PROFESSIONAL ETHICS

The American Horticultural Therapy Association (AHTA) is dedicated to promoting all levels of interest in the development of horticulture and related activities as a therapeutic and rehabilitative medium. The AHTA strives to improve the performance of programs utilizing horticultural activities in human development through communications, coordination, knowledge dissemination, and promotion of education and training. Furthermore, AHTA is dedicated to enhancing the professionalism of horticultural therapists. This service is predicated on a basic belief in the intrinsic worth, dignity and potential of each human being. Respect for this belief shall guide the member's professional conduct.

Responsibility §: A member's primary responsibility is to the client served. A member shall be continuously aware of the relationship between responsibilities to the clients served, to the profession and to the employer. A member shall model, encourage and expect ethical and competent behavior from colleagues, whether or not they are members of this association, and shall attempt to rectify behavior which is contradictory.

Professional Competence §: A member shall demonstrate competency gained through education and experience with particular emphasis on the utilization of the horticultural environment in therapy and rehabilitation. A member shall be able to comprehend and interpret information acquired during the therapeutic horticultural activity process, and be familiar with studies of human and plant interaction.

Confidentiality §: A member shall respect the privacy of the client served and shall safeguard all information and materials obtained during the therapeutic horticultural activity process. To the extent consonant with employer methodologies, a member shall inform the client served, or the client's agent, about the purpose of any professional service being contemplated. Information generated during the administration of such services shall only be communicated to those with an essential need to know as part of the therapeutic horticultural activity process; the client served shall be made aware of who is in receipt of such information. If such information is used in teaching, research or writing, the identity of the client served shall be protected.

Interprofessional Relationships §: A member shall be cognizant of the relationship to other professionals involved in the service to clients and shall be aware that the welfare of clients receiving services depends on the capacity of all professional personnel to integrate their efforts. As part of this awareness, a member shall avoid practicing in areas not within the member's professional competence. Likewise, a member shall assist other professionals to comprehend this role and shall be prepared to respond when others seek assistance.

Publications §: A member shall communicate to other members, through publications, seminars, workshops or other appropriate means, information the member believes will contribute to improve or expand the existing knowledge. In such instances, the member shall adequately acknowledge the contributions of all individuals or organizations.

Consultation §: A member shall insure and be able to document competency in all areas related to acceptance of consulting assignments. Information regarding either the organization or clients served shall be regarded as confidential and shall not be used for unethical purposes or personal advantage.

Chapter 2

People-Plant Relationship

Paula Diane Relf

INTRODUCTION

Horticulture has traditionally been defined very narrowly as the production of specific, high-value plants for the commercial market and, in more recent years, the subsequent services related to the installation and maintenance of landscape plants. However, the fields of horticultural therapy, community gardening, and school gardening focus on the value of the process of growing plants rather than the end product. This has resulted in the exploration of a new definition for horticulture.

According to Webster's *New World Dictionary* (Second College Edition), we see that "horticulture" is derived from the root words *hortus* (a garden) and *cultura,* for which the dictionary refers us to the word "culture." Under "culture" we find "cultivation of the soil; the development, improvement or refinement of the mind, emotions, interests, manners, tastes, etc.; the ideas, customs, skills, arts, etc. of a given people in a given period; civilization." Combining *hortus* with the other definitions of "culture" brings us to a more inclusive definition and expands the horizons of educators, researchers, and practitioners in this field:

> *Horticulture:* The art and science of growing flowers, fruits, vegetables, trees, and shrubs resulting in the development of the minds and emotions of individuals, the enrichment and health of communities, and the integration of the "garden" in the breadth of modern civilization.

By this newly developed definition, horticulture encompasses PLANTS, including the multitude of products (food, medicine, O_2) essential for

---

Some of this material appeared in a related form in Relf, D. (1992). Human Issues in Horticulture. *HortTechnology* 2(2):159–171.

human survival, and PEOPLE, whose active and passive involvement with "the garden" brings about benefits to them as individuals and to the communities and cultures they comprise.

The purpose of this chapter is to provide a broad overview of the role of horticulture in human life quality. The learning objectives are to

1. recognize the theoretical basis for explaining the response of people to plant;
2. identify psychological and physiological responses of individuals to plants;
3. identify the psychological, economic, and social value of plants in community development;
4. contrast a child's perception of a garden with an adult's;
5. identify plants role in physically modifying the environment for greater health and comfort; and
6. identify the role of plants in cultural continuity and in the evolution of philosophy and fine art.

BACKGROUND THEORIES

Overload and Arousal

There are several theories discussed by Ulrich and Parsons (1992) to explain how and why being around plants can be beneficial. The simplest theories, the *overload* and *arousal* theories, maintain that in the modern world, we are bombarded constantly with so much noise, movement, and visual complexity that our surroundings can overwhelm our senses and lead to damaging levels of psychological and physiological excitement. Environments dominated by plants, on the other hand, are less complex and have patterns that reduce arousal and, therefore, reduce our feelings of stress.

Learning

Another theory maintains that people's responses to plants are a result of their early *learning experiences* or the cultures in which they were raised. According to this theory, those individuals, for example, who grow up in western Texas will have a more positive attitude toward flat lands with sparse, natural vegetation and cultivated crops, such as sorghum and cotton, than someone from the mountains of Virginia. Along the same line, this theory could be used to explain why Americans seem to prefer foundation plantings in their front yards even though the style of architecture has changed and these plants are no longer needed to hide unattractive foundations, or why Americans desire broad expanses of lawn that urban water systems cannot readily maintain. According to Ulrich, this theory also holds that modern, Western cultures condition people to like nature and plants and to have negative feelings about cities. However, this theory does not take into account the similarities in responses to nature found among people from different geographical and cultural backgrounds, or even those from different historical periods.

Evolution

The evolution theory maintains that our responses to plants are a result of evolution; that is, since we evolved in environments comprised primarily of plants, we have a psychological and physiological response to them. This evolutionary response is seen in an unlearned tendency to pay attention and respond positively to certain combinations of plants and other natural elements, such as water and stone. The most positive types of responses researchers found have been to the settings resembling those most favorable to survival for early humans. For example, one researcher has linked preference for certain tree forms to a high probability of finding food and water in nature near similarly shaped trees (Orians, 1986). Balling and Falk (1982) interpret their research with individuals from elementary school through senior citizens as providing limited support for the hypothesis of an evolutionary preference for savanna-like settings. Another researcher has shown that many features we particularly enjoy in the modern landscape, such as pathways that gently curve into the woods, were important to early man in terms of safety and exploration (Kaplan and Kaplan, 1989). The Kaplans' (1982) evolutionary perspective links settings high in vegetation with intuitively and cognitively based preferences and restorative influences. Ulrich (1983) puts forth a theory that the first level of response to natural scenes including vegetation is emotional. His "psychoevolutionary" perspective holds that this emotional response to nature is central to all subsequent thoughts, memory, meaning, and behavior as related to human environments.

PLANTS AND THE INDIVIDUAL

Physiological Responses to Plants

Ulrich's work strongly supports the idea that our immediate responses to plants are evolutionary with an affective or emotional basis and physiological response. In one study of college students under stress from an exam, views of plants increased positive feelings and reduced fear and anger (Ulrich, 1979). Another of his studies documented physiological changes related to recovery from stress, including lower blood pressure and reduced muscle tension (Ulrich and Simons, 1986). With a view of nature, recovery from stress was reported by physiological indicators within four to six minutes, indicating that even brief, visual contacts with plants, such as in urban tree plantings or office parks, might be valuable in restoration from mild, daily stress.

Psychological Responses to Plants

Steven Kaplan (1992) attributes the restorative value of participation with nature, particularly wilderness experiences, to the ability to fulfill several criteria:

- *Being away,* that is, providing a setting so different from the stressful setting that there is a feeling of escape and an increased likelihood of thinking about other things.
- *Extent,* which implies that the setting is large enough in scope to experience without exceeding its boundaries, and that the various parts of the setting are connected or belong to the whole. Extent is not defined by physical size, but rather by conceptual size; thus a miniature garden, a terrarium, or a vegetable plot may provide for one person what acres of wilderness provide for another.
- *Fascination* elicits involuntary attention; that is, you do not have to focus your attention consciously on what you are doing, as is often required by stress-inducing jobs, in order to avoid distraction or day-dreaming. Fascination allows you to recover from the efforts of the directed attention given to more stressful work.
- *Compatibility* is established by an environment that is conducive to meeting your personal goals; that is, in a compatible environment, what you want to do and are inclined to attempt are needed and feasible.

Participation in restorative experiences meeting these criteria may be essentially passive (sitting in a park) or active (maintaining a vegetable garden). A significant amount of research has been done with regard to experiences outside of populated areas where the participant (hiker, camper, fisher) is, in fact, simply passing through an environment controlled and directed by the U.S. Forest Service. Few studies have been conducted in which participation requires the commitment of caring for the environment necessitated by gardening. However, the results of one study of the garden experience (R. Kaplan, 1973) indicate that this model for restorative experiences would hold true.

Preference as an Indicator of Perception

Each individual brings accumulated knowledge and history to the perception of an environment, thereby influencing how it is experienced. These perceptions are very difficult to identify and interpret as they may be on a subconscious level. However, the Kaplans use an intermediate concept, *preference,* which is easy to elicit. By analyzing the patterns of preference within given populations, it has been possible to learn about perceptions and categories of environments. These have been divided into two major types of environmental categories: those based on content and those based on spatial configuration or arrangement. Content categories are divided based on the amount and kind of human influence; for example, scenes dominated by buildings would form a discrete category as would scenes with vegetation and no buildings, roads, or other human artifacts. Both Ulrich and Kaplan have demonstrated that scenes of nature/ vegetation are preferred significantly over scenes of buildings, and Ulrich and Simons (1986) have demonstrated that recovery from stress, based on physiological measurements, is faster when viewing scenes of nature. Honeyman (1987) expanded on these studies to include scenes with buildings and plants. Her findings suggest that even in an urban environment with buildings, the presence of vegetation may produce greater restoration than settings without vegetation. Preference judgments for categories based on spatial configuration or arrangement suggest an underlying criterion related to presumed possibilities for action, as well as potential limitations. In addition, spatial configuration categories can be distinguished in terms of openness with low differentiation (predominantly sky with farmland, bogs, marshes, etc.); lack of openness with low differentiating characteristics, but with the view blocked; and strong spatial definition often characterized as "parkland" (relatively open with distinct trees to enhance depth). The "parkland" settings tend to be among the most highly preferred kinds of settings (R. Kaplan, 1992).

Physiological Responses Compared to Preferences

Wise and Rosenberg (1988) measured both physiological response and aesthetic preference in a study on the role of nature decor in alleviating the symptoms of stress created by work-productivity demands in a simulated space station. The bulkhead of the simulated crew cabin had one of four scenes: savannah-like, mountain waterscape, hi-tech abstract, or blank control. The mountain waterscape was the most aesthetically preferred and was highly successful in stimulating remembered and imagined outdoor experiences. However, the savannah-like scene was significantly more effective in producing measurable physiological stress reduction. The effect was just as strong for participants who expressed a preference for the scene as for those who disliked it. Particularly interesting is the fact that these results were found whether the subject was looking at the scene or not.

Views from a Window Impact Health

Studies related to the view from a window have given interesting results. Office workers with essentially no outside view were more likely to decorate their work spaces with scenes of nature than workers with windows (Heerwagen and Orians, 1986). Several studies of interiors with windows have documented higher preference for views with vegetation or nature than alternatives that were "visually impoverished" (Markus, 1967; Verderber, 1986). Another study (Kaplan, Talbot, and Kaplan, 1988) reports that workers with a view of natural elements, such as trees and flowers, experienced less job pressure, were more satisfied with their jobs, and reported fewer ailments and headaches than those who either had no outside view or could only see built elements from their windows. Particularly interesting in this study was the finding that simply the knowledge that the view was available was important to the employees, even if they did not take advantage of it. Health-related benefits of window views of vegetation have been documented in several studies. Moore (1982) reported that inmates who had a view of nearby farmlands and forests had fewer sick call reports than those with a view of the prison yard. West (1985) found a lower frequency of stress symptoms, such as headaches, among inmates with natural views than those with views of buildings and prison walls. In a study of gall-bladder surgery patients, Ulrich (1984) reported shorter, post-operative hospital stays, fewer potent pain drugs, and fewer negative staff evaluations about patient conditions among those with a view of trees than those viewing a wall.

Physical Benefits of Gardening

Work also is being done to evaluate the impact of active participation in gardening on general physical health from the perspective of exercise. Taylor (1990) cites several sources to illustrate the physical value of gardening, reporting that an individual can burn as many calories in forty-five minutes of gardening as in thirty minutes of aerobics. One hour of weeding burns 300 calories (the same as walking or bicycling at a moderate pace), and manual push-mowing of the lawn burns 500 calories per hour (the same rate as playing tennis).

Summary and Implications for Health and Wellness

Views of nature have positive, physiological impacts on individuals whether or not they are consciously aware of them. These effects include lower blood pressure, reduced muscle tension, and lower skin conductance. In addition, documentation shows that views through a window produce a reduction in the need for medical treatment. Finally, the availability of views of nature, whether or not individuals take advantage of the views, has been demonstrated to play a role in worker satisfaction. It would appear from this limited research that appropriate configuration of vegetation (or stated horticulturally, a properly conceived landscape based on knowledge of human responses to plants) can have positive physiological effects on individuals without their awareness and additional, positive psychological effects on people who are aware of them. Actual participation, either active or passive, in a nature experience (e.g., gardening) can enhance further the value of plants on an individual's mental and physical health.

Individuals appear to benefit significantly from access to views of nature/vegetation and opportunities for passive encounters with plants and/or active participation in gardening experiences on a continuous basis for optimum physical and psychological functioning. The type and configuration of the vegetation may influence its effectiveness in this regard. We need to better understand from both a health perspective and preference perspective what is most effective, then apply that knowledge to the selection of plants that can be sustained within the urban setting, while taking into consideration the increased constraints on water resources, available space and light, adequate soil conditions, and decreased use of chemicals on plants. Since it appears that physiological responses to vegetation can differ from aesthetic or culturally acquired preference responses, we need to pursue actively an understanding of a healthful landscape. With sufficient information, horticulturists may play a role in altering culturally based or learned responses to vegetation by strongly reenforcing more environmentally sustainable and humanly healthful landscapes. The production, installation and maintenance of appropriate landscapes will continue to be major goals of horticulture, but the actual plant content and configuration involved may be altered by further studies.

PLANTS AND THE COMMUNITY

Physical Impact of Plants

Plants provide a positive physical surrounding in which it is more comfortable to live and work by purifying the air, moderating temperatures through shade or windblocks, reducing glare and noise, removing pollutants from the air, screening unattractive sights, and increasing relative humidity (Nighswonger, 1975). In urban microclimates, Herrington (1980) reports that plants are useful in moderating the temperature effects of solar and infrared radiation, thus increasing comfort levels. However, psychological factors, such as expectations and desires for certain environmental conditions, have a greater impact on perceived comfort than do actual temperatures.

Psychological Impact of Plants Within a Community

A "community" is defined as a group of people living in close proximity and sharing similar interests and values, usually implying friendly association. The interaction and collective values of the members of the community give it the uniqueness that defines it as a community. A community may be formed from any grouping of people; thus a neighborhood, a retirement village, a school, an office complex, or a housing project can become a community. Plants play a role in the development of healthy communities in three distinct ways:

- By providing a physical condition or appearance that makes people proud to be considered part of the community and by enhancing the economic and social condition of the community.
- By providing opportunities for the sharing of values, interests, and commitments that open the door to friendly association and lead to further cooperation, which has the impact of demonstrating the individual's ability to have control and responsibility for changes in the community.
- By providing a surrounding that is more comfortable physically in which to live and work.

The physical condition of an area, be it a neighborhood or an office complex, provides a measure of the self-worth of the area, defines the value of the individuals within that area, and projects that definition to outsiders. Thus, if an area is dilapidated or vandalized, has trash-filled vacant lots, or is sterile steel and concrete, it sends messages that those in charge (the city government, the owner, the employers) do not place value on the area and the people there; it implies that the people have no intrinsic worth and no control over their environment; and it tells outsiders that this is not a good place to be. A study in Atlanta (Brogan and James, 1980) examined the association between psychosocial health of the community and the physical environment (e.g., landscaping and nearby land use) and sociocultural environment (e.g., population density and income). The results indicated that the characteristics of physical and sociocultural environments were about equally important in explaining the variations in the psychosocial health of the community. Groups, such as the Partners for Livable Places, maintain that plants are the fastest, most cost-effective agents for changing negative perceptions of an area, enhancing the economic and social conditions, and improving the psychosocial health.

Creating a Positive Community Atmosphere

To explore the value of plants in creating a positive community atmosphere, researchers have looked at the role of nature/vegetation/plants in several related areas: environmental preferences and perceptions; neighborhood satisfaction; and the choice of place to live and economic impacts, such as residential property value and value to recreation and tourism.

Environmental Preferences and Perceptions

Research has shown that people prefer scenes of nature over urban scenes with buildings and man-made features; and among urban scenes, those with vegetation are preferred to those without (Kaplan, Kaplan, and Wendt, 1972; Herzog, Kaplan, and Kaplan, 1982; Herzog, 1989). Environmental perception studies seek to understand the qualities that make vegetation a preferred element. Schroeder (1990) reports that trees and forested areas, water, good maintenance, and peace and quiet were among the most preferred features of urban parks. Schroeder and Cannon (1987) also found that yard and street trees both were effective in enhancing the aesthetic quality of residential areas. No studies have been identified that explore the role of flowers, shrubs, groundcovers, or smaller vegetation. In a study of students' responses to faculty offices (Campbell, 1979), the presence of plants and wall posters led to positive ratings and clutter led to strong negative ratings.

Neighborhood Satisfaction

Fried (1982) found that the strongest indicator of local residential satisfaction was the ease of access to nature, and that this was the most important factor (after marital role) to life satisfaction. Frey (1981) likewise found that the availability of natural elements in the surrounding area strongly affected neighborhood satisfaction. Based on a survey questionnaire of residents in Detroit, Michigan, Getz (1982) reported that parks and street trees were second only to education in the perceived value of municipal services offered. They also were an important factor in determining where people would choose to live and in residential neighborhood satisfaction. Rachel Kaplan (1985) reported that the most important factors in neighborhood satisfaction among the multiple-family housing complexes she studied were the availability of trees, well-landscaped grounds, and places for taking walks.

Browne (1992) determined that among residents of retirement communities, pleasant, landscaped grounds were important (48.5 percent) or essential (50.5 percent) to 99 percent of the residents. A window view of green, landscaped grounds was three times as important as a view of activity areas. The configuration and natural elements of the grounds were given as the most important reasons for selecting the particular retirement community.

Perceived security and personal safety play a role in neighborhood satisfaction. Two studies document the importance of design and maintenance in perceived security. In a study of urban parking lots (Shaffer and Anderson, 1985), security was rated high only when vegetation was well maintained and appeared to be installed as part of a landscape design. The results of interviews with African-American residents with low and moderate incomes in Detroit by Talbot and Kaplan (1984) indicated that well-maintained areas incorporating built features were preferred over unkept, densely wooded areas that often elicited concern of physical danger.

Economic Value of Plants to a Community

People clearly are willing to pay more to have plants in their immediate surroundings. Residential property values are enhanced by their proximity to urban parks and greenbelts (Correll and Knetson, 1978; Hammer, Coughlin, and Horn, 1974; Kitchen and Hendon, 1967). Using professional appraisers' estimates, unimproved, residential land was determined to have higher value if there were trees on the land, and a scattered arrangement was determined to have a higher value than concentrated arrangements of trees with the same percentage of tree coverage (Payne and Strom, 1975). Individual home owners estimate that a well-maintained landscape increases the market value of their homes by 15 percent, while real estate professionals attribute 7 percent of the value of a residence to an attractive landscape (Weyerhauser, 1986). However, only 34 percent of the professionals include a dollar figure specifically for the landscaping when appraising residential property.

The willingness to pay for the use of urban forests is another method to determine economic value of urban vegetation. Travel-cost models were developed for three urban forest sites in the Chicago area, and willingness to pay up to $12.71 per visit was established (Dwyer, Peterson, and Darragh, 1983).

To determine the value of interior plants to the hotel/tourism industry, Evans and Malone (1992) conducted a study at Opryland. The twelve acres of indoor space has approximately 18,000 plants valued at over $1 million. The annual horticultural budget is approximately $1.2 million. The study attributes several positive impacts to the "greatscapes": the unusually high occupancy rate of 85 percent, numerous awards, and continued expansion. Most importantly, the higher rate ($30/night) for those rooms overlooking the gardens and the high occupancy rate of those rooms translate into $7 million in additional room revenue annually.

Social Value of Plants in a Community

Opportunities for the sharing of values and becoming a community are created when people participate actively toward one issue of concern to them all. Lewis (1992) states,

> Matthew Dumont, a community psychiatrist, has looked at the city to try to understand it in terms of the mental health needs of the city dweller. He states that the city dweller has a need for *stimulation,* to break the monotony of daily life; for a *sense of community,* which arises not because people are forced to live together, but rather from some spontaneous action such as creating a garden; and for a *sense of mastery of the environment,* reassuring him that he is not a helpless cog in the overwhelming machinery of living.

Community gardening, landscaping, and tree-planting projects provide excellent settings in which all of these needs can be met. Lewis describes changes that have taken place in communities as a result of people working together in gardening projects sponsored by the New York City Housing Authority, and similar results from gardening programs sponsored by the Chicago Housing Authority. In Philadelphia, the Pennsylvania Horticultural Society has played a leading role in helping communities help themselves through gardening (Bonham, 1992).

Sociological Impact of Community Greening

Results of these projects include the conversion of vacant lots into playgrounds and gardens, the cleaning of streets around planted areas, and the formation of neighborhood groups to supervise the gardens. According to Bonham, the most important factor in the success of the gardens is the development of neighborhood leadership—gardeners who coordinate the gardening projects and provide the initiative to continue and expand. A study to understand what motivates community gardeners (Clark and Manzo, 1988) revealed differences between gardeners and non-gardeners in terms of previous environmental experience, "rootedness" in the community, social interaction with people in the community, and values placed on growing things and access to nature. USDA Cooperative Extension gardening programs have socioeconomic (Patel, 1991; 1992) and community development (Grieshop, 1984) consequences that enrich the people and communities in which they are conducted.

In urban, tree-planting programs, sociological factors may be more important than biological factors in tree survival (Ames, 1980). Public works plantings with no involvement from the community in planning or installation may lack grassroots support and be open to local action to subvert the effort. For example, as part of one model city's program, officials decided to plant 2,000 trees, few of which were standing two years later. However, Ames reported that with community involvement from the initial conceptualization through planting and maintenance, tree survival increased and many human benefits resulted, such as enhancement of the sense of community among participants, a positive social identity for the participants, increased personal identification with the neighborhood, and allowance for personal control over the neighborhood.

Children and the Garden

Children represent a subset of the larger community in which they live, since they perceive their environment from a unique perspective. How children perceive and experience their environments has been a focus of research (Eberbach, 1987). Studies also have documented the differences in perception among different age groups (Zube, Pitt, and Evans, 1983). These studies have focused on the larger environment of nature, cityscapes, or playgrounds. However, a few studies have looked at the child as a participant in the garden and the perceptions children hold of the natural world in the limited context to which they respond and are able to understand (Bunn, 1986; Jessee et al., 1986). Eberbach (1992) presents three observations of children and gardens based on her research with elementary school children and their art interpretations of a garden:

- *Children understand what a garden is and have aesthetic preferences.* While 19 percent of the children illustrated their gardens exclusively with functional plants, such as fruits and vegetables, 47 percent used ornamental plants chosen for aesthetic purposes.
- *Perceptions of gardens are shaped by a child's cognitive development.* Younger children's concepts of a garden were limited to a few environmental elements (plants, soil), while older children linked the garden into a whole picture that included people, paths, tools, animals, etc. These perceptions were felt to be functions of the cognitive or developmental levels of the children.
- *Activity is used to understand a garden and one's place in the garden.* The children's drawings were saturated with elements implying activity—paths, bridges, swings, and tools. Touching and interacting with the elements of the garden are essential to the values a child gains from the experience.

Creating a garden or natural environment that meets a child's requirements for understanding and responding will provide an atmosphere for encouraging curiosity and motivating learning.

Summary and Implications for Healthy Communities

The degree to which plants create a positive, community atmosphere is measured, in part, by people's perceptions of and preferences for plants and the economic investment they are willing to make to have plants around them. Scenes with plants are highly preferred over those without, and plants play a significant role in neighborhood satisfaction. The proper maintenance of plants is also a factor in the positive perception of plants. By working together in tree plantings, community gardens, and beautification projects, people get to know each other, thus creating a true community with inhabitants who have a sense of allegiance to and responsibility for their surroundings. In addition to providing these sociocultural benefits, plants are extremely important in mediating environmental factors, such as temperature, noise, and pollution.

To maximize the benefits plants have on our communities, we need to support increased urban horticultural activities. Achieving this requires documentation of the impact of these activities in forms that will be accepted by budget officials in government and private industry, as well as by taxpayers. It is as important to understand the role of social and emotional factors in the survival of urban plants as it is to understand the botanical and physical factors. Horticulturists must be as actively involved in the research that determines the social and emotional factors relating to urban plant survival as they traditionally have been involved in the botanical and physical factors.

There is clearly a need to teach children an appreciation of plants/nature through active participation in gardening. Sponsoring school and youth gardens needs to be a high priority for adults in all areas of horticultural activities, including commercial horticulturists, communicators, educators, and nonprofessionals, such as master gardeners and garden club members.

PLANTS AND HUMAN CULTURE

Plants traditionally are associated with food production and are seen as key factors in the evolution of civilization from the perspective of agricultural domestication of food crops. Additional consideration of the impact of plants on our culture recognizes their value in terms of clothing, shelter, medicine, and other economic goods. Domestication of plants and animals has had an equally significant impact on their evolution. As Jules Janick (1992) expresses it,

> The end result of the agricultural revolution has been a fundamental change in the human condition. The interaction of humans, crops, and domestic animals has resulted in the fused genetic destinies. An abundance of food causes changes in selection pressure and alterations of human evolution equivalent to those wrought by the domestications of plant and animal species.

However, an understanding of the role of plants in our culture cannot be limited to the view of meeting primarily physical and economic needs. Sociologists, anthropologists, artists, historians, and other professionals are beginning to explore the people-plant relationship to gain a better understanding of our humanity.

Plants are an integral part of our rituals from birth to death. Despite the prevalence of plants in celebrating and grieving, little is understood about their significance in these roles. Recently, researchers have begun to explore how plants are valued and interpreted when used as a celebration or ritual element. McDonald and Bruce (1992) report that in their study, 82 percent of the respondents associated a horticultural descriptor with Christmas, and subjects tended to rate the description of Christmas scenes including horticultural elements as more meaningful and enjoyable than those without such elements. The discovery of clusters of different kinds of pollen in the grave of a Neanderthal at Shanidar cave, Iraq, (Solecki, 1975) indicates that flowers have been part of the funeral process since earliest man. Shoemaker and Relf (1990) found that flowers are an important part of the bereavement process as a source of comfort and warmth and to help deal with grief. Their function in brightening up the somber environment and providing a conversational diversion also were appreciated highly. The primary reasons for sending flowers are to comfort survivors and show respect for the deceased. Eisuke Matsuo (1992) explored how flowers are used on grave sites in Japan.

In addition to providing a cultural continuity through our traditions and ceremonies, plants have strong linkages to the evolution of our philosophies and fine arts. Rosenfield (1992) maintains that the courtly gardens of the Italian Renaissance incorporated design elements for achieving "epideictic aims in the task of civilizing man;" that is, these display gardens were intended to impress, in order to reinforce the civilizing influence of rhetoric. "The visitor came to the garden as a spectator, to celebrate the glories of human being . . . the garden shared with pageants and festivals the ability to display spectacle while the beholder's point of view shifted so as to absorb a richer array of visual impressions than words alone could offer . . . to grasp more readily the symbolic meaning than would ritual or formal instruction . . . epideictic highlighting led the visitor toward a more personal realization. When we 'come home' to Nature, we rediscover our own nature."

Levi (1984) traces the relationship between natural and urban settings from classical to modern times as reflected in philosophy, literature, and art. He argues that this relationship expresses fundamental shifts in the human experience. The landscape is no longer a medium of emotional involvement and sensuous enjoyment. According to him, "modern industrial and technical objects and modes of living receive a stronger human allegiance than nature, and the consequences are apparent in current views, values, and modes of perception."

Cremone and Doherty (1992) looked at the symbolism of the flowers in paintings from the same period to understand the role of flowers in awakening moral consciousness. The symbolisms conveyed by plants include ethical and religious messages, flowers symbolic of marriage, concern over death and transience, greed and speculation. Zeven and Brandenburg

(1986) used paintings from the sixteenth to the nineteenth centuries to study the history of domesticated plants, providing an example of a practical bond between the fine arts and horticulture. Shearer (1992) has studied plants in art in terms of their reflections of the cultures and philosophies represented at the time they were painted. Further, she views plants "not as beautiful, sentimental or decorative objectives, but as universal forms whose very structure offers a window into the underlying vital principles of nature itself." She describes Leonardo Da Vinci's studies of plants as part of a process to understand the fundamental truths in nature. Of Mondrian, the father of geometric abstractions, Shearer states that it is clear that he rejected the human figure as his muse and sought out the "hidden dynamic in plants" and that trees "are the single most important subject matter for his evolvement into abstraction . . . in his search for the 'universal.'" Mondrian's "transmutation of the natural form from realism to geometric abstraction" is seen as a reflection of the early twentieth century when the conflict between nature and technology was in central focus.

The spiritual aspects of interaction with nature are explored and clarified by Schroeder (1991). To understand what he is writing, it is useful to consider first his point that the human psyche functions in two different modes: the rational, analytical mode associated with science and technology, and the intuitive mode manifested in the "ambiguous language of nonverbal imagery and symbolism," which is more the realm of art, music, and poetry and the source of spiritual phenomena. He further emphasizes that spiritual phenomena can be conceptualized in psychological terms rather than in supernatural terms and, as such, are a legitimate topic for scientific inquiry. He uses the following statement to sum up the use of the term "spiritual" in relation to nature: "'Spiritual' refers to the experience of being related to or in touch with an 'other' that transcends one's individual sense of self and gives meaning to one's life at a deeper than intellectual level."

Based on his study of the concepts of depth psychology by C.G. Jung, which concerns itself with the unconscious mind outside the awareness and/or control of the conscious ego, Schroeder (1991) discusses the application of the archetypes to nature. Archetypes are basic, instinctive patterns of behavior, emotion, and imagery common to all humans. To identify and understand the spiritual significance of nature, one might turn to mythology, literature, religious traditions, and art of various cultures in order to seek out archetypical responses. Through such studies, one identifies a rich tradition associated with trees, such as "World Tree," "Tree of Life," and "sacred groves," as well as symbolisms associated with the garden as an origin of humanity. Jung has described several archetypes derived from psychological analyses of his patients. The three considered most relevant by Schroeder are the Great Mother with both positive, nurturing aspects and negative, destructive aspects that are projected onto nature and personified as Mother Nature; Anima, considered to be the unconscious, feminine side of a man's personality and associated with creative, intuitive, and spiritual aspects of life; and the Self, representing a movement toward wholeness and a unique, integrated personality. An understanding of these archetypes, how they developed through our culture and how we interpret their projection onto nature can have significant value in exploring people-plant interaction.

Summary and Implications for Plants and Human Culture

The role of plants in the evolution of civilization reaches far beyond food, fiber, and medicine. The domestication of plants and animals allowed for massive changes in human culture. The act of cultivation brought intellectual, psychological, and social rewards that are reflected in our folklore, literature, and art. Plants and gardens have been used as havens for reflection by philosophers, as teachers for those who would learn by example and examination, and as sources of inspiration and symbols of virtue/vice by artists and poets. Plants and nature are woven into the unconscious mind of humans and serve as a source of spiritual renewal.

All aspects of human culture are rich with references and meanings regarding plants since they have played an integral role in the development of our civilization. Knowing more about human interaction with plants, from our food to our folklore, will help us better understand ourselves and our roles in the "grand scheme of things." While study in these areas is primarily the domain of social scientists, relatively few studies have focused on the importance of people-plant relationships in the development of our culture and the application of that information to modern life. Social scientist have given several explanations for the dearth of studies in the current literature ranging from "taking plants so much for granted that it never occurred to them" to a belief that work in this area was so "fundamental and accepted that all important information had been acquired at some unspecified time in the past." Through interdisciplinary efforts, horticulturists can supply the insight about plants and the guidance needed by social scientists to explore further and communicate this critical area.

ADDITIONAL RESOURCES

Books

The Healing Dimensions of People-Plant Relations: A Research Symposium. Mark Francis, Patricia Lindsey, and Jay Stone Rice, Editors. Davis, CA: University of California. 1994. 498 pages. $39.

People-Plant Relationships: Setting Research Priorities. Joel Flagler and Raymond P. Poincelot, PhD, Editors. 1993. Binghamton, NY: Food Products Press. 368 pages. $54.

The Role of Horticulture in Human Well-Being and Social Development. Diane Relf, Editor. 1992. Portland, OR: Timber Press. 254 pages. $54.

Computerized Bibliographies

People-Plant Interaction (1,305 citations) and *Horticulture Therapy* (1,184 citations) bibliographies are available on 3.5-inch, DS/HD diskettes containing the citations in WordPerfect 5.0. The material also can be ordered on 3.5-inch diskettes as DOS text files. $15 for each bibliography. For updates, return original diskette with $5.

Videotapes

The Art of Rhonda Roland Shearer. $15.

Role of Horticulture in Human Well-Being and Social Development: Reflections of Jules Janick, Charles Lewis, Roger Ulrich, Russ Parson, and Diane Relf. $15.

Web Page Address

A wealth of information is available on our Virginia Cooperative Extension Consumer Horticulture Web site at *http://www.hort.vt.edu/.*

REFERENCES

Ames, R.G. 1980. The sociology of urban tree planting. *The Journal of Arboriculture* 6(5):120–123.

All prices include shipping and handling. Make checks (in U.S. dollars payable on a U.S. bank) payable to **Treasurer, Virginia Tech,** and mail to PPC/Office of Consumer Horticulture, Virginia Tech, Blacksburg, VA 24061-0327.

Balling, J.D. and J.H. Falk. 1982. Development of visual preference for natural environments. *Environment and Behavior* 14(1):5–28.

Bonham, B. 1992. Philadelphia Green's Greene Countrie Towne model as an agent for community development: Findings of case studies. In D. Relf (ed.), *The Role of Horticulture in Human Well-Being and Social Development: A National Symposium.* Portland, OR: Timber Press.

Brogan, D.R. and L.D. James. 1980. Physical environment correlates of psychosocial health among urban residents. *American Journal of Community Psychology* 8(5):507–522.

Browne, C.A. 1992. The role of nature for the promotion of well-being of the elderly. In D. Relf (ed.), *The Role of Horticulture in Human Well-Being and Social Development: A National Symposium.* Portland, OR: Timber Press.

Bunn, D.E. 1986. Group cohesiveness is enhanced as children engage in plant-stimulated discovery activities. *Journal of Therapeutic Horticulture* 1:37–43.

Campbell, D.E. 1979. Interior office design and visitor response. *Journal of Applied Psychology* 64(6):648–653.

Clark, H. and L. Manzo. 1988. Community gardens: Factors that influence participation. In *Proceedings of the Nineteenth Annual Conference of the Environmental Design Research Association* 19:57–61. Pomona, CA.

Correll, M.R. and J.L. Knetson. 1978. The effects of greenbelts on residential property values: Some findings on the political economy of open space. *Land Economics* 54(2):207–217.

Cremone, J.C., Jr. and R.P. Doherty. 1992. Vita brevis: Moral symbolism from nature. In D. Relf (ed.), *The Role of Horticulture in Human Well-Being and Social Development: A National Symposium.* Portland, OR: Timber Press.

Dwyer, J.F., G.L. Peterson, and A.J. Darragh. 1983. Estimating the value of urban forests using the travel cost method. *The Journal of Arboriculture* 9(7):182–185.

Eberbach, C. 1987. Gardens from a child's view: An interpretation of children's artwork. *Journal of Therapeutic Horticulture* 2:9–16.

Eberbach, C. 1992. Children's gardens: The meaning of place. In D. Relf (ed.), *The Role of Horticulture in Human Well-Being and Social Development: A National Symposium.* Portland, OR: Timber Press.

Evans, M.R. and H. Malone. 1992. People and plants: A case study in the hotel industry. In D. Relf (ed.), *The Role of Horticulture in Human Well-Being and Social Development: A National Symposium.* Portland, OR: Timber Press.

Frey, J.E. 1981. Preferences, satisfaction, and the physical environments of urban neighborhoods. Unpublished doctoral dissertation for the University of Michigan, Ann Arbor, MI.

Fried, M. 1982. Residential attachment: Sources of residential and community satisfaction. *Journal of Social Sciences* 38(3):107–119.

Grieshop, J.I. 1984. Serendipity and community development: A study of unplanned community development consequences in a community service program. *Journal of the Community Development Society* 15(2):87–103.

Hammer, T.R., R.E. Coughlin, and E.T. Horn. 1974. The effect of a large urban park on real estate. *AIP Journal* 40(4): 274–277.

Heerwagen, J.H. and G. Orians. 1986. Adaptations to windowlessness: A study of the use of visual decor in windowed and windowless offices. *Environment and Behavior* 18(5):623–639.

Herrington, L.P. 1980. Plants and people in urban settings. Proceedings of the Longwood program seminars 12:40–45. Longwood Gardens, PA.

Herzog, T.R. 1989. A cognitive analysis of preference for urban nature. *Journal of Environmental Psychology* 9(1):27–43.

Herzog, T.R., S. Kaplan, and R. Kaplan. 1982. The prediction of preference for unfamiliar urban places. *Population and Environment* 5(1):627–645.

Honeyman, M. 1987. Vegetation and stress: A comparison study of varying amounts of vegetation in countryside and urban scenes. Unpublished master's thesis from the Department of Landscape Architecture, Kansas State University, Manhattan, KS.

Janick, J. 1992. Horticulture and human culture. In D. Relf (ed.), *The Role of Horticulture in Human Well-Being and Social Development: A National Symposium*. Portland, OR: Timber Press.

Jessee, P., M.P. Strickland, J.D. Leeper, and C.J. Hudson. 1986. Nature experiences for hospitalized children. *Children's Health Care* 15(1):55–57.

Kaplan, R. 1973. Some psychological benefits of gardening. *Environment and Behavior* 5(2):145–162.

Kaplan, R. 1985. Nature at the doorstep: Residential satisfaction and the nearby environment. *Journal of Architectural Planning Research* 2:115–127.

Kaplan. R. 1992. The psychological benefits of nearby nature. In D. Relf (ed.), *The Role of Horticulture in Human Well-Being and Social Development: A National Symposium*. Portland, OR: Timber Press.

Kaplan, R. and S. Kaplan. 1989. *The experience of nature*. Cambridge, New York: Cambridge University Press.

Kaplan, S. 1992. The restorative environment: nature and human experience. In D. Relf (ed.), *The Role of Horticulture in Human Well-Being and Social Development: A National Symposium*. Portland, OR: Timber Press.

Kaplan, S. and R. Kaplan. 1982. *Cognition and environment*. New York: Praeger.

Kaplan, S., R. Kaplan, and J.S. Wendt. 1972. Rated preference and complexity for natural and urban visual material. *Perception & Psychophysics* 12(4):354–356.

Kaplan, S., J.F. Talbot, and R. Kaplan. 1988. Coping with daily hassles: The impact of nearby nature on the work environment. Project Report of the USDA Forest Service, North Central Forest Experiment Station, Urban Forestry Unit Cooperative Agreement 23-85-08.

Kitchen, J.W. and W.S. Hendon. 1967. Land values adjacent to an urban neighborhood park. *Land Economics* 43: 357–360.

Levi, A.W. 1984. Nature and art. *Journal of Aesthetic Education* 18(3):521.

Lewis, C.A. 1988. Hidden value. *American Nurseryman* 168(4):111–115.

Lewis, C.A. 1992. Effects of plants and gardening in creating interpersonal and community well-being. In D. Relf (ed.), *The Role of Horticulture in Human Well-Being and Social Development: A National Symposium*. Portland, OR: Timber Press.

Markus, T.A. 1967. The function of windows: A reappraisal. *Building Science* 2:97–121.

Matsuo, Eisuke. 1992. Cut flower usage for ancestral tombs in Kagoshima Japan. *HortTechnology* 2(2):236–238.

McDonald, B.G. and A.J. Bruce. 1992. Can you have a merry Christmas without a tree? In D. Relf (ed.), *The Role of Horticulture in Human Well-Being and Social Development: A National Symposium*. Portland, OR: Timber Press.

Moore, E.O. 1982. A prison environment's effect on health care service demands. *Journal of Environmental Systems* 11(1):17–34.

Nighswonger, J.J. 1975. Plants, man and environment. Cooperative Extension Service Publication C-448. Kansas State University, Manhattan, KS.

Orians, G.H. 1986. An ecological and evolutionary approach to landscape aesthetics. In E.C. Penning-Rowsell and D. Lowenthal (eds.), *Meanings and Values in Landscape*.

Patel, I.C. 1991. Gardening's socioeconomic impacts. *Journal of Extension* 29(4):7–8.

Patel, I.C. 1992. Socioeconomic impact of community gardening in an urban setting. In D. Relf (ed.), *The Role of Horticulture in Human Well-Being and Social Development: A National Symposium*. Portland, OR: Timber Press.

Payne, B.R. and S. Strom. 1975. The contribution of trees to the appraised value of unimproved residential land. *Valuation* 22(2):36–45.

Rosenfield, L.W. 1992. Gardens and civic virtue in the Italian Renaissance. In D. Relf (ed.), *The Role of Horticulture in Human Well-Being and Social Development: A National Symposium*. Portland, OR: Timber Press.

Schroeder, H.W. 1990. Perceptions and preferences of urban forest users. *Journal of Arboriculture* 16(3):58–61.

Schroeder, H.W. 1991. The spiritual aspects of nature: A perspective from depth psychology. In Proceedings of Northeastern Recreation Research Conference, Saratoga Springs, NY.

Schroeder, H.W. and W.N. Cannon. 1987. Visual quality of residential streets: Both street and yard trees make a difference. *Journal of Arboriculture* 13(10): 236–239.

Shaffer, G.S. and L.M. Anderson. 1985. Perceptions of the security and attractiveness of urban parking lots. *Journal of Environmental Psychology* 5(4):311–323.

Shearer, R.R. 1992. Beyond romanticism: The significance of plants as form in the history of art. In D. Relf (ed.), *The Role of Horticulture in Human Well-Being and Social Development: A National Symposium*. Portland, OR: Timber Press.

Shoemaker, C.A. and P.D. Relf. 1990. The role of flowers and plants in the bereavement process. Final report of the Department of Horticulture, Virginia Polytechnic Institute and State University, Blacksburg, VA.

Solecki, R.S. 1975. Shanidar IV, a Neanderthal flower burial in northern Iraq. *Science* 190(28):880–881.

Talbot, J.F. and R. Kaplan. 1984. Needs and fears: The response to trees and nature in the inner city. *Journal of Arboriculture* 10(8):222–228.

Taylor, M.K. 1990. The healthy gardener. *Flower and Garden* March/April: 46–47.

Ulrich, R.S. 1979. Visual landscapes and psychological well being. *Landscape Research* 4(1):17–23.

Ulrich, R.S. 1983. Aesthetic and affective response to natural environment. In I. Altman and J.F. Wohlwill (eds.), *Behavior and the Natural Environment.* New York: Plenum 85–127.

Ulrich, R.S. 1984. View through a window may influence recovery from surgery. *Science* 224:420–421.

Ulrich, R.S. and R. Parsons. 1992. Influences of passive experiences with plants on individual well-being and health. In D. Relf (ed.), *The Role of Horticulture in Human Well-Being and Social Development: A National Symposium.* Portland, OR: Timber Press.

Ulrich, R.S. and R.F. Simons. 1986. Recovery from stress during exposure to everyday outdoor environments. In J. Wineman, R. Barnes, and C. Zimring (eds.), *The Costs of Not Knowing: Proceedings of the Seventeenth Annual Conference of the Environmental Design Research Association.* Washington, DC: Environmental Design Research Association.

Verderber, S.F. 1986. Dimensions of person-window transactions in the hospital environment. *Environment and Behavior* 18(4):450–466.

West, M.J. 1985. Landscape views and stress response in the prison environment. Unpublished master's thesis from the Department of Landscape Architecture, University of Washington, Seattle, WA.

Weyerhaeuser (Staff). 1986. *The value of landscaping: Ideas for today.* Vol. IV. Tacoma, WA: Weyerhaeuser Nursery Product Division.

Wise, J.A. and E. Rosenberg. 1988. The effects of interior treatments on performance stress in three types of mental tasks. CIFR Technical Report No. 002-02-1988. Grand Rapids, MI: Grand Valley State University.

Zeven, A.C. and W.A. Brandenburg. 1986. Use of paintings from the 16th to 19th centuries to study the history of domesticated plants. *Economic Botany* 40(4):397–408.

Zube, E.H., D.G. Pitt, and G.W. Evans. 1983. A lifespan developmental study of landscape assessment. *Journal of Environmental Psychology* 3(2):115–128.

Chapter 3

Vocational, Social, and Therapeutic Programs in Horticulture

Rebecca Haller

INTRODUCTION

Horticultural therapy includes a broad array of services, settings, and populations served. With programs in nearly every type of health care and social service setting—from adult day care to psychiatric to corrections—treatment approaches to meet the needs of patients or clients are also varied. As shown in Table 3.1, these may be categorized as (1) vocational, (2) therapeutic, and (3) social, with each reflecting a distinct type of programming with corresponding purpose, goals, and program design.

Learning Objectives

For a comprehensive understanding of horticultural therapy practice, these three main types of programming will be outlined in this chapter. For each, we will look at concepts, program settings, service recipients, funding, benefits to clientele, treatment teams, and treatment goals.

TABLE 3.1. Types of HT Programs

HT Program Types	Models	Focus/Goal for Patient/Client
Vocational	Rehabilitation	Employment
Therapeutic	Medical	Recovery from illness or injury
Social	Wellness	Quality of life, wellness

Program Types

Vocational horticultural therapy programs are primarily concerned with enhancement of skills and/or behaviors that lead to employment for those clients served. Based on a medical model (involving the treatment of someone who is ill or injured), therapeutic programs may include physical rehabilitation, psychiatric, or long-term care, for example. Social programs enhance leisure skills and quality of life and are often activity based. Before we explore these three types of horticultural therapy programs further, let's look at some common themes found in all of them.

Commonalities

Horticultural therapy programs seek to improve the well-being of their participants. Wellness involves an absence of illness and also maximal health and self-actualization of the client. By focusing on the whole person, the therapist attempts to encourage the development of a fully functioning individual. For example, in providing vocational training to a woman who has had a brain injury, the horticultural therapist is concerned with more than just an outcome of employment. Life satisfaction, community integration, self-esteem, and health are all considered to improve the overall well-being of the woman served.

Techniques that enable and empower participants to achieve maximum independence are common to every type of program in horticultural therapy. This may involve the use of adaptive or modified tools to enable persons with physical disabilities to participate more fully in horticultural activities. It may require breaking down a task to its simplest components or providing illustrated instructions so that someone with a cognitive disability can work independently. The attitude and approach of the therapist is critical to empower clients/patients to be less dependent on professionals for support. By allowing appropriate levels of "risk taking" and encouraging the development of a community support system, the horticultural therapist lends dignity, independence, and enhanced skills to the individual.

By giving support and encouragement, horticultural therapy programs help individuals to function in the least restrictive environments possible. Within a particular program or facility there may be a continuum of degrees of supervision or structure that can be chosen. For example, employment in an enclave (a work setting in which a crew of workers with disabilities works within industry with a specially trained supervisor) is more restrictive than competitive employment because of the extra staff supervision present in an enclave. Residence in a nursing home or long-term care facility with twenty-four-hour care is more restrictive than adult day care where the individual returns home at night. Again, the goal of horticultural therapy is to help prepare the individual for the least intervention, structure, or support necessary to meet his or her needs. Through successful horticultural therapy experiences, clients are often motivated to participate fully in this and in other therapies that help them to maximize achievements.

Horticultural therapy programs have much in common, but also differ in their approaches based on the type of programming offered. Now, let's explore each of these three main types: vocational, therapeutic, and social.

VOCATIONAL PROGRAMS

Conceptual Basis for Treatment

A vocational horticultural therapy program focuses on employment. Services may include prevocational and vocational training, job placement and/or supported employment (see Table 3.2). The desired outcome for individuals in the program is placement in the least restrictive environment in which the individual functions successfully, such as sheltered, supported, or competitive work settings. Within this range of work settings an individual may progressively move over months, or years, toward more independence. For example, an employee/trainee at a sheltered workshop may progress by increasing productivity on a particular job and thus earn higher wages. Improving work skills, such as initiating tasks and staying on task more consistently, may lead to placement in supported employment within the industry. In supported employment, the individual receives whatever professional support is necessary for success, such as a job coach, an enclave supervisor, or other ongoing vocational services. At the most independent end of the employment scale is competitive employment. In this case, the individual works in industry without professional intervention. Supports may be used to secure or train for the job initially, but are subsequently removed as the individual demonstrates independent work skills. See Table 3.3 for types of vocational placement.

Although historically this range of employment options has been considered a continuum through which a trainee moves, the current trend is to skip steps in the continuum and train people directly on the job at the site of employment. This trend results from work with more severely disabled clients for whom the skills learned in training facilities do not necessarily transfer to a new work setting. An increasingly less common approach is to train individuals in a special facility, such as a sheltered workshop or school prior to placement in industry jobs.

TABLE 3.2. Vocational Services

Services	Description
Prevocational training	Training in vocationally related personal and social adjustment behaviors or skills in readiness for work training
Vocational training	Training in specific skills to enhance ability to work independently or semi-independently
Job placement	Services geared toward finding, securing, and maintaining employment
Supported employment	Services that provide necessary supports to enable employment in industry, such as job coaching or enclave supervision

TABLE 3.3. Types of Vocational Placements

Placement	Duration	Staff	Client Wages	Location	Other
Sheltered	short or long-term	crew supervisors, therapists, trainers	generally below minimum wage, based on time studies	within facility	most intensive support to client
Enclave	long-term	crew supervisors	may be below minimum wage, based on time studies	industry within the community	contract with industry to perform a job, may replace one or more positions with a supervised crew
Supported	short or long-term	job coaches, vocational counselors	generally industry wages (but not always)	industry within the community	supports vary depending on individual needs
Competitive	long-term	job placement specialists, may have a job coach for initial training period of employment	industry wages	industry within the community	regular employment

Regardless of type of vocational setting, placements, or programs, all are based on a rehabilitation model. Rehabilitation has been defined as "restoration, after a disease or injury, of the ability to function in a normal or near-normal manner" (Punwar, 1994). Clients/patients are assisted with restoring physically, psychologically, socially, and vocationally in order to live a happy and productive life. This model also includes the concept of "habilitation," defined as "the acquisition of new skills that the client did not previously possess" (Punwar, 1994). Habilitation may especially describe the process of working with people with developmental disabilities or socioeconomic disadvantages. New skills are stressed over restoration of those previously held.

Settings for Programs

Vocational horticultural therapy programs are found in many settings.

Public Schools

Educational programs for special education students in public high schools typically include a vocational or preoccupational component. Teachers may work on or off site with students or may develop partnerships with the local horticulture industry to provide training. A horticultural environment is used to teach basic job skills as well as skills specific to that industry.

Sheltered Workshops

Sheltered workshops provide an opportunity for productive work to individuals with a need for significant supervision and support. The work setting is in a school or training center, not in industry. Wages are based on a percentage of the prevailing wage for similar work in industry and are calculated according to individual productivity. Workers typically earn less than minimum wage. In the United States, a number of sheltered workshops serving people with developmental disabilities operate nurseries or greenhouse businesses.

Correctional Facilities

Prisons have effectively used horticulture as a work-training medium for many years in the United States. Programs may focus on education, therapy, or production, all using work as a primary activity.

Special education students learn basic work skills through indoor and outdoor gardening tasks. (Photo by Andris K. Walter)

Industry

Supported employment services for people with disabilities can be centered within places of employment in horticulture. A horticultural therapist may supervise a crew of workers with disabilities employed in a greenhouse or nursery facility, for example. A job coach may train and/or support an individual to work in a retail floral shop. Job placement of people with disabilities in the horticulture industry may also be provided by a horticultural therapy program.

Service Recipients

People with Developmental Disabilities

People with developmental disabilities are the most frequently served population by vocational horticultural therapy programs. Programs include vocational training centers that utilize horticulture, special education classes, and job placement programs.

Injured Workers

Horticultural therapy may be used to rehabilitate or retrain workers who have become disabled, either within or following treatment in a rehabilitation hospital. Cooperative efforts with rehabilitation agencies can enhance efforts.

Legal Offenders

Incarcerated individuals, juvenile offenders, and those on work parole have participated in many types of work programs that use horticulture.

People with Psychosocial Disorders

People with chronic and acute mental illness have benefitted from work training and/or placement in horticulture. Services occur in training facilities, industry, or mental health hospitals.

People with Traumatic Brain Injuries and Spinal Cord Injuries

Services for people who have had traumatic brain or spinal cord injuries may include sheltered employment, training, support, or placement. An initial physical rehabilitation process usually precedes vocational horticultural therapy programming.

Socially/Economically Disadvantaged

In programs that focus on youth who are socially and/or economically disadvantaged, horticulture is used for summer job training, "school-to-work" transitional programs, or community-supported agriculture programs.

Funding

Results of a 1994 survey of AHTA members (Laminack and Haller, 1995) indicate that operational budgets of programs reported as having a vocational component averaged $70,495 annually. (Note that the assumption of the authors that operational budgets included salaries may not have

A child with a visual impairment is encouraged to explore the garden with minimal support from a horticultural therapist. (Photo by Andris K. Walter)

been made by all respondents. If salaries were not included in all of the figures reported, the total average budgets seen here may be lower than in reality.) The largest source of funding was self-earned income (from sales of products and services). Thirty-seven percent of annual operating budgets came from this source. Likewise, correctional programs, which tend to be vocationally oriented, reported that an average of 35 percent of the operating budget was self-earned. Because of the production emphasis of vocational programs, self-earned income can be a significant source of funding.

Benefits to Clientele

The use of horticulture in vocational programs with diverse populations and settings has many benefits. For individuals it offers a chance to complete tasks geared to their capabilities. The varied and adaptable nature of horticulture allows for successful performance as well as challenge. Work

---

AHTA Vocational Programs

Horticulture Hiring People with Disabilities (HHPD), Rural Horticulture Employment Initiative (RHEI), and Enterprise Community Horticulture Employment Initiative (ECHEI) are service projects of the American Horticultural Therapy Association. They are partially funded through federal grants under the Projects with Industry (PWI) program of the U.S. Department of Education, Rehabilitation Services Administration.

The HHPD project was established in October 1982 to provide full-time permanent employment for workers with disabilities in the horticulture industry. It has been funded through a variety of resources since that time. The RHEI program was established in 1992 to extend the effort to rural areas, and the ECHEI project was established in 1995 to carry this initiative into selected Enterprise Communities.

Although the horticulture industry offers numerous and varied job opportunities, workers with disabilities have not been recruited for these jobs in significant numbers. It is the purpose of HHPD/RHEI/ECHEI to

A. enhance, expand, and increase private and public employment opportunities for workers with disabilities in the horticulture industry.

B. increase the interest, involvement, and participation of horticulture businesses and their representatives in the training and employment of workers with mental and physical disabilities.

---

with indoor plants, flowers, landscaping, and crop production produces tangible results, easily seen by the participants as well as valued by society for their beauty and utility. Pride and satisfaction in accomplishments and enhanced self-esteem result. Landscape maintenance and retail nursery work facilitate integration into the community. Such a large and diverse industry as horticulture offers choices and opportunities to find appropriate work for the trainee. Unlike the static nature of many vocational training centers or sheltered employment settings, horticulture offers a seasonally changing environment in which to work on cognitive and functional skills. The fact that the work is "real"—in that plants require care—seems to motivate participants to do what it takes. This may lead to improved skills such as problem solving, attention to task, or ambulation, to name a few.

Treatment Teams

In vocational programs, treatment teams generally include the horticultural therapist or trainer, job coach, vocational counselor, job placement specialist, and case manager. The adult individual receiving services (referred to as "client," "trainee," or "consumer") is usually also part of the

The plant-filled spaces created by horticultural therapy programs can be a showcase for an institution. (Photo by Rebecca Haller)

team. In addition, residential staff, occupational therapists, family members, medical personnel, mental health counselors, employers, and others may be included as needed. This interdisciplinary team of professionals and advocates cooperatively establishs goals, objectives, methods, and evaluation of services with each individual in the program.

Treatment Goals

Treatment goals in vocational programs in horticultural therapy focus on employment. Goals are individualized and developed by the treatment team to reflect the client's needs, desires, and abilities for an appropriate level of employment. Long-range goals describe the desired outcome or placement, while short-range goals and objectives are developed to specify particular actions or behaviors necessary to achieve the long-range goals. Let's look at some examples.

For Dan, a twenty-two-year-old man with Down's syndrome, current placement in a sheltered workshop or vocational training center is considered by the treatment team to be appropriate for the immediate future. His long-range goal is to be competitively employed in a production greenhouse. Because he is physically fit, has well-developed social skills, and is motivated, it is expected that he will be able to work successfully without the need for professional support. He will, however, most likely need support with finances, housing, and other activities of daily living.

In order to reach his long-range goal, objectives for Dan might include those related to specific greenhouse tasks and general work performance. For example, Dan may have an objective to "correctly pot up rooted cuttings," "place drip irrigation emitters in pots," or "add wires to hanging baskets." Behavioral objectives might include "ask for new assignment," "continue working in distracting environment," or "put tools away when finished." These objectives are based on formal evaluation and observation of Dan in a work setting, with priority needs for training identified.

In another scenario, Dan might be placed immediately in the desired eventual job setting with a job coach to support the skill and behavioral development necessary for independence. The job coach's level of support is gradually reduced as Dan takes on more and more responsibility and initiative to complete the work assigned. In either case, Dan's interests, aptitudes, and challenges are taken into account at each step.

Julie, a high school student who is blind, contemplates a job after graduation "working with plants." Placed in a summer program of vocational training in horticulture, she works with a job coach and other trainees to gain experience in the field. For her, the eventual long-range goal is to find meaningful employment after high school graduation. Her goals while in the summer training program are to try out different tasks in horticulture, learn more about the different job possibilities, gain specific plant-related skills, and improve general work habits and skills. Specific objectives identified by her trainer are to "ask for clarification of instructions when necessary" and to "establish the physical scope and boundaries of assigned tasks prior to beginning to work." Both of these objectives are related to a broader need to take initiative and to compensate for blindness in effective ways.

Additional Goals

In addition to individualized treatment goals, vocational horticultural therapy programs also have programmatic goals that might include serving a specified number of individuals annually, finding job placements, or providing a realistic work setting based on industry methods.

Vocational horticulture programs may be asked to meet goals for the organization of which they are a part. For example, the training program may be expected to earn revenue through sales, to maximize opportunities for community integration, to provide plants for landscaping, to grow food, or to be a showcase for tours and public relations efforts.

Raised beds at Denver Botanic Gardens are accessible and user-friendly because they bring the soil level up to a comfortable height. (Photo by Rebecca Haller)

With these diverse goals, striking a balance between meeting the needs of the client, the organization, and the program itself is important to the success of a vocational horticultural therapy program. This can be a challenge. The wise horticultural therapist uses production demands to enhance training effectiveness by emphasizing the "value" of the work and utilizing efficient equipment and techniques as found in industry. If food production is expected, finding charitable ways to give away the excess may motivate trainees to higher output. Whatever solutions are found, the balance is essential and often requires adjustment and fluidity on the part of the therapist.

THERAPEUTIC PROGRAMS

Conceptual Basis for Treatment

A therapeutic program in horticultural therapy is designed to assist individuals in recovery from illness or injury. Reentry into the mainstream of society (in work, place of residence, and leisure activity), to whatever extent possible, is the ultimate aim for those served. Professional services may include physical exercise, cognitive development, counseling, mental health services, communication skills development, and adaptations of structures, tools, and/or techniques.

The medical basis for treatment in therapeutic programs focuses on disease and its remediation or cure. To be therapeutic is to be curative or to serve to preserve health. Broad application of this model is found in horticultural therapy programs.

Settings

Rehabilitation Hospitals

Therapeutic horticulture programs are found in rehabilitation hospitals nationwide. They often include both indoor and outdoor horticulture facilities for year-round programming. In some cases, horticultural therapy is brought to the patient who is unable to leave his or her room via a rolling cart containing pots, soil, tools, and even grow-light units. The focus is on rehabilitation, following an injury, illness, or addiction.

Psychiatric Hospitals

The first sites of horticultural therapy programs in the United States were mental health institutions. Horticultural therapy is typically incorporated into broader activity-therapy departments and patients receive treatment in groups both on an inpatient or outpatient status. As health care costs have risen, the average length of hospitalization in these facilities has dramatically declined in recent years, impacting the nature of horticultural therapy programming offered. Treatment services for people with mental illness are now available in community mental health centers, where outpatient therapy is the standard fare (Punwar, 1994.)

Long-Term Residential Care Facilities

Serving primarily older persons needing skilled nursing care, horticultural therapy programs are frequently found in long-term care facilities. Physical and mental health of residents is promoted and maintained through gardening and plant-related activity therapies. Facilities typically include outdoor gardens with raised beds and containers and indoor windowsill or artificial light gardens. As in rehabilitation hospitals, horticultural therapy services may be taken to the residents in their rooms when appropriate.

Service Recipients

Within these various medically focused settings, the range of populations served by horticultural therapy programs is broad.

People Who Have Had Spinal Cord Injuries

Possibly the first population that comes to mind when one thinks of utilizing horticultural therapy for rehabilitation, people who have had spinal cord injuries often are assisted by the horticultural therapist in learning new ways of gardening independently.

People Who Have Experienced Traumatic Brain Injuries

Usually seen by horticultural therapists as patients in rehabilitation hospitals, people with traumatic brain injuries may alternatively receive services as residents of long-term care facilities or other rehabilitation centers.

People Who Have Had Orthopedic Injuries

Horticultural therapists in rehabilitation settings provide activities that strengthen and promote use of injured joints and provide modified tools and gardens to enable participation by patients.

Aging Adults

Older adults who have medical diagnoses that require treatment and/or long-term care are served by horticultural therapy programs with a therapeutic emphasis.

Stroke Survivors

Again, usually served by horticultural therapists in rehabilitation hospitals or long-term care, people who have had strokes are helped to regain physical and cognitive functioning and/or psychological health through participating in horticultural therapy activities.

People with Chronic or Terminal Illness

Horticultural therapists may also work in hospitals, rehabilitation centers, or long-term care facilities with people who are chronically or terminally ill. Patients or residents may have diseases such as cancer, AIDS, and multiple sclerosis.

People with Mental Illness

Mental illness, as previously stated, was one of the first medical concerns to be addressed by horticultural therapy programs. It is still widely used to help these populations today.

Funding

According to the horticultural therapy funding survey described earlier (Laminack and Haller, 1995), physical rehabilitation programs among AHTA members had average annual budgets of $28,829, with 39 percent of that budget coming from grants. Psychiatric programs averaged $38,529, with the largest source of income coming from self-earned sources (29 percent). Sales of the products of horticultural therapy programs are important for those with a therapeutic as well as vocational focus. With increased concern for cutting costs within health care, horticultural therapy programs are in the unique position of being able to generate revenue in this manner.

Benefits to Clientele

In therapeutic programs, horticulture provides benefits to participants through physical exercise, social interaction, and opportunities to relearn skills such as sequencing, memory, and problem solving. The activities are geared to build self-confidence, social skills, and self-esteem. Perhaps most importantly, horticultural therapy provides a location and attitude where the patient can feel well. The plant-filled horticultural therapy setting is dramatically different than the rest of the hospital, with smells, appearance, and ambiance that are more natural and relaxing. In a medical facility, the focus on illness and its treatment is pervasive. For patients, horticultural therapy can give respite from this focus and room for renewed hope.

Treatment Teams

Because of the diversity of horticultural therapy settings within the therapeutic model, the composition of treatment teams are equally varied.

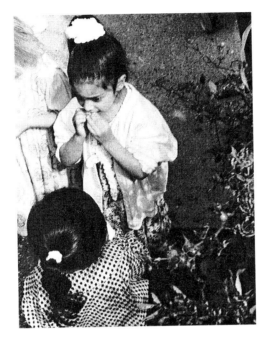

Plant materials must be carefully chosen to avoid toxicity, but tasting can be an important feature of a program. (Photo by Andris K. Walter)

In a physical rehabilitation setting, a team may consist of physical therapists, occupational therapists, nurses, therapeutic recreation specialists, speech therapists, horticultural therapists, counselors, and others. In a psychiatric setting, the horticultural therapist may be part of a treatment team that includes a psychiatrist and/or psychologist, mental health counselor, activity therapist, and others. In long-term care, medical personnel such as nurses may work closely with the horticultural therapist in developing and implementing individual client/patient programs.

Treatment Goals

To identify goals addressed by horticultural therapy programming with a therapeutic emphasis, it is again useful to look at the three settings of service: rehabilitation, psychiatric, and long-term care facilities. In general, a long-range goal is to regain or maintain functioning following an accident or illness.

In rehabilitation programs, long-range goals are based on the expected eventual placement or level of recovery. The therapist needs to have answers to these questions: Is the individual expected to return home? With what level of support, if any? Will the individual require a modified environment or tools for work or home? Short-range goals or objectives address the steps needed to reach long-range goals. Let's look at Lisa, for example, who is a sixty-year-old who has had a stroke. With an expectation that she will return to her own home, Lisa wishes to continue to garden outdoors independently. Earlier in her rehabilitation program, she may work on stamina or maneuvering her new wheelchair, or on regaining hope for the future during horticultural therapy sessions. Because she now has functional use of her right side only and uses a motorized wheelchair, she receives help from the horticultural therapist to locate new, and adapt existing, hand tools that enable her to complete gardening tasks with one hand. She will be able to use the same movements practiced in the horticulture environment to feed herself independently. Clumsiness and spills are much less embarrassing when the medium is soil rather than food, so the practice is less stressful and grueling for her. The horticultural therapist also makes recommendations to Lisa for adapting her own garden at home. By continuing to garden at home, she gains an avenue for reintegration into her community. She will be outdoors with more chance for interaction with neighbors and to regain self confidence in her ability to engage in a satisfying activity.

In psychiatric programs, types of therapies are also quite varied. Short-range objectives may include increasing frustration tolerance, interacting appropriately in a group activity, strengthening assertiveness, dealing with grief or addictions, focusing on beauty, managing stress or coping with activities of daily living, or reducing hallucinations and delusions, to mention a few. All of these, and many more, positive changes can be supported through the use of horticultural therapy. In many mental health settings the horticultural therapist uses the plant medium and environment as a tool to overcome the resistance or fear of the patient and to provide a "grounding" connection to reality and life cycles. Stress management and relaxation are keys to successful programming. To cope with, manage or heal the illness is the general object.

In long-term care facilities the desired outcome of the horticultural therapy program is usually to maintain functioning levels of participants, although improvements in functioning levels are also possible. An elderly nursing home resident's quality of life and overall health may improve through consistent participation in outdoor gardening activities with a group horticultural therapy program. Compared with days of inactivity and isolation, the required exercise of mind, body and social skills can have positive results. In a long-term care facility residents must relinquish control over many aspects of their lives, including choice of meals, schedules, and decor and space. Horticultural therapy activities can offer residents a chance to actively participate in choices again. For instance, they may be given plants or sections of the garden to call their own with some options for what and when to plant. Objectives for residents might include "choosing plant materials to use," "participating in activities," "standing with the support of a raised bed," "using both hands," "initiating conversation," "identifying flower names," or "following instructions." The wealth of experience of the elderly resident is considered and incorporated into activities by the therapist.

Special Considerations

Concerns about pathogens and allergens in horticultural therapy programs may be raised in medical facilities. In some programs, therapists use rock wool or other artificial growing media to avoid dusts and soil particles. Patients may need to be draped with waterproof aprons to protect them from exposure to soils during activities. Flowers that produce pollen that commonly cause allergies are usually avoided. For safety reasons, when working with individuals who may be disoriented or confused (e.g., those with Alzheimer's, dementia, mental illness) plants that are toxic to eat or touch are prohibited and sharp tools are kept locked away and used only under close supervision.

Although not usually as big a focus in therapeutic programs as in vocational ones, there may be a need to balance production demands with the needs for individual treatment. This is reflected in the statistic that an average of 29 percent of the annual budget for psychiatric programs comes from self-earned income. Some level of production must be maintained to achieve this income.

SOCIAL PROGRAMS

Conceptual Basis for Treatment

Horticultural therapy programs with a social emphasis serve to improve the general well-being of participants, often through the use of gardening as a leisure activity. Horticulture is seen as a valuable recreational pursuit. Interaction with others is encouraged as part of a process to increase life satisfaction and general well-being (Rothert and Daubert, 1981).

Based primarily on the wellness model, social programs are more concerned with the growth of the whole person than with treatment of a specific disease or injury (Austin, 1991). Activities that promote health of body, mind, and spirit are offered by the therapist.

Settings

Although social programs that use horticultural therapy are found in both public and private facilities, they are more likely to be public than are vocational or therapeutic horticultural therapy programs.

Retirement Homes

With a growing older population, retirement living centers are a burgeoning business. Many encourage gardening activities in common areas, on balconies, patios, and indoors. Garden clubs may be organized under the direction of an activities coordinator to bring residents with similar interests together.

Community Gardens

Community gardens are found all across the country in cities both large and small, reflecting a desire (and perhaps need) to cultivate and grow. When horticultural therapists are involved, they usually work with disadvantaged populations to improve social integration, personal satisfaction, mental health, and overall well-being. Additionally, they may design community gardens to enable seniors to more fully participate. Intergenerational programs that pair elders with youth engaged in gardening offer mutual health benefits to age groups that are increasingly isolated from each other in modern society.

Residential Group Homes

Group homes and apartments that house people with developmental disabilities, mental illness, or other disabilities also offer gardening opportunities for their residents. In addition to providing exercise, the possibility of improved nutrition, and outlets for stress, well-tended gardens can help the home and its residents be more accepted in a neighborhood. Gardens not only enhance the appearance of a property, but also can be an easy, nonthreatening topic of conversations with neighbors. What a great way to build bridges!

Community gardens can be productive sites for social integration and personal satisfaction. (Photo by Andris K. Walter)

Senior Centers

Senior Centers may offer plant-care classes, garden clubs, and/or gardening space. Community programs facilitated by a horticultural therapist usually have a social emphasis, encouraging seniors to lead healthy, active lives. Gardens provide seniors with nutritious food and help reduce grocery bills for those on fixed retirement incomes.

Service Recipients

Elderly

Socially oriented horticultural therapy programs serve older adults in retirement living centers, day-care, long-term care, or community gardens. Again, the purpose is to support general well-being through better nutrition, feelings of control over parts of their lives, and a sense of purpose.

Socially Disadvantaged

Community gardens in low-income or immigrant neighborhoods can be the site of horticultural therapy services that empower residents to belong to a community with pride and to grow healthy food. Improved nutrition and lower food bills result.

Homeless

The homeless person is also usually served by the horticultural therapist in a community garden setting, allowing him or her to regain a sense of place and control.

People with Disabilities

In the residential settings just described, horticultural activity can be an achievable healthy leisure alternative to a sedentary lifestyle. Neighborhood acceptance of the residents may be enhanced by seeing them engaged in this "normal" activity and by the more beautiful property that results.

Funding Sources

From the 1994 survey of horticultural therapy funding (Laminack and Haller, 1995), social programs range from an average annual budget of $21,994 for recreational programs to $71,248 for programs that include community development. (Again, note that budget figures are thought to include professional salaries.) Grants provided 33 percent of the funding for recreational programs and 30 percent for community development programs. The largest source of income for community development was fees for services (39 percent). Fees for services may include charges for community garden plots or activities in retirement centers, for example.

Benefits to Clientele

As previously stated, participants in social horticultural therapy programs can benefit in a number of ways, all enhancing their well-being. Exercise and fresh air, healthy food production, a chance to belong to a community, having an object of focus and a rewarding hobby, reducing stress, and connecting with the land are a few of the gains possible.

Children with disabilities gain social skills and discover the wonders of plants through well-designed activities. (Photo by Andris K. Walter)

Composition of Treatment Teams

In social programs, horticultural therapists may work independently or with gardeners, community organizers, activity directors, or therapeutic recreation specialists. Unlike vocational or therapeutic programs, the approach is not clinical or treatment oriented. Volunteers often play a critical role in these types of programs. In fact, some are initiated and implemented entirely by volunteers from garden clubs or master gardener programs.

Treatment Goals

Goals for participants in these programs tend to be for the group, rather than for the individual. The therapist or volunteer leader may help clients to increase interaction with others, reduce stress, improve nutrition, develop a hobby, exercise, or cultivate a sense of belonging. Again, enhanced well-being is sought.

Special Considerations

Although social programs have been separately categorized here for point of discussion, they may be combined with therapeutic or vocational programs. For example, an activity department of a long-term care facility may develop a garden club for residents (social), while the horticultural therapist works with a stroke survivor on adaptations or physical rehabilitation (therapeutic) in the same facility. Outside garden clubs or other volunteers may be able to facilitate the garden club, but may lack the training and skills necessary to work in the therapeutic program. Of course, they can provide valuable assistance to the horticultural therapist in both settings!

CONCLUSION

Program Design in Actual Horticultural Therapy Practice

Many different people participate in horticultural therapy at all stages of life, conditions, and circumstances. Because programs are so diverse, each particular population can receive services in many different settings and within any of the three major program types that have been described.

Let's look at seniors to illustrate this point. Table 3.4 identifies settings and program types for horticultural therapy services for seniors. A similar range of settings and program types could be outlined for many other populations.

TABLE 3.4. Settings and Types of Horticultural Therapy Programs Serving Seniors

Settings	Program Type(s)
Retirement living	Social
Vocational program	Vocational
Physical rehabilitation hospital	Therapeutic
Community garden	Social
Adult day care	Social, Therapeutic
Mental health center	Therapeutic
Long-term care	Therapeutic, Social

Overlap of Program Types

In actual practice, a horticultural therapy program may include more than one program type. For example, within a mental health or rehabilitation hospital, a program could have many components—vocational, therapeutic, and social–with activities designed to meet specific patient goals. Services for each group may be structured to include only a therapeutic emphasis or to combine elements of vocational, therapeutic, and social programs in the treatment plans.

Prevalence of Program Types

Members of the American Horticultural Therapy Association are asked annually to identify the populations with whom they work and the types of programs used. In 1995, 294 members reported involvement with therapeutic programs, 154 with vocational programs, and 145 with social programs (see Table 3.5). (Other program categories that could be chosen were "administration," "education," "production," and "other." A total of 273 members included one or more of these categories.) Note that one or any number of program types could be selected. Out of 294 members with therapeutic programs, 175 also represented vocational and/or social programs, indicating the considerable overlap that exists.

Implications of Program Diversity

Education

The horticultural therapy profession includes programs and practitioners of all three types: vocational, therapeutic, and social. Partly because of these variations, existing literature about horticultural therapy can be confusing. Students may encounter articles that describe each of these program types on separate occasions and wonder which model is "true." It is important to know the conceptual differences and overlap of program designs.

TABLE 3.5. Prevalence of Program Types

Types of Programs	# of Members Working in Program Type
Therapeutic	294
Vocational	154
Social	145
Other (includes administration, education, production)	273

In order to be qualified to practice horticultural therapy in all three types of programs, students must understand the models upon which they are based. Program design, staffing, and goals emanate from them. The models discussed in this chapter are summarized in Table 3.1.

There are advantages and disadvantages to both specialized and general educational approaches. Educational programs that allow for specialization within one or more areas may better prepare the student for employment in a specific setting. However, with rapidly changing health care delivery and funding systems, comprehensive study that includes all three models gives the graduate more flexibility and opportunity to respond to these changes.

Purpose

To be an effective horticultural therapist, not only must the professional know and use the components of the program types described in this chapter, but also employ good therapeutic techniques. A client-centered approach that respects the desires and individuality of each person served is essential. In practice, this may mean asking opinions, tranferring control to the client, encouraging creativity, allowing risks and mistakes, celebrating successes, listening actively, and understanding the whole person. Those served by a therapist who employ these practices are often motivated and empowered to better their lives.

It is important to remember that horticultural therapy exists as a profession generally to improve human well-being. It emphasizes the process over the outcome or product and truly celebrates the person served. That is the essence of all programming in horticultural therapy.

BIBLIOGRAPHY

Mattson, R. and Shoemaker, J. 1982. *Defining horticulture as a therapeutic modality. Part 2: Models in horticultural therapy.* Manhattan, KS: Kansas State University.

Olszowy, D. 1978. Horticulture for the disabled and disadvantaged. Springfield, IL: Charles C Thomas.

U.S. Department of Health and Human Services. 1980. *The Melwood manual: A planning and operations manual for horticultural training and work co-op programs.* Washington, DC: U.S. Dept. of Health and Human Services.

REFERENCES

Austin, D. 1991. *Therapeutic recreation: Processes and techniques.* Champain, IL: Sagamore Publishing.

Laminack, J. and Haller, R. 1995. Horticultural therapy program funding survey. Proceedings of twenty-third annual conference. Gaithersburg, MD: American Horticultural Therapy Association.

Punwar, A. 1994. *Occupational Therapy: Principles and Practice*. Baltimore, MD: Williams and Wilkins.

Rothert, E. and Daubert, J. 1981. *Horticultural therapy for senior centers, nursing homes, retirement living*. Glencoe, IL: Chicago Horticultural Society.

PART TWO:
SPECIAL POPULATIONS
FOR HORTICULTURAL THERAPY
PRACTICE

Chapter 4

Stroke, Spinal Cord, and Physical Disabilities and Horticultural Therapy Practice

Matthew Wichrowski
Nancy K. Chambers
Linda M. Ciccantelli

INTRODUCTION

Statement of Challenges and Issues

People must continually adapt to changes in their lives. Changes such as accidents or disease that cause disabilities may be the most challenging for individuals to confront. Nearly all of us will have to face a disability at least once in our lives. The potential for adaptation and healing within the human condition has led to the practice of physical medicine and rehabilitation (PM&R). This discipline uses therapeutic activities to maximize the potential of an individual's personal and societal capacities by improving functional abilities.

Therapists who assist disabled individuals toward maximizing their potential are faced with many issues and challenges:

- *Attitudes.* Therapists must examine their own attitudes toward disabilities, become aware of how they affect the therapeutic interaction, and adopt healthy attitudes in creating a positive therapeutic milieu for the patient.
- *Attributes.* Personal qualities that are helpful in achieving this end are: empathy or sensitivity to others needs; nonjudgmental acceptance of others; and a desire to aid an individual in adjusting to a new

and very challenging set of circumstances in their lives by helping and supporting maximum functioning while allowing the individual to take risks to gain independence.

- *Diversity.* Due to the diversity of physical disabilities, their manifestations, and their effects on human functioning, the therapist encounters complex assessment issues, a range of goals, numerous treatment modalities, and medical precautions to learn, such as cardiac patients not reaching above shoulder level to take cuttings from hanging baskets, or extra care taken with diabetic patients when working with scissors or thorny plants. The most important attribute the therapist needs to be effective in working with such diversity is flexibility and good problem-solving skills.

Purpose and Objective of Chapter

The purpose of this chapter is to introduce the reader to the practice of horticultural therapy as applied to the field of physical medicine and rehabilitation (PM&R). The chapter will do the following:

1. Introduce the range of disabilities encountered with their etiology, incidence, and physical, social, cognitive, and emotional manifestations.
2. Introduce the rehabilitation team, which performs assessments and sets appropriate goals.
3. Introduce treatment issues facing the therapist in implementing a horticultural therapy program.
4. Discuss adaptation of horticultural activities to meet clients' treatment needs.
5. Address specific assessment and documentation tools specific for physical disability programs.

Significance

Many issues make physical medicine and rehabilitation—the third phase of medicine after prevention and acute care—an important area of study and training for the horticultural therapist. Despite the shortening of hospital stays, the field of PM&R itself is growing. Our aging population means that people are living longer but not necessarily healthier. There is an increased ability to save lives at birth and from accidents and disease. Often this leaves major changes in a person and the need for rehabilitation services, which horticultural therapy can address. Horticultural therapy is very versatile, can meet many needs in a physical rehabilitation setting, and integrates easily into the team approach to meet different patient, family, and medical goals.

THE CLIENT POPULATION

Physical disabilities affect people regardless of age, gender, culture, or economic status. Genetic makeup, the environment, and interaction between the two all can affect the occurrence of a disabling condition. Trauma before, during, or right after birth can cause conditions in children that will endure for their entire life. Injuries caused by accidents involving cars, sports, industrial machines, falls, or weapons usually occur in the young to middle-aged population, and cause trauma such as spinal cord injuries, brain injuries, fractures, nerve injuries, and amputations (Spencer, 1983). Neurological conditions such as stroke, Guillain-Barré Syndrome, and polio affect some individuals. Progressive diseases such as multiple sclerosis and Parkinson's disease begin to affect people in the middle of their lives. Older adults tend to be more prone to lower extremity amputation as a result of poor circulation and fractures from falls, and are more likely to have a cerebral vascular accident (stroke) or cardiac problem.

In most cases, patients in a rehabilitation setting have been referred directly from an acute-care hospital after being stabilized from the cause for admittance. Causes could include orthopedic surgery, stroke, spinal cord injury, amputation, cardiac surgery, or brain or spinal cord tumor surgery. Although the underlying purpose for referral to the PM&R division may not be related to an individual's premorbid status, such as Parkinson's disease or spina bifida, these preexisting conditions become important factors during treatment.

This section presents disabilities/diagnoses frequently encountered in a rehabilitation setting. It discusses physical, psychological, and social issues and some of the effects these disabilities may have on the individual.

Parkinson's Disease

Parkinson's disease is a slowly progressing neurological condition caused by degeneration of neurons in the substanta nigra and globus pallidus resulting in damage to the basal ganglia of the brain. This may be caused by carbon monoxide and manganese poisoning, encephalitis, senile brain changes, and arteriosclerosis. It usually occurs between the ages of forty and eighty years, and has marked effects on one's lifestyle (Spencer, 1983).

Physical symptoms of Parkinson's disease include a gradual increase of muscular rigidity and weakness, slowing of response in voluntary movements, and characteristic shuffling gait. General reduction of range of motion and tremors reduce fine motor skills. Tremors may shift from one muscle group to another and increase with emotional excitement, yet can be inhibited temporarily by conscious effort. There may be periods of nonprogression of the symptoms. Reduction of motor control also affects vocal muscles, which causes a slurring of speech, decreased volume, and monotone sound.

Psychological effects usually center around loss of functional ability. The outlook for the future may seem hopeless, which may affect motivation toward therapy and life in general; depression and withdrawal may occur. Such symptoms have obvious effects on social abilities, and there is a tendency to become isolated.

Parkinson's disease in itself is not a cause for admission to a PM&R division, which generally is intended for short-term rehabilitation before discharge home. Rather it becomes a confounding condition in a patient who might have had a stroke but who also suffers from Parkinson's disease.

Multiple Sclerosis

Multiple sclerosis (MS) is a progressive disease of the nervous system and is of an unknown origin. It is characterized by a slow decomposition of the integrity of the myelin sheath insulating the nerve fibers. Their compromised sheath short circuits nerve signals and results in fatigue. The degenerated sheath is replaced by plaques which affect the white matter of the brain and spinal cord (Spencer, 1983). The disease is characterized by intermittent exacerbations and remissions and runs an unpredictable though unrelenting course.

Physical symptoms depend on the location of the plaques and can include motor weakness, sensory loss, visual difficulties, tremors, and ataxia. Progression of the disease causes loss of range of motion, paraparesis, and eventually paraplegia.

The decline in cognitive function, memory, and abstract reasoning mirrors the decline in physical function (Fink and Houser; Spencer 1983). Attention span and judgment may also be affected.

Social skills decline along with physical abilities. Disturbances of vision make it difficult to focus and give others eye contact. Anxiety, depression, and irritability effect motivation to participate in social activities. Dysarthria may affect the ability to speak and in severe cases may make speech unintelligible.

Patient working on plant propagation activity with a volunteer. (Glass Garden, Rusk Institute, New York University Medical Center)

MS has many psychological implications as well. Individuals with MS may have mood swings spanning from euphoria to depression in quick fluctuations. Low self-esteem and lack of motivation toward therapy may affect the rehabilitation process.

Cardiac Rehabilitation

Cardiac rehabilitation is used after an individual suffers a heart attack or myocardial infarction. Chances of a myocardial infarction (MI) occurring increase with age, improper diet, degree of obesity, and type-A behavior. Most commonly caused by a clogging of the arteries (arteriosclerosis), an MI occurs when the blood supply is cut off from a portion of the heart muscle, killing those cells. After a heart attack, pumping efficiency is reduced and the individual must make adjustments to their lifestyle and rebuild their strength and endurance.

A heart attack has many psychological effects on the individual. Depression may be caused due to tension and anxiety related to changes in roles and lifestyle. There are fears concerning death, reoccurrence, and lost abilities. Self-esteem may be affected by lack of motivation created by these emotional issues (Trombly, 1983). Since many social activities also require a degree of physical stress (e.g., sports, hiking, exercising in health clubs), the individual who has had a heart attack may have to redefine his or her social role and choice of activities. The change in activity potential and fear of a reoccurrence can lead the individual to withdraw from social situations. Withdrawal can lead to depression or lack of motivation in rehabilitative efforts or health maintenance upon the return home.

Amputations and Orthopedic Rehabilitation

Amputations often occur as a result of accidents usually involving automobiles or other heavy machinery. Later in life, amputations may be the result of poor circulation to extremities, as is the case with diabetes. Most amputations (70 percent) involve the lower extremities.

A cardiac rehabilitation patient learning energy conservation techniques during plant propagation activity. (Glass Garden, Rusk Institute, New York University Medical Center)

The extent of the amputation defines the rehabilitative issues and effects on the individual's lifestyle. The use of orthotic devices helps to restore some of the functional use of the affected area, but weakness in function is an unfortunate result. This causes a redefinition of abilities and strategic use of prosthetics to keep functional losses at a minimum.

Psychological issues have a profound affect on the individual and center around the loss of a body part. The individual may not feel like a whole person. These feelings can impact on a person's self-image, self-esteem, and the ability to return to activities enjoyed before the loss. Readjustment to life afterward can be very difficult. Individuals with prosthetic devices may feel self-conscious in group situations. There is a risk of social withdrawal and depression, which in turn may affect their motivation toward rehabilitation and readjustment to a fulfilling lifestyle.

Individuals who lose a limb due to complications of diabetes or circulatory disorders have, in addition, a larger array of medical issues including skin breakdown, which in turn interferes with their ability to use a prosthesis.

Orthopedic rehabilitation includes those individuals, generally elderly, who have had hip or knee fractures or replacements due to falls or accidents. These patients experience impeded functioning after surgery, fear of recurrence, pain, stiffness, and medical precautions that slow their recovery. Hospital rehabilitation is short term and includes education on weight-bearing and hip precautions, safety, and adaptive techniques. Patients also must increase their endurance, strength, and range of motion in order to rebuild affected joints. These conditions do not usually affect cognitive functions or the ability to socialize, aside from the loss of opportunity associated with limitations in mobility, but care must be taken to avoid withdrawal, isolation, and possible depression, which would interfere with motivation toward rehabilitation.

Spinal Cord Injury

Spinal cord injuries occur when the spinal cord is traumatized or severed, usually in an accident, assault, fall, or by tumors, disease, or congenital abnormalities. Roughly 10,000 spinal cord injuries occur annually (Trombly, 1983). Rehabilitation for spinal cord injuries can last many months and have a vast impact in all areas of an individual's life.

Because of this immense impact, there are many psychological variables associated with spinal cord injury. Loss of mobility, extended length of rehabilitation, and permanent loss of function cause anxiety, fear, and depression. Hopelessness and lack of motivation can have strong effects on an individual's attitude toward rehabilitation and future quality of life.

Persons with a spinal cord injury will generally experience weakness or loss of function in all areas innervated by the cord at and below the point of injury. This creates a variable course of treatment depending on location of the lesion (see Figure 4.1).

Cerebrovascular Accident

A cerebrovascular accident (CVA) or stroke are terms used to describe a condition where there is an interruption of blood flow to a portion of the brain. Blood flow can be interrupted because of a blockage or occlusion. A thrombosis is a clot that originates at a certain point and prevents oxygenated blood from proceeding further down the artery. An embolism is a clot that originates somewhere else in the body, breaks loose, and travels to a point where it gets stuck and prevents blood from flowing further. A hemorrhagic stroke occurs when a blood vessel pops, typically due to an aneurysm. An aneurysm occurs when there is a weakness in a portion of a blood vessel. Pressure in the vessel causes part of the vessel wall to balloon out. Eventually the ballooned portion becomes so weak that it pops, leaking blood into the brain cavity instead of supplying it to brain cells.

Strokes typically affect between 170 and 190 per 100,000 individuals, with probabilities increasing with age. Roughly 500,000 people a year have strokes in America (Bond-Howard, 1993). Contributing factors may include hypertension, arteriosclerosis, and congenital vascular weakness.

When blood flow is interrupted to a certain area of the brain, the cells in that area die and affect the performance of the activities controlled by that area. One consequence of this is paralysis (hemiplegia), or weakness (hemiparesis) on the opposite side of the body from which the damage in the brain has occurred.

As well as right and left CVAs, stroke affecting the brainstem occurs in some individuals. This type of stroke is usually more serious because of the importance of the functions governed by the brainstem. Control of many involuntary life-sustaining functions are carried out in this area. Occasionally, mild symptoms of a stroke may occur and later go away. These transient ischemic attacks (TIAs) are caused by temporary vascular insufficiency and often indicate that a more serious CVA may occur in the future. Other causes of hemiplegia include traumatic brain injury, heart attack, and tumors of the head or spinal cord. Symptoms from these causes are often very similar to those caused by a CVA, and similar rehabilitative methods and goals can be used.

Due to the vast complexity of human nature and notably the human brain, the consequences of a CVA on the functional ability of an individual

FIGURE 4.1. Functional Ability Associated with Level of Spinal Cord Injury

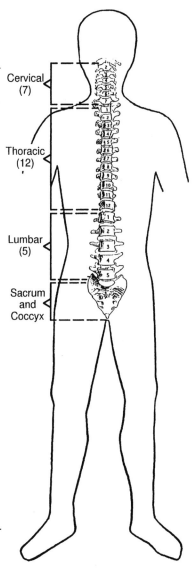

C_{1-3} Limited neck control; dependent on respirator; mobility limited to power wheelchair with chin or breath controls; environmental systems, computers, page turners managed with mouth stick, eye, or voice control.

C_4 Neck control; scapular elevation; weakened respiration; power wheelchair mobility with chin control; limited self-feeding possible with assistive devices, mouth-stick activities, page turner, computer, environmental controls.

C_5 Deltoids; fair to good shoulders; biceps; elbow flexion; manual wheelchair with quad tips for short, smooth surfaces, electric wheelchair for outside or long distances; needs assistance with transfers; may be able to do weight shifts; self-feeding, writing, telephone, computer with splints or assistive devices.

C_6 Dorsiflexion; wrist extension; able to propel wheelchair with quad tips, may need electric wheelchair for long distances, sliding board transfers with or without assist; able to feed self with assistive devices; bathe and dress upper extremities; may need assistance with lower extremities; may drive with hand controls; able to write with splints.

C_7 Wheelchair independent; limited thumb extension and abduction; limited grasp, release, and dexterity; independent with transfers, feeding, bathing, dressing, bowel and bladder care, and skin inspection; can sit up, move in sitting position, and roll over.

T_{1-2} All muscles of upper extremities fully innervated; independent with fine motor skills; weakness in trunk control.

T_{3-10} Intact upper extremities; partial to good trunk balance; independent wheelchair skills including ramps and curbs; independent in all transfers; able to drive with hand controls; better respiration and endurance.

$T_{12}-L_2$ Hip flexors; able to ambulate short distances with braces and crutches.

L_{2-4} Hip abductors; total ambulation with short leg braces with or without crutches or canes.

L_5-S_3 Movement of ankles and toes; total ambulation without assistive devices.

Cervical (7)

Thoracic (12)

Lumbar (5)

Sacrum and Coccyx

vary greatly. The effects of a lesion depend on the functional responsibilities of the part of the brain damaged. Control centers in the brain are usually interrelated, and changes in function affect many faculties. This dramatically increases the complexity of the rehabilitative process for stroke victims and presents a challenging case for the therapist.

A major physical consequence of a stroke is paralysis or weakness in the side opposite of the side of the brain affected by the CVA. Passive range of motion is complete initially, but gradually spasticity affects the muscles and contractures may occur, limiting motion. Some people begin to regain voluntary control in a few days while some never regain their premorbid abilities. Hypertonia of the shoulder muscles can cause a stretching of the tendons, resulting in subluxation of the shoulder.

While persons with a right CVA or left CVA may have very similar physical needs and therapeutic goals, the psychological effects are greatly varied and differ depending on the side affected and the specific area of the brain damaged by the stroke, e.g., depression is often a consequence of stroke. Even though deficits will vary between individuals, there are usually clusters of symptoms and disabilities that are similar among people with right or left CVAs (see Figures 4.2 and 4.3). This leads to similar methods and goals for each type of stroke.

A stroke can greatly disrupt one's lifestyle and produce many strong emotional reactions. Individual coping styles vary greatly from individual to individual and have a strong influence on rehabilitation and recovery.

There are many social implications for stroke survivors as well. Individuals with degrees of aphasia find it difficult to communicate with others, especially in group situations. These people feel isolated and often avoid social situations. Many of the usual social situations are confusing and cause anxiety. In general, a disability makes it difficult to get around and represents a loss of opportunity for social interaction, often leading to increased isolation.

HORTICULTURAL THERAPY AND TREATMENT

Therapeutic Environment/Rehabilitation Team

The field of physical medicine and rehabilitation (PM&R) as developed by Dr. Howard A. Rusk uses an integrated team approach in patient diagnosis, assessment, and treatment, and emphasizes the whole person, with emotional, psychological, and social as well as physical needs.

The PM&R team understands that physical trauma or the onset of chronic illness, may leave the patient with multiple disabilities disruptive

FIGURE 4.2. Effects of Right Cerebrovascular Accidents (CVA)

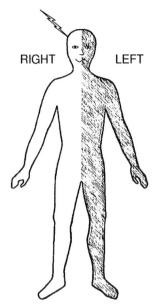

RIGHT CVA

Characteristics include the following:

- Left-side paralysis or weakness from head to toe
- Spatial/perceptual deficits
- Time/space deficits
- Left neglect/field cut
- Impulsive behavior
- Decreased attention
- Poor insight/judgment/reasoning
- Lability
- Denial

FIGURE 4.3. Effects of Left Cerebrovascular Accidents (CVA)

LEFT CVA

Characteristics include the following:

- Right-side paralysis or weakness from head to toe
- Aphasia/communication disability–receptive, expressive, or global
- Decreased attention
- Decreased verbal learning
- Memory deficits
- Reading/writing/math disabilities
- Anxiety, depression
- Lability
- Isolation/withdrawal

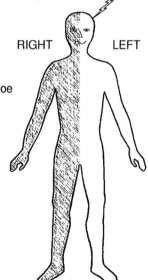

to family life and future plans, alter his body image, terminate work roles, and consequently reduce self-esteem, security, and independence. In addition, opportunities for social contacts are reduced, psychological integrity is threatened, and the patient may lose control over intimate physical functions (Versluys, 1983). Normal reactions to changes of this nature include anxiety, denial, depression, mourning, and regression (Versluys, 1983). These reactions, in addition to those of the individual's family and support group, greatly influence the motivation of the individual in treatment, and their effects should be considered in developing a plan of treatment.

The typical treatment team includes the physiatrist (physician whose specialty is PM&R), nurse, occupational therapist, physical therapist, social worker, dietician, psychologist, speech therapist, and vocation counselor.

Patients working with a horticultural therapist on dried flower arrangements. (Glass Garden, Rusk Institute, New York University Medical Center)

All prescriptions for treatment come from the physiatrist, and team members' roles within each facility are usually clearly delineated, though always focusing on promoting health and wellness, increasing independence, and improving the quality of life for individuals with disabilities by improving physical, cognitive, and social functioning (Table 4.1).

Treatment Issues

There are some treatment issues that must be understood when implementing a horticultural therapy program.

Group Programming

Horticultural therapists can treat people using both individual and group modalities. Individual therapy allows for one-on-one attention and is well suited for very physically or cognitively impaired individuals. Medically involved patients who are confined to their beds or homebound are also good candidates for individual horticultural therapy; however, recognizing that being in a garden has intrinsic values and benefits, the goals are generally focused on increasing the individual's ability to work in the garden with other people.

Horticultural therapy lends itself best to group programming. Working in a group provides a wider array of benefits for the participants, with a lessened need for paid staff. Help offered by assistants and volunteers can increase the amount of personal attention, thus allowing more individuals to participate safely and benefit from the program. Group activities allow for physical aspects of a treatment plan to be met while addressing psychosocial needs pertinent to recovery. Studies have shown that self-esteem is positively influenced by group participation and that there is a strong relationship between group work, patient self-concept, and response to treatment (Trombly, 1983; Versluys, 1983).

Group work including people with different disabilities, levels of function, and at varying stages of rehabilitation allow the individual to learn strategies from watching others. Seeing others at more advanced stages of recovery instills hope for the future and facilitates adjustment to one's condition. Support from the group provides confidence as well as a forum to develop social skills and assist in community reintegration.

TABLE 4.1. The Roles of Each Treatment Team Member

Member	Primary Focus
Nursing	Patient Training: • bathing and showering • bowel and bladder control • self-medication education • skin care/pressure sores • family education on home management
Occupational Therapy	• treatment for restoration of physical function (e.g., increase muscle strength and coordination, fine motor control, follow directions, focus) to allow activities of daily living (ADL), e.g., brushing teeth, self-grooming, dressing, eating, writing, returning to work or school • work capacity and work tolerance development • skills increase with adaptive equipment and prosthesis in ADL function • help with readjustment to home management routines, e.g., adapt equipment and work simplification (e.g. one-handed cutting boards) • prevocational training, work skills, and habits • barrier-free design consultants—home and workplace • electronic technical aids for home environmental controls and computer access systems
Social Work	• discharge planning • match with available social service networks • liaison with insurance, etc. • family counseling
Therapeutic Recreation	• improving quality of life and independent lifestyle • community excursions, practice in public transportation, pools, parks, theaters • leisure resources in the community, travel consultant • stress reduction through play and recreational activities • redirect avocational interest
Speech and Language Therapy	• remediation of communication disorders—aphasia, dysarthria, swallowing, recall, and memory
Physical Therapy	• gross motor movement, e.g., ambulation, standing balance • individual exercises to maximize strength, endurance, mobility, and balance • hydrotherapy • ambulation therapy (for stroke and amputations) • exercises for relaxation • vestibular problems—diagnosis and treatment

TABLE 4.1 (continued)

Driver Education	• automobile modifications • retraining
Psychology	Adjustment • evaluation of all patients because of complexity of problems with physical disabilities, including those associated with motor impairment, brain injury, and pain • stress-related issues • sexuality issues • cognitive retraining
Orthotics/ Prosthetics	• research into newer and lighter weights • biofeedback and electrical impulse functioning orthoses including splints, devices, and wheelchairs
Vocational Services	• evaluation • counseling • referrals • training • placement • determining maximum level of vocational activity and economic independence
Teachers	• regular school classes so children do not fall behind with extended hospital stays
Dietician	• education relating to special dietary needs
Pastoral Service	• counseling and service
Horticultural Therapist	• activities assisting in the development of fine motor skills, range of motion, hand-eye coordination, strength, balance, and perception • assistance in achieving OT goals • assistance in PT goals • activities that assist in the psychological adjustment and well-being of the patient • providing assistance to psychologist in cognitive retraining • opportunities for socialization in group • avocational pursuits for after discharge

Focus on Outcomes

Adaptation of horticultural activities should be aimed at enabling the individual throughout his or her stay at the rehabilitative facility as well as after they go home. Assistance should be provided to ensure successful completion of activities. Techniques should be taught in an effort to help compensate for disability and to help the individual to conserve energy. Raised beds and other barrier-free techniques can be used after discharge from the hospital to ensure positive gardening experiences in vocational and avocational pursuits.

Cultural Differences

Another issue concerns cultural awareness. Certain activities and plant materials are more desirable within various cultures and may serve to peak interest and improve motivation. Different plant materials may elicit various emotional responses both helpful and harmful to therapeutic gains.

Appropriate Technology

Another factor to consider in the choice of horticultural activities concerns equipment and material recommendations for post discharge. It is useless to recommend equipment or tools to people to assist in their gardening if they are not available to them or if they are unaffordable. It can be very frustrating to try to grow a plant under conditions that are not conducive to its growth. This sets the person up for failure and undermines therapeutic gains. It is important to choose activities and materials that suit an individuals lifestyle and resources.

Identification of Goals

Assessment is the first step in evaluating an individual to determine what goals and objectives are appropriate for them. Assessment should be done formally at periodic intervals and informally within each treatment session to fine tune the treatment plan and maximize the therapeutic benefits for the individual. Activities that are too easy cause boredom and activities that are too challenging cause frustration and may aggravate a person's symptoms. Assessing and providing for the specific treatment plan for each individual can go a long way in motivation toward therapy and therapeutic gains. This is especially important in situations where a person's individuality is reduced through disability and hospitalization and the person has become dependent on others. Accurate assessment and adaptation for individual goals are extremely important aspects of the treatment plan.

Goals of a horticultural therapy program may include some or all of the following:

1. Provide and facilitate activities that stimulate cognitive development and sensory enrichment and allow opportunities for decision making, problem solving, sequencing, orientation, judgment, following directions, and increased attending skills.
2. Provide, encourage, and facilitate horticultural activities that promote the use of verbal and/or augmentive communication among peers, cooperation and sharing, and leadership skills when appropriate.
3. Provide the opportunity to practice and reinforce new functional skills, e.g., change of dominance, one-handed techniques, strengthening the use of the affected hand, joint protection, use of adapted tools, and fine motor coordination. Increase strength, range of motion, activity tolerance, and dynamic sitting or standing balance.
4. Introduce new horticultural projects and related community activities that can enhance physical activity levels, increase family and community integration, and decrease isolation post discharge. Educate and help problem solve home plant and garden care using safe and/or adaptive techniques and energy conservation practices.

The next section will cover issues relating to the implementation of horticultural activities, precautions, and specific activities that can be used to achieve the aforementioned goals.

Once goals and objectives have been determined, the next stage is for the patient to achieve them. The following presents some helpful hints for working with individuals with certain disabilities and ways to use horticultural activities and materials to help in achieving patient goals.

Parkinson's Disease and Multiple Sclerosis

Parkinson's disease and multiple sclerosis are degenerative diseases that present the patient with similar physical challenges. Many of the goals and objectives of patients with these conditions will be similar, often, the psychological (self-esteem) goals are most important.

When working with patients who have either of these diseases, it is very important to assess the cognitive and physical status in order to choose effective activities. The level of activity should be enough to challenge but not strain or frustrate the patients' physical or cognitive abilities; watch for fatigue. Also, tremors and disturbances of vision may create safety issues with sharp objects, in which case maximum assistance would be appropriate so the patient can complete the project and thus feel a sense of success and fulfillment.

Cognitive and social stimulation are very important for the individual. Patients should be included in the group conversation even if prompting is

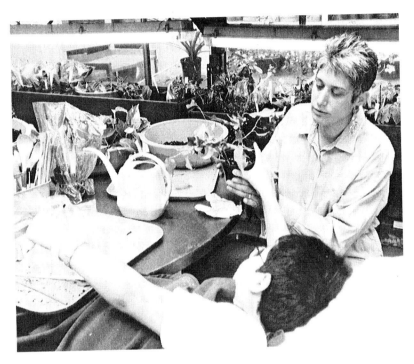

Staff assisting pediatric patient with propagation task. (Glass Garden, Rusk Institute, New York University Medical Center)

needed in advanced cases. Anxiety, moodiness, and depression may also affect motivation and should be monitored by the therapist. Provide encouragement and point out successes. A sense of humor and sense of fun are very effective motivators.

Horticulture as an avocational pursuit can be very therapeutic, especially for higher-functioning individuals. Also, plan more complex projects for cognitively intact individuals who stay in the program for extended periods of time so that they continue to be stimulated and interested. Beneficial activities include watering, planting seeds, propagating with cuttings, light pruning, transplanting, and use of sensory stimulating materials.

Cardiac, Orthopedic, and Amputation Rehabilitation

Other clusters of disabilities encountered in a rehabilitation setting have similar goals, objectives, and precautions. Cardiac and orthopedic rehabilitation, as well as rehabilitation for individuals with amputations, will be grouped together in this section.

Persons who have had a heart attack, need to be aware of the specific changes in their cardiac function when choosing activities, and many can perform a wide range of activities. They should not, however, perform activities that cause too much strain on their compromised cardiovascular systems. Heart rate and blood pressure should not exceed levels set by the doctors. Increases in strenuous activity should be gradual and carefully monitored. At the same time, it is important to help the patient overcome his or her fear of activity and to gradually increase self-confidence.

Therapists working with individuals who have sustained bone fractures or joint replacement face a number of treatment issues. Their rehabilitative focus is on regaining functional mobility and independence, gradual weight bearing and strengthening, and safety awareness. Individuals recovering from bone fractures should be careful to increase task difficulty gradually. Individuals with hip replacements should not cross their legs or bend at the waist.

Individuals with amputations should exercise unaffected parts of their body and learn to use a prosthetic device where applicable. The stump area around the amputated limb is very sensitive, especially just after surgery. Take care not to bump or otherwise injure the stump. Learning to cope with the loss of a body part and self-esteem issues should be assessed and dealt with. Natural acceptance by group members, staff, and volunteers and high expectations for task performance can help begin the patient's recovery process.

Patients with these disabilities are usually cognitively intact, faced with extreme stressors, often in pain, and have major adjustments to make in their lives. Opportunities for social interaction, relaxation, stress reduction, and activities providing for distraction from pain should be readily offered. Examples of work simplification and energy conservation strategies should be discussed. The possibility of horticulture as an avocational pursuit also can be assessed, and appropriate education and resource lists can be offered. The following goals and objectives are applicable for individuals in the above-mentioned groups.

Spinal Cord Injury

When an injury to the spinal cord causes damage to or severs the cord, the psychological, emotional, and physical consequences are enormous.

The level of injury to the cord directly affects the individual's ability to physically function. If the spinal cord is only partially severed or damaged (incomplete), resulting in paraparesis or quadriparesis, the individual may have remaining coordination, motion, strength, or sensation, which are important to increase and maintain.

Good communication skills are vital to establishing the therapeutic relationship with the person with a spinal cord injury (SCI). The patient will most often be the best source in defining his or her needs and in articulating therapy goals. Others with SCIs, particularly those who have already been through rehabilitation, can be invaluable in providing feedback, support, and suggestions.

There is often a great deal of denial, frustration, anger, and anxiety following such a drastic sudden change in one's abilities and life such as a SCI imposes. It is important to provide empathetic support, assistance, and feedback, which is essential to therapeutic gains. Success in horticultural activities can help boost confidence and increase independence. The indi-

Spinal injury patient displaying her cactus garden. (Glass Garden, Rusk Institute, New York University Medical Center)

vidual may discover for the first time, or rediscover, the range of horticultural tasks they are capable of achieving despite their physical limitations. Problem-solving skills learned from horticultural activities can be applied in other areas of rehabilitation.

Relearning and redefining physical levels of function are primary to the SCI patient, who will often be able to verbalize their strengths, weaknesses, medical precautions, as well as identify areas where they will require assistance. Increasing strength and mobility; improving range of motion and balance; and learning energy conservation techniques are important areas to focus upon.

Medical considerations and precautions are paramount in the treatment of the SCI patient, and may include increased sensitivity to heat or cold; sun sensitivity; fluid restrictions; bowel and bladder programs; lack of sensation in arms, legs, or other parts of the body; skin breakdown; and sudden drops in blood pressure from standing or exertion. The therapist and patient must be vigilant in noting any marks on the skin from tools or devices, which can lead to pressure sores and skin breakdown.

Goals for SCI patients might include the following:

- *Goal:* Improve fine motor skills
  Objective: Plant various sizes of seeds
- *Goal:* Increase strength
  Objective: Lift watering can to water plants (various graded sizes should be available, e.g., one quart can = two pounds, one gallon = eight pounds)
- *Goal:* Improve dynamic sitting balance, range of motion, strength, coordination
  Objective: Repot large pot at table
- *Goal:* Improve problem-solving skills
  Objective: Plant a small cactus garden, using precautions and assistance when necessary
- *Goal:* Increase or develop horticultural knowledge and accessible gardening techniques
  Objective: Practice home gardening skills
- *Goal:* Improve writing ability
  Objective: Write label for each plant with patient's name, plant name, and date, using adaptive devices and techniques as necessary
- *Goal:* Increase standing tolerance and endurance
  Objective: Work on horticulture projects while standing

Stroke

Individuals who have had strokes present a vast array of abilities and disabilities. Some of the many complex issues in the treatment of individuals with a stroke include level of function and degree of return of the affected side, strength, gross and fine motor skills, visual perceptual and cognitive changes, communication abilities, and level of independent functioning upon recovery. Accurate assessment in these areas is essential for positive therapeutic outcomes. Choice of therapeutic activity can vary greatly from case to case depending on the part of the brain involved. Horticulture provides many opportunities to exercise different levels of exertion in physical and cognitive activities, which is beneficial in the rehabilitation of stroke patients. Many people have enjoyed gardening in the past and this sometimes makes for a significant difference in the motivation for therapy for the patient.

Left CVA

For individuals with a left CVA, assess for aphasia to determine how much a person understands and how well they can communicate. Speak slowly, using a normal tone and simple declarative sentences. Do not use a "special" voice. Recognize nonverbal cues and body language to understand nonverbal patients' communications. Use nonverbal cues, demonstrations, and body language to help communicate your intentions. Being animate may assist in getting the message across. Try not to talk too much in detail or chatter as this "static" can confuse aphasic individuals.

The therapist should assess for cognitive status and learning style. Visual prompts and modeling are helpful in teaching procedures of various activities. Complex tasks can be broken down into simple steps to aid in comprehension and learning. Learning is maximized when feedback is given often, even after each step in some cases, instead of waiting until the end of the activity. This offers encouragement and develops self-confidence. Although somewhat anxious and cautious in their approach to activities, the individual with a left CVA is considered easier to retrain than a person with a right CVA.

Right CVA

Individuals with a right CVA have different issues relating to the cognitive aspects of their condition. They often are quite articulate, superficially, in discussing their stroke, yet they often have little understanding or insight regarding the depths of their disabilities. They are impulsive in their functioning and will complete the task inaccurately before the therapist has demonstrated the steps. The use of task analysis for complex tasks and repetitive prompting is necessary to keep the patient focused on the activity without distraction from his own talk.

Individuals with a right CVA also tend to have more visual difficulties. If the person has a field cut or visual neglect, work on the person's right side initially. As therapy progresses, the therapist can place more and more materials toward the left side. Prompting may be needed to get the patient to scan in order to see more of their complete visual field. This is often best achieved with a gradual shaping process depending on the severity of the visual disturbance.

Safety issues are a primary concern. Use of sharp tools should be closely monitored. Many patients are confused and may try to eat soil or plant materials. Precautions should be taken regarding body positioning and special care should be taken when repositioning a patient in their wheelchair or transferring them to or from their chairs.

The therapist should be aware that the patients, during the course of the session, will often cry and become quite depressed. It is appropriate to stop, take a short break, acknowledge the sadness, and continue on with the session. This lability and depression will decline over time.

Documentation is also an extremely important part of the treatment plan. It allows for the various members of the team to coordinate their efforts and see how their patients are performing in other areas. Discrepancies or consistencies in various areas of performance can provide valuable information in assisting in an individual's recovery. Forms such as those in Figures 4.4 and 4.5 are used in rehabilitative settings for documentation purposes.

CASE STUDIES

Bruce: Spinal Cord Injury

Background

Bruce was an eighteen-year-old high school senior from an intact, very supportive upper-middle-class family from the suburbs of New York City. He was the driver in a single-car accident in which he sustained a spinal cord injury—C 4-5 incomplete.

Bruce underwent his rehabilitation at the Rusk Institute of Rehabilitation Medicine in the pediatric unit. During the course of a typical day,

FIGURE 4.4. Sample Form for Documentation

NEW YORK UNIVERSITY MEDICAL CENTER
The Rusk Institute of Rehabilitation Medicine
HORTICULTURAL THERAPY
GROUP ACTIVITY TREATMENT PROCEDURE

Name: _____ Age: _____
Chart No.: _____ Room: _____
Diagnosis: _____
Disability: _____
Physician: _____

KEY
5 – independent
4 – consistent/w/supervision
3 – minimum assist
2 – moderate assist
1 – maximum assist
N/A – non-applicable

Attendance:

M		T		W		TH		F	
½	1	½	1	½	1	½	1	½	1

M		T		W		TH		F	
½	1	½	1	½	1	½	1	½	1

TREATMENT GOALS:

	5	4	3	2	1	N/A
MOBILITY						
Comes independently to HT session						
Comes on time to HT session						
Able to maneuver within the greenhouse						
PHYSICAL/PERCEPTUAL ABILITIES						
Performs horticultural tasks: bilaterally						
left-handed						
right-handed						
Able to grasp/release tools during activity						
Able to manipulate tools						
Able to manipulate plant and non-plant materials						
Able to fill pot accurately						
Able to center cuttings in pot						
Able to place cuttings in dibble holes						
Able to place plant in upright position						
Able to place scissors in proper position for cutting						
Selects correct stem when cutting from multi-stem plant						
Able to space cuttings in whole pot						

	5	4	3	2	1	N/A
SOCIAL INTERACTION						
Hearing impairment limits socialization						
Foreign language limits socialization						
Aphasia limits socialization						
Initiates interaction w/therapist						
Interacts once approached						
Interacts with peers						
Interacts appropriately with others.						
Fits easily into group						
Responds accurately/appropriate to questions						
Able to make self understood						
Able to discuss physical conditions realistically						
COGNITIVE ABILITY						
Follows verbal &/or written directions: 1 step						
2 step						
more						
Follows demonstrated directions: 1 step						
2 step						
more						
Able to remember task sequence between sessions						
Able to focus on task						
Able to maintain attention span for 1 hour session						
Able to shift from one task to another						
Able to control behavior to complete tasks accurately						
Follows safety precautions						
Understands basic horticultural concepts						

Able to work with plant materials: in front . . . . .
                                   above head . . . . .
                                   to sides . . . . .

Able to water plants with: 1 lb. can filled . . . . .
                              2 lb. can filled . . . . .

Able to water plants accurately . . . . .
Able to wash hands &/or nails in sink . . . . .
Endurance permits completion of horticultural tasks . . . . .
Able to work on specimen plants (over 6″ pot) . . . . .
Able to complete horticultural tasks correctly . . . . .
Able to find all materials on table for task . . . . .
Able to control physical problems/pain during task . . . . .

WRITING ABILITY
Writes own name on plant label . . . . .
Writes date on label . . . . .
Writes plant name on label . . . . .
Handwriting is legible . . . . .

Understands purpose of HT treatment . . . . .
Able to adhere to time schedule . . . . .
Aware of seasons, weather, whereabouts . . . . .
Able to overcome problems encountered during tasks . . . . .

EMOTIONAL STATUS
Is willing to try new activities . . . . .
Seeks assistance when appropriate . . . . .
Has confidence in horticultural tasks attempted . . . . .
Perseveres on difficult tasks . . . . .
Able to control emotional status during activity . . . . .

AVOCATIONAL INTERESTS
Selects own plant material for propagation . . . . .
Shows interest in learning cultural practices of plants . . . . .
Maintains own plants propagated during treatment . . . . .
Anticipates taking plants home . . . . .
Visits greenhouse on own time at least 3x/week . . . . .

CURRENT FUNCTIONAL STATUS:

PROJECTED TREATMENT PLAN:

Therapist

FIGURE 4.5. Sample Form for Documentation

HORTICULTURAL THERAPY FACT SHEET

HT TIME: Daily M T W Th F

Ref. by: _____ Ext._____

NAME _____ AGE _____

Sex_____Room # _____

Married Single Widow Divorced Children:

Occupation: _____

PHYSICIAN_____

DIAGNOSIS_____

DISABILITY_____

AMBULATION STATUS: Ambulates
Cane/Ws   Walker/Ws   Crutch/Ws   Wheelchair   Needs Escort

SPEECH STATUS: Normal
Foreign Language   Aphasia   Dysanthia   Expressive   Receptive
Hearing:
Writing / Reading:

PHYSICAL STATUS: Dominance: R L
A. UEF: Present   Weak   Absent   Arthritic
B. COORDINATION: Tremors   Spasticity   Ataxia
C. PERCEPTUAL STATUS:
  Normal Glasses Double Vision Visual Field Cut Visual Neglect

---

MENTAL STATUS:
    Oriented   Memory   Attention Span   Labile   Flat
    Follows directions: Verbal   Visual   Both   Multi-Step

PRECAUTIONS (Meds):

GOALS:

COMMENTS:

Bruce would attend school classes, occupational therapy, physical therapy, psychological and vocational counseling, and special nursing classes for spinal cord injury patients, which included bowel and bladder management, skin care, exercise, medications, and other special topics. He and his family also had extensive meetings with the barrier-free design specialist, in order to build Bruce his own separate accessible apartment at their home. Bruce also worked with the electronic assistive device specialist, who helped Bruce adapt to a computer for school use and adapt to home independence (e.g., environmental controls such as telephone, lights, television, radio, air conditioner).

Six months posttrauma, Bruce had neck, diaphragm, and shoulder control and elbow flexion. He was able to use an electric wheelchair independently for mobility (except transfers) and was able to move his arm laterally. He had no pinch, grip, or fine muscle control in his wrists or hands, which meant he was not able to write or feed himself unless he had assistive devices. He was not able to dress, bath, or transfer unassisted, or able to manage bowel and bladder self-care. He would forever need assistance in his basic functioning.

Bruce was referred to the horticultural therapy program by the occupational therapist to offer him an additional modality in which to practice and improve his fine motor skills (using his hand/wrist splints provided by the occupational therapists), to improve his dynamic sitting balance, and to maximize functional performance and overall well-being. During his lengthy course of treatment his goals included the following:

- Increase strength and stamina
- Increase awareness of safety issues
- Increase dynamic sitting balance
- Increase self-esteem and confidence by trying new activities

Horticultural Therapy

Bruce attended the greenhouse daily for his one-hour session. He established a very close, warm, and fun relationship with the horticultural therapist (male) and was most interested in the fish pond and bird life in the greenhouse. He was not especially interested in the plants, even though he did participate in the daily horticultural therapy program with other patients, where he worked on horticulture activities with his hand splints. Bruce often selected his own plants from the greenhouse collection for propagation and learned the steps for propagation quickly so he could tell others. He interacted appropriately with others and was an asset to the group because of his sense of humor.

Bruce was able to hold a spoon using a C-clamp and, with a volunteer's assistance, was able to fill his pot with dry soil by using lateral motion of his arm. He lacked strength to tamp it down. A complicated orthotic device from the occupational therapist allowed him to pinch. This meant he could hold his cutting, use the dibble with a flat washer to stop his hand from sliding down, and place the cutting in a hole, using his arm laterally. Assistance was needed to hold the watering can and for most other tasks. Bruce also wrote his own label with assistance and a splint to hold the pen.

Bruce's friendship with the horticultural therapist increased. They traveled after work to purchase some interesting fish for the conservatory pond. These trips increased Bruce's self-confidence and self-esteem. He was accepted, like a "normal" person, in the horticultural environment, and often visited with his family and friends in the greenhouse.

Postdischarge

Bruce graduated from high school and went to a local college to major in computers. He always stopped by when going for a doctor's appointment and would report on his and his plants' status. (He had a full-time aid and discarded the use of the hand splints devised by the occupational therapists. At the university, he took notes on a tape recorder and used a computer for his work).

A year postdischarge, while visiting the hospital for a checkup, Bruce brought in his plants for repotting into larger containers. He often called to chat, and once, to ask advice on size, materials, and soil for raised beds outside his apartment for a vegetable garden. He called at summer's end to brag about his watermelons.

Bruce called again after he graduated from college. He was not that interested in a computer career, and decided to go back to school to become a landscape architect. This is a viable and realistic career choice because of the current capabilities of computers in design work. Bruce also told us he had his own greenhouse, built for him to house tanks to grow aquatic plants. He had developed a small business and was selling the plants to local nurseries.

Horticultural therapy helped to change Bruce's life by offering and opening to him an achievable and emotionally and intellectually satisfying vocational future.

Carol: Stroke Victim

Background

Carol is a fifty-three-year-old African-American woman who was admitted to physical rehabilitation three weeks after a left cerebrovascular accident resulted in right hemiparesis with mild expressive aphasia. Carol was employed as a nurse practitioner and a teacher at an acute-care university hospital. She is divorced with no children and has a history of alcohol abuse. Prior to this stroke, she was in recovery from her alcoholism but she smoked heavily, and her busy schedule allowed little time for exercise or the pursuit of leisure interests. Her work was her life.

Diagnosis and Treatment

The admitting physical medicine and rehabilitation physician predicted that Carol might see the return of movement and function in her right arm. The prognosis for walking was less promising. The psychologist evaluated Carol for depression as well as adjustment issues related to her disability. Speech therapy was prescribed for her as well as intensive occupational and physical therapy. The physicians agreed that Carol was an excellent candidate for horticulture therapy as well.

Goals of Therapy

Carol's functional goals in occupational and horticultural therapy were

- wheelchair mobility,
- strengthening and increasing the function of her right arm,
- using the left arm as an assist, and
- overall endurance.

Carol's personal goals in horticultural therapy were to

- explore the possibility of returning to her vocation as a nurse practitioner and teacher by simulating work-related activities in the greenhouse,
- explore gardening as a leisure interest, and
- assess whether her own home environment was accessible for indoor or outdoor gardening.

Therapy Results

Carol became very animated and somewhat overwhelmed when she first entered the greenhouse because she had previously had a strong interest in horticulture. Carol grew up in a home that adjoined a field, a wooded area, and a large vegetable garden. The horticultural therapist encouraged Carol to spend free time and her "open" leisure groups in the greenhouse or just "hang out" and help with the plants.

As Carol's physical and occupational programs intensified, she made considerable physical and functional gains. She attained functional use of her right arm with a slight weakness being the only inhibiting factor to function. She was able to ambulate for short distances with a quad cane. Her aphasia diminished and she only had slight word-finding problems when she was under stress or in a hurry to express a thought. Her speech therapist sat in on one horticultural therapy session and observed as Carol identified many of the plants by name from memory.

Carol practiced ambulation in the greenhouse with her physical therapist and shared her knowledge of the various plants and their care. During her free time, Carol also assisted in making flower arrangements for the patient dining room. She received glowing compliments on her beautiful designs. Carol socialized easily with other patients and was very supportive of those patients fearful of medical issues or who lacked motivation to participate. Carol became empowered in her own rehabilitation as she taught others how to care for plants and served as a support person.

Carol's training as a nurse enabled her to develop an understanding of her own condition and to explore her vocational potential. Many of Carol's friends, students, and co-workers visited the hospital and encouraged her to return to work when she felt strong enough. It was agreed that Carol would come back part-time in her teaching capacity after her discharge. Her return to clinical work would be evaluated further down the line as she explored her areas of strengths and weakness. Carol realized that she would need to change her lifestyle: she must focus on conserving her energy and having more "fun."

Carol expressed an interest in continuing horticulture as a leisure activity. With the help of the horticultural therapist, she purchased an indoor artificial light unit that provided adequate growing space. This was necessary as her apartment had poor lighting and her outdoor garden was very small. Container gardening was also discussed as a future option.

After Carol returned to work, she continued to visit the greenhouse whenever she had a follow-up hospital visit. She later volunteered once a week for the evening horticultural therapy sessions. Carol decided to continue teaching and expand her responsibilities in that area. Her clinical work continued only in a supervisory capacity.

FUTURE PRACTICE

Current and future change in the U.S. health-care delivery system has created a climate of flux for hosticultural therapists within the field of physical disabilities. While the need for rehabilitation services increases with a growing disabled population, both the length of hospital stay and types of services covered by third-party payers have decreased significantly due to new reimbursement and managed-care controls by outside carriers. Our roles as service providers will need to evolve along with the changing structure of health care in the future. Various roles within inpatient, outpatient, and community-based care will present opportunities and challenges for future practice.

Becoming part of a managed care package where horticultural therapists work as part of a team treating certain types of disabilities (e.g., cardiac, orthopedic, stroke) appears to be a positive step. Programming for children presents many opportunities to explore in different clinical applications. Inpatient stays of children are significantly longer on the average than adults, and children benefit greatly from the stimulation and education that horticultural therapy provides. Benefits from therapy with subacute patients at bedside can be assessed for practice in this area. However, we must remember our roots are in the garden, and wherever feasible, the patient should participate in a garden setting.

Health maintenance for outpatients is another area to explore within the realm of managed care. Horticultural therapists can be utilized in the role of prevention and choice of healthy avocational pursuits. Assessment of work capacity, work readiness, and work hardening training can be performed at horticultural therapy sites in conjunction with vocational assessment and training.

GLOSSARY

active movement: Unassisted movement of a muscle or joint using minimal resistance

aphasia: Communication disorder caused by brain lesions, which results in a loss or decrease in the ability to express oneself through speech, comprehend the spoken word, read, or write can be "expressive," "receptive," "global," or varying degrees of involvement

apraxia: Inability to carry out, on request, purposeful coordinated movements, without impairment of muscles or senses

ataxia: Inability to coordinate voluntary muscular movement, which is also symptomatic of some nervous disorders

bilateral coordination: Ability to perform movement with both arms and hands simultaneously

coordination: Ability to perform smoothly and at will movements of the limbs simultaneously or alternately

CVA: Cerebrovascular accident, which is the interruption of the blood supply due to a thrombus (clot), embolus (a clot that has traveled from its site of formation), or a hemorrhage, resulting in a specific pattern of symptoms depending on the area of the brain affected

DIP: Distal interphalangeal joint, nearest the finger tip

dysarthria: Difficulty in articulating words caused by weakness of the muscles used in speech

dysphagia: Difficulty swallowing, usually in patients with a left CVA

eye-hand coordination: Ability to perform activity with the hand while using a visual control, with one hand used as a stabilizer while the other manipulates (e.g., taking cuttings)

eye, hand, and foot coordination: Ability to perform activity involving all three (e.g., spading, planting shrubs, moving flats or pots)

heavy palmar grip: A grip characterized by fingers fully flexed to the palm with the thumb firmly positioned over the fingers to give added strength; used for handling heavy tools and when strength is needed; utilizes ulnar deviation as a very important component of this grip and actively uses radial deviation for activities such as hammering

hemianopsia (field cut): Loss of the half field of vision in one or both eyes following a stroke. In the hemiplegic, it occurs in the half field of the affected side (e.g., the left hemiplegic has loss of vision to the left)

hemiparesis: Weakness of one side of the body

hemiplegia: Paralysis of one side of the body and extremities resulting from injury to the motor centers of the brain; manifests in paralysis on the side of the body opposite the side of the affected brain following a CVA

IP: Interphalangeal joint, a joint in the finger (the thumb has only one IP, whereas the other fingers have two)

lability: Emotional instability that manifests in inappropriate laughing or crying

light palmar grip: Flexion of all the fingers but not fully down to the palm; used when holding large objects such as a drinking glass, telephone receiver, etc.

MCP: Metacarpophalangeal joints, between the finger and hand, the joints nearest the wrist

neglect: Lack of awareness of one side of the body and space, usually following a right CVA, in which the individual will show neglect to the left of the midline

opposition: Touching of the fingers, in turn, to the tip of the thumb

paraparesis: Weakness in the lower extremities

paraplegia: Symmetrical paralysis of both lower extremities

passive movement: Movements of an individual's joints through the full range done by a therapist or mechanical device

perseveration: Recurrence or repetition of speech or activity

pinch grip: Triangle pinch formed by the approximation of the thumb and finger pads of the index and middle fingers; used in picking up and holding small, fine objects such as pencils and buttons

PIP: Proximal interphalangeal joint, joint nearest the mcp joints

proprioception: Joint position sense; the ability to tell where the joints and limbs are without looking directly at them

quadriparesis: Weakness of all four extremities

quadriplegia: Paralysis of all four extremities

ROM: Range of motion, usually measured in degrees, through which the normal, unaffected joint can be expected to move, both actively and passively

spasticity: Increased resistance to sudden passive movements

stereognosis: Ability to identify objects placed in the hand without using visual clues (tactile recognition); pain and touch sensation intact, though one may be unable to interpret nerve impulses sent to the brain to identify objects

tolerance: Ability to sustain activity over a specified period of time; also endurance

BIBLIOGRAPHY

Agness, Phyllis J. (1985). *Learning Disabilities and the Person with Spina Bifida.* Lisle, IL: Spina Bifida Association of America.

American Heart Association. (1989). *How Stroke Affects Behavior.* Dallas, TX: American Heart Association.

Bond-Howard, Barbara. (1993). *Introduction to Stroke.* Seattle, WA: Idyll Arbor, Inc.

Fink, S.L. and Houser, H.B. (1966). An Investigation of Physical and Intellectual Changes in Multiple Sclerosis. *Arch. Phys. Med. Rehabil.* 47(2), 56-61.

Hopkins, Helen and Smith, Helen (ed). (1983). *Willard and Spackman's Occupational Therapy*. Philadelphia, PA: J.B. Lippincott Co.

Mclone, David. (1986). *An Introduction to Spina Bifida*. Chicago, IL: Children's Memorial Hospital.

Rothert, Eugene. (1994). *The Enabling Garden*. Dallas, TX: Taylor Publishing Co.

Rothert, Eugene and Daubert, James R. (1981). *Horticultural Therapy at a Physical Rehabilitation Facility*. Glencoe, IL: Chicago Horticultural Society.

Spencer, A.S. (1983). Functional Restoration—Theory, Principles, and Techniques. In Helen Hopkins and Helen D. Smith (eds.), *Willard and Spackman's Occupational Therapy*. Philadelphia, PA: J.B. Lippincott Co.

Trombly, Catherine. (1983). *Occupational Therapy for Physical Dysfunction*. Baltimore, MD: Williams & Wilkins, p. 5.

Versluys, H.P. (1983). Psychosocial Adjustment to Physical Disability. In Catherine Trombly (ed.), *Occupational Therapy for Physical Dysfunction*. Baltimore, MD: Wilkins & Wilkins, 12.

Chapter 5

Traumatic Brain Injury and Horticultural Therapy Practice

David Strauss
Maria Gabaldo

INTRODUCTION

Every fifteen seconds, one person in the United States sustains a traumatic brain injury. More than 56,000 of these Americans die each year, while another 373,000 are hospitalized. Nearly 100,000 of those who are hospitalized have sustained such severe brain injuries that they will never again be the same (Brain Injury Association, 1995).

Traumatic brain injury (TBI) is caused by an injury to the brain through an external physical force. Victims of TBI now have a greater chance to survive than ever before, due to the advances in trauma and emergency medicine over the last two decades. Their life spans are estimated to be only five years shorter than individuals who have never sustained a brain injury.

The TBI survivor will not live the life he or she experienced before the injury. The survivor must learn to cope with lifelong cognitive, physical, behavioral, and emotional changes—changes that forever affect the survivor, family, and friends.

The objectives for this chapter are to define the causes and affects of traumatic brain injury, and to explore how horticultural activities are used to assess and improve an individual's social, behavioral, physical, and cognitive capabilities after TBI. Individual case studies will illustrate horticultural rehabilitation strategies and show how horticultural therapy helps a TBI survivor engage in activities that encourage and enhance rehabilitation efforts. As this chapter documents, horticultural therapy helps a TBI survivor develop not only valuable work habits but rebuild the self-confidence and life skills that a traumatic brain injury took away (Straus and DiDonato, 1994).

THE CLIENT POPULATION

Injury is the leading cause of death for Americans under forty-five years of age. Traumatic brain injury is responsible for the majority of these deaths, and TBI is the leading cause of disabilities in both children and young adult males (Brain Injury Association, 1995).

Injuries that cause TBI include (see Figure 5.1) the following:

- Motor vehicle accidents, which account for 50 percent of all TBI injuries
- Falls, accounting for 21 percent
- Assaults and violence, accounting for 12 percent
- Accidents through sports and recreation, accounting for 10 percent

In all injury categories, alcohol plays a significant factor. More than 50 percent of adults with brain injuries were intoxicated at the time of injury (Brain Injury Association, 1995).

The National Pediatric Trauma Registry reports that more than 30,000 children per year sustain permanent disabilities as a result of brain injuries. Children are especially at risk in the afternoon hours when they are dismissed from school, when the majority of accidents occur on the roads, at home, and in recreation areas. Males aged fourteen to twenty-four have the highest rate of TBI, and are twice as likely as females to sustain a brain injury. Accidents that result in TBI are more likely to occur in mid-afternoon to early evening and on weekends, especially during the summer months.

The Brain Injury Association, Inc., (1995) has called TBI the "Silent Epidemic" because every person has a 1-in-10 chance of suffering from a significant brain injury during his or her lifetime. Regardless of when or how a brain injury takes place, each injury is unique. No two survivors of TBI are alike. The severity and long-term effect of an individual's disabilities from TBI are primarily determined by which particular area of the brain is damaged and the extent of the injury.

Each part of the brain controls specific functions of the body, ranging from vision to emotional behavior. The frontal lobe controls the individual's executive functions—from planning and problem solving to sexual disinhibition to impulse control. The occipital lobe controls vision, while the temporal lobe is the brain's primary memory area.

FIGURE 5.1. Fact Sheet on Traumatic Brain Injury

Brain Injury Association, Inc.
(formerly National Head Injury Foundation, Inc.)

FACT SHEET

TRAUMATIC BRAIN INJURY

DEFINITION

Traumatic brain injury (TBI) is an insult to the brain, not of degenerative or congenital nature but caused by an external physical force that may produce a diminished or altered state of consciousness, which results in an impairment of cognitive abilities or physical functioning.

SCOPE

Injury is the leading cause of mortality among Americans under forty-five years of age, and traumatic brain injury (TBI) is responsible for the majority of these deaths. It is estimated that TBI claims more than 56,000 American lives annually.[1]

Each year, about 373,000 Americans are hospitalized as a result of TBI. Of these, 99,000 individuals sustain moderate to severe brain injuries resulting in lifelong disabling conditions.[2]

In the fifteen seconds it takes to read these statistics, one person in the U.S. sustains a traumatic brain injury.[3]

After one traumatic brain injury, the risk for a second injury is three times greater; and after a second TBI, the risk for a third injury is eight times greater.[4]

CAUSES OF TRAUMATIC BRAIN INJURY

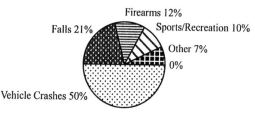

Firearms 12%

Falls 21%   Sports/Recreation 10%

Other 7%

0%

Vehicle Crashes 50%

Vehicle crashes are the leading cause of TBI, accounting for 50% of all injuries.

Falls are the second leading cause, accounting for more than 20% of all traumatic brain injuries.

Alcohol is a significant factor in the occurrence of TBI. More than 50% of persons with brain injury have been intoxicated at the time of injury.[5] [6]

1776 Massachusetts Avenue, NW, Suite 100, Washington, DC 20036
(202) 296-6443 FAMILY HELPLINE (800) 444-6443

FIGURE 5.1 (continued)

WHO SUSTAINS TRAUMATIC BRAIN INJURIES

Males aged fourteen to twenty-four years are at highest risk, followed by infants and the elderly.[7]

Males are twice as likely as females to sustain TBI due to differences in risk exposure and lifestyle.

According to the National Pediatric Trauma Registry, more than 30,000 children sustain permanent disabilities as a result of brain injuries.[8]

WHEN TRAUMATIC BRAIN INJURY OCCURS

Mid-afternoons to early evenings, weekends, and the summer months are critical times during which TBI is most likely to occur.[9]

Children are especially at risk in the afternoon hours after they are dismissed from school; 42.6% of children's injuries occur on roads, 34.3% at home, and 6.6% in recreation areas.[10]

CONSEQUENCES

Cognitive: may include short- and long-term memory loss; difficulties with concentration, judgment, communication and planning; spatial disorientation.

Physical: may include seizures; muscle spasticity; vision, hearing, smell and taste loss; speech impairment; headaches; reduced endurance.

Psychosocial/Behavioral/Emotional: may include anxiety and depression, mood swings, denial, sexual difficulties, emotional liability, egocentricity, impulsivity and disinhibition, agitation, isolation.

COST

The cost of traumatic brain injuries in the United States is estimated to be $48.3 billion annually. Hospitalization accounts for $31.7 billion, whereas fatal brain injuries cost the nation $16.6 billion.[11]

SOURCES

[1]Kraus JF, McArthur DL. *Epidemiology of Brain Injury.* Los Angeles: University of California Los Angeles, Department of Epidemiology, Southern California Injury Prevention Research Center, February 1995. In press.

[2]Kraus J, Sorenson S. Epidemiology. In Silver J, Yudofsky S, Hales R (eds), *Neuropsychiatry of Traumatic Brain Injury.* Washington, DC: American Psychiatric Press, Inc. 1994.

[3]FDA consumer. *Head Injuries.* Health ResponseAbility Systems, 1993. (File downloaded from America Online).

[4]Annegers JF, Garbow JD, Kurland LT, et al. The incidence, causes and secular trends of head trauma in Olmstead County, Minnesota. 1935-1974. *Neurology* 1980; 30:912-919.

[5]Ruff RM, Marshall LF, Klauber MR, Blunt BA, Grant I, Foulkes MA, et al. Alcohol abuse and neurological outcome of the severely head injured. *Journal of Head Trauma Rehabilitation* 1990; 5:21-31.

[6]Kreutzer JS, Doherty KR, Harris JA, Zasler ND. Alcohol use among persons with traumatic brain injury. *Journal of Head Trauma Rehabilitation* 1990; 5:9-20

[7]Kraus JF, Epidemiology of head injury. In Cooper, PR (ed), *Head Injury.* Baltimore: Williams & Wilkins, 1993.

[8]Research and Training Center in Rehabilitation and Childhood Trauma. *National Pediatric Trauma Registry.* Boston: Tufts University School of Medicine, New England Medical Center, Spring 1993.

[9]Rimel RW, Jane JA. Characteristics of the head-injured patient. In Tosenthal M, Griffith ER, Bond MR, Miller JD (eds), *Rehabilitation of the Head Injured Adult.* Philadelphia, FA Davis, 1983.

[10]Research and Training Center in Rehabilitation and Childhood Trauma. *National Pediatric Trauma Registry.* Boston: Tufts University School of Medicine, New England Medical Center, April 1995.

[11]Lewin-ICF. *The Cost of Disorders of the Brain.* Washington, DC: The National Foundation for the Brain, 1992.

The extent of any individual brain injury is determined by many factors. An open brain injury involves a fractured skull; a closed brain injury does not. The brain is usually protected by the protective covering of the skull and by cerebrospinal fluid (CSF) that allows the brain to float with the skull and be cushioned against trauma. With TBI, the brain receives a blow so severe that any protection offered by the skull and the CSF is compromised. Consider this description:

> The brain is soft and is surrounded by a hard bony skull—somewhat like Jell-O in a bowl. A hard blow to the bowl will shake the Jell-O inside, pulling parts away from each other. Similarly, with a blow to the head, the brain bounces and swirls within the skull causing widespread trauma—a little bit of damage everywhere. Billions of nerve cells are compressed, stretched, twisted or torn by these violent forces. . . .
>
> Now imagine that half of the inside surface of the bowl of Jell-O is rough like a cheese grater. The same hard blow noted before will cause more injury to the Jell-O next to this rough area because of the rough surface. Similarly, the inside surface of the skull next to the frontal and temporal lobes of the brain is rough. In most cases of traumatic brain injury, there is more damage to the frontal and temporal poles—areas involved with attention, memory, and emotion.
>
> Now, imagine that all of the Jell-O is free-floating in the bowl except for a half-dollar-sized portion stuck to the bottom. Twisting the Jell-O in the bowl will cause more damage to this fixed area. Similarly, the brain is fixed by the brain stem . . . twisting of the upper brain on the fixed brain stem causes injury to the pathways to and from the brain stem and spinal cord. It often causes loss of consciousness. Extreme injury of this type causes severe paralysis and spasticity of the arms and legs (Calub, DeBoskey, and Hecht, 1991).

Brain injuries do not always involve loss of consciousness, and the length of a coma is often used to define the severity of a brain injury: under one hour, mild; from one to twenty-four hours, moderate; and over twenty-four hours, severe. Diseases like strokes cause "focal" injury to the brain—an injury that leads to deficits in one area of the brain but leaves other functions completely unaffected. TBI causes what is known as "diffuse injury" throughout the brain. Due to various degrees of movement of the brain within the skull and often a lack of blood flow to the brain, diffuse injury causes multiple problems throughout all areas of the brain. Consequently, the TBI survivor must cope with problems in three distinct areas: (1) physical, (2) cognitive, and (3) behavioral and emotional (see Figure 5.2).

The combination of physical, cognitive, behavioral, and emotional problems burden the TBI survivor with yet another disability—a lack of social acceptance. It can be extremely difficult for TBI survivors to reenter the social community. Inappropriate speech, awkward physical appearances, overt actions, and subtle yet incongruous gestures often make successful, daily interactions outside a therapeutic setting difficult to achieve. TBI survivors exhibit ataxia (a muscular incoordination characterized by tremors and unsteady gait) or aphasia (an inability to understand and/or communicate through speech, reading, writing), as well as dysarthria (neurologically based speech difficulty resulting from the inability to coordinate musculature for speech production). Survivors who also exhibit traits of disinhibition will lack a sense of control over impulses, in thoughts, as well as actions.

FIGURE 5.2. Problems That Confront Survivors of Traumatic Brain Injury—Three Distinct Areas

Physical	• speech	• vision
	• hearing	• lack of coordination
	• paralysis	• seizure disorders
	• vertigo	• headaches
	• spasticity	• contractures
Cognitive	• concentration	• short-term memory
	• attention	• long-term memory
	• perception	• judgment
	• planning	• sequencing
	• communication	• reading skills
	• writing skills	• orientation
Behavioral and Emotional	• fatigue	• excessive emotions
	• anxiety	• depression
	• low self-esteem	• sexual dysfunction
	• restlessness	• lack of motivation
	• agitation	• inability to cope
	• mood swings	• self-centeredness

A TBI survivor often loses the ability to monitor his or her own performance. The individual has tremendous difficulty coping, performing tasks, and achieving personal goals in a social setting. This situation effects self-identity and self-worth, which in turn can lead to loneliness, isolation, and poor self-esteem for the individual with TBI.

An average acute-care hospitalization stay for a person with TBI is twenty-seven days; acute rehabilitation in another facility now lasts an average of thirty-six days, which is down from 118 days just five years ago. Additional rehabilitation is often required on an outpatient basis, in a post-acute program. It is the rule rather than the exception that the TBI survivor faces five to ten years of intensive rehabilitation services, followed by a lifetime of intervention and rehabilitation assistance that will eventually cost an average of $6 million (ReMed, 1996). Progress may occur throughout this lifelong recovery process. Survivors are known to show substantial improvements within a year after their injury, with additional progress coming slowly, sometimes as much as five or ten years post-injury. Throughout this recovery process, horticultural therapy is uniquely suited to address the treatment and rehabilitative needs of the TBI survivor.

IDENTIFICATION OF TREATMENT ISSUES

As previously mentioned, neurological damage caused by a traumatic brain injury can manifest itself in many ways. A survivor may experience physical, cognitive, behavioral, and emotional symptoms. Most often, there is a combination of symptoms that cause a general decrease in a survivor's level of functioning. Also, the brain damage that affects cognition impacts all other areas because memory, concentration, relearning, and new learning are all important in the recovery process (Dow, 1989). Because of these widespread effects, the survivor interfaces with many rehabilitation professionals during recovery.

Typically, an interdisciplinary treatment team is assembled to help the survivor regain skills in his or her areas of loss (see Figure 5.3). The coordinated efforts of this team are the key to successful rehabilitation. Not only must the survivor work to regain skills in each area of loss, but he or she must also work to cognitively reintegrate these skills. The treatment team must facilitate this integration by planning treatment that reinforces skills and helps the survivor generalize compensatory strategies.

A horticultural therapist can be a valued team member because horticultural activities can be adapted to meet many different needs and can meet many needs simultaneously, which is important when a survivor is

FIGURE 5.3. Interdisciplinary Team Members

Neurologist
Physiatrist
Psychiatrist
Neuropsychologist
Behavioral Psychologist
Rehabilitation Nurse
Speech Pathologist
Physical Therapist
Occupational Therapist
Horticultural Therapist
Recreational Therapist
Music Therapist
Art Therapist
Nutritionist
Cognitive Therapist
Vocational Counselor
Social Worker
Case Manager
Family Members
Survivor
Personal Care Attendants

trying to relearn how to integrate skills. For example, planting seeds is an activity that has physical, cognitive, and visual components. A survivor may have deficits in one or all of these areas, and the brain must coordinate, sequence, and integrate all the skills required to accomplish this activity. The horticultural therapist must know what is being done in other disciplines in order to structure the activity in a way that will emphasize skills and strategies in a like manner. Consistency is important to the process of relearning and new learning for an individual with neurological injury.

IDENTIFICATION OF TREATMENT GOALS AND OBJECTIVES

Recovery from a traumatic brain injury is a long process. Treatment team members and the treatment focus will change along the way based on the survivor's progress and level of functioning (Dow, 1989). The Rancho Scale, developed by Rancho Los Amigos Hospital is commonly used to assess the level of cognitive functioning (see Figure 5.4).

FIGURE 5.4. Rancho Los Amigos Scale of Cognitive Levels and Expected Behavior

Level I	No Response	Patient is unresponsive to all stimuli.
Level II	Generalized Response	Patient demonstrates inconsistent, nonpurposeful, nonspecific reactions to stimuli. He or she responds to pain, but the response may be delayed.
Level III	Localized Response	Patient has an inconsistent reaction directly related to the type of stimulus presented. He or she responds to some commands and may respond to discomfort.
Level IV	Confused, Agitated Response	Patient is disoriented and unaware of present events, with frequent bizarre and inappropriate behavior. His or her attention span is short and ability to process information is impaired.
Level V	Confused, Inappropriate, Nonagitated Response	Patient exhibits nonpurposeful random or fragmented responses when task complexity exceeds abilities. Patient appears alert and responds to simple commands. He or she performs previously learned tasks but is unable to learn new ones.
Level VI	Confused, Appropriate Response	Behavior is goal-directed. Responses are appropriate to the situation with incorrect responses due to memory difficulties.
Level VII	Automatic, Appropriate Response	Patient exhibits correct routine responses, which are robotlike. He or she appears oriented to setting, but insight, judgment, and problem solving skills are poor.
Level VIII	Purposeful, Appropriate Response	Patient correctly responds and carries over new learning. He or she needs no required supervision, and has poor tolerance for stress, and some abstract reasoning difficulties.

Source: Adapted from Hagen, C. and Malkmus, D. *Intervention strategies for language disorders secondary to head trauma.* American Speech-Language-Hearing Association, short courses, Atlanta, Georgia, 1979.

Horticultural therapy is most effective for individuals who are at Levels VI-VIII of the Rancho Scale. The focus of horticultural therapy treatment will depend on how the therapist and other team members feel it can best address the survivor's needs, and how it is utilized in treatment programming. Horticultural therapy is used in some programs to address specific areas such as vocational training or recreational/social skills. Other programs use horticultural therapy as a tool to address global rehabilitation goals such as generalization of compensatory strategies or psychological adjustment to life after the injury. The following goals and objectives demonstrate the diverse ways that horticultural therapy can be used in the rehabilitation of a brain injury survivor.

The functioning levels of the survivor will determine the complexity of the goals and objectives. Horticultural therapy can address a wide range of skill levels. Two examples of objectives will be given under each goal listed in the areas of physical, cognitive, behavioral, and emotional skills to illustrate how activities can match survivor needs.

Physical

• *Goal:*	To improve general physical conditioning and endurance.
Basic Objective:	Survivor will tolerate standing for ten minutes at potting bench while assisting therapist with transplanting activity.
Advanced Objective:	Survivor will work in greenhouse for two-hour block daily without a break, five consecutive days for two weeks.
• *Goal:*	To improve bilateral integration of upper extremities.
Basic Objective:	Survivor will repeatedly cross midline (a minimum of four times) when filling pot with soil, while holding trowel in nondominant hand, with therapist providing physical guiding.
Advanced Objective:	Survivor will take hanging baskets down from the suspended bar, water them with a watering can, and return them.

COGNITIVE

• *Goal:*	To improve ability to correctly sequence multistep task.
Basic Objective:	Survivor will correctly plant four-pack with seeds by following step-by-step verbal instruction given by therapist.
Advanced Objective:	Survivor will correctly complete daily closing routine of greenhouse by following written checklist.
• *Goal:*	To increase expression of organized thoughts.
Basic Objective:	Survivor will verbally state in a concise manner what action he or she just completed after each step of the chosen task.
Advanced Objective:	Survivor will verbally explain and demonstrate to the therapist how to take a stem cutting. Success will be judged by the therapist's ability to follow the survivor's directions.

BEHAVIORAL

• *Goal:*	To decrease impulsive actions.
Basic Objective:	Survivor will count to ten while applying water to each pot before moving on to the next pot on the bench.
Advanced Objective:	Survivor will consult a completion checklist for each task assigned when he or she feels the task is done and before moving on to the next task to determine if it was done completely.
• *Goal:*	To improve initiation.
Basic Objective:	Survivor will start the task independently after the therapist gathers and sets up necessary materials to make started mix.
Advanced Objective:	Survivor will independently consult job board upon entering greenhouse and begin the number-one-ranked task.

EMOTIONAL

• *Goal:*	Improve ability to self-monitor anger.
Basic Objective:	Survivor will discuss with therapist what caused an outburst during their session after he or she has regained composure.
Advanced Objective:	Survivor will self-initiate a five-minute time-out when working in the greenhouse if he or she anticipates an outburst, and will later process the situation with the therapist.
• *Goal:*	Develop relaxation techniques.
Basic Objective:	Survivor will sit in greenhouse for fifteen minutes while listening to music of his or her choise two times each week.
Advanced Objective:	Survivor will identify one to three horticultural activities that he or she finds relaxing and will schedule one into his or her weekly calendar no less than two times each month.

ADAPTATIONS OF ACTIVITIES

Cognitive impairments will be the most limiting barrier to independent functioning (Dow, 1989; Bray, 1987). Horticultural therapy provides opportunity for early successes because the work is relatively simple and may be highly structured. The therapist must present information to the survivor in the modality that is best received. Depending on where the brain has been damaged, this may be in a written, spoken, pictorial, or demonstrated form. It may also require a combination of any or all of these forms. Multisensory learning is more easily retained. The survivor may not indicate that he or she did not understand or is a bit confused. The therapist must be prepared to try another approach if a problem arises.

General disorientation and/or memory deficits may require massive repetition of a particular task or routine before the survivor can do it independently. Verbal cues, checklists, and pictures provided initially can then be gradually faded as the survivor gains competency. Distractibility requires that the environment be made as quiet and still as possible. As the survivor's attention to task improves, distracters can be introduced to increase tolerance of environmental stimuli. External structure is a compensatory technique that helps the survivor make sense of and put into order the world around him or her. A clean, well-organized workspace is essential. A tool board with traced outlines of where each tool belongs provides cues to a survivor with poor visual memory. Job cards with sequenced step-by-step directions provide cues to a survivor with poor verbal memory. Items always located in the same place and a starting routine that is always the same provide order for a survivor with consistency or orientation problems.

All horticultural activities can be graded up or down in complexity and structure by the therapist through alterations in the amount and type of assistance that is given. Problem solving is always encouraged. Missing materials or a broken tool planned by the therapist can present an opportunity for the survivor to practice problem-solving steps independently or with the help of the therapist.

Physical impairment is common for a survivor. The extent and the recovery of functional loss depends on the cause and seriousness of the impairment. Some physical impairments are caused by physical injury suffered in the accident that also caused the brain injury. These may be temporary, such as broken bones, sprains, and lacerations. However, some physical injuries may also cause long-term deficits such as peripheral nerve damage, soft-tissue damage in the back or at joints, and partial or total loss of limbs (Bray, 1987).

Other physical impairments, which may be temporary or permanent, are caused by neurological injury and damage that effect the body's motor function. Loss of gross and fine motor control may improve as brain swelling goes down and the brain recovers. Neuronal firing and the internal muscle mechanisms responsible for controlling the amount and length of muscle contraction can be impaired due to brain-tissue damage. As a result, survivors may have permanent problems with flaccid muscles, spastic muscles, ataxia, or contractures. A survivor may become a wheelchair user as a result of extensive neuromotor damage to the central nervous system. Activities and tools must be adapted to accommodate physical impairments. Consultation with the physical or occupational therapist on the treatment team provides direction for the horticultural therapist when it comes to positioning of the survivor and the required adaptation of tools.

Visual and perceptual impairments are another consequence of brain injury. Physical damage to the optic nerve may cause permanent vision loss. Damage to parts of the brain responsible for interpreting what the eye sees may cause visual field losses or diplopia. Perceptual deficits may include problems with depth perception, figure and ground, visual closure, and scanning (Dow, 1989). The therapist must be aware of any visual problems and adapt the workspace so that items are within the survivor's visual field. Perceptual skills can be addressed by using sharply contrasting colors and by providing compensatory tools to mark depths, and visual cues to mark fill lines. Consultation with the survivor and other members of the treatment team is always an important source of information when adapting activities (see Figure 5.5).

FIGURE 5.5. Special Tools and Equipment

- Lightweight tools for persons with decreased strength or use of only one hand

- Weighted cuffs for forearms of persons who have tremors in their hands (can increase control)

- Attachable extra handles for long-handled garden tools for persons with decreased strength, coordination, or use of only one hand

- Built-up handles for persons with weak grips, cone-shaped ones for persons with hands in spastic flexion

- Plastic-covered job cards outlining each discrete step in simple terms, printed in large type, with healthy spaces between the lines so it is easy to read

- Posters and pictures of important things to remember posted throughout work area

- Gloves for people who are tactile defensive or can't or won't get their hands dirty

- Chalkboard or dry-erase board for use as job board

- Flags to use as memory markers when persons have poor visual memory, which can be placed to help them remember where they left off when watering, which benches have been done, etc.

- Kitchen timers for persons with memory deficits or time disorientation to help keep track of time deadlines

- Labels on everything

PRECAUTIONS

Be aware that environmental conditions in a greenhouse or garden may be problematic for some survivors. High temperatures and humidity may trigger negative physical or behavioral reactions. Sunlight may also be problematic due to medication, vision, or neurological sensitivities (may trigger headache).

Severe stress can cause deterioration in the performance of cognitive and/or physical skills because it can affect attention and increase muscle tone. The therapist should monitor the complexity of the cognitive and physical components of each activity because of the demands they place on the integrated system. A complex cognitive component may require coupling with a simple motor response and visa versa in order to allow for success (Bray, 1987).

Medications commonly prescribed for survivors may have side effects or contraindications to be aware of when planning activities. Antidepressants may affect fine motor control, balance, vision, fatigue level, and heart rate. Antianxiety agents may cause drowsiness and decreased coordination in initial phases. The most commonly prescribed mood stabilizer, lithium, causes increased thirst and urination and a mild tremor. Perspiration causes concentration of lithium levels in the blood. Be sure to provide adequate hydration and salt replacement, and be careful of strenuous activity in warm weather. Anticonvulsants may affect cognitive abilities and cause irritability and restlessness (O'Shanick and Zasler, 1990). Survivors may experience posttraumatic epilepsy (PTE). Be trained in how to respond to a seizure.

Injury to the frontal lobe is believed to have an affect on a survivor's ability to control emotions (Sbordone, 1990). Some survivors easily become frustrated, agitated, or angry. When working with a survivor known to be emotionally explosive, be thoughtful in planning the work area and the tools to which he or she will have access. Be trained in the use of de-escalating techniques.

OBSERVATION/DOCUMENTATION
AND CLIENT FEEDBACK

Observations made in horticultural therapy sessions can be of great value to the treatment team. The structure and atmosphere of horticultural therapy will be unlike many of the other therapies. Horticultural therapy provides goal-directed activity with visible consequences. It allows the survivor to be a caregiver after a long period of being a care receiver. The environment allows for relaxation and improved performance. This engagement in purposeful, in vivo activity offers opportunity to observe things such as spontaneous problem solving, contextual learning, compensation, and decompensation. When writing a session summary, consider commenting on the following areas:

- Quality of movements
- Visuospatial skills
- Sensory skills
- Visuomotor coordination
- Perceived level of stress
- Affect and conversation

No matter how mild or severe the brain injury, the survivor can lose sight of why he or she is participating in horticultural therapy. Concrete thinking does not allow the survivor to see a relationship between current activities and ultimate goals. It is very important to explain the purpose of each and every session before you start. Discuss what objective in the survivor's treatment plan is being addressed in the session. Review what gains you have seen and what skill or behavior you are hoping to observe. At the end of the session, evaluate the survivor's performance. Include positive observations as well as shortfalls. Brain injury can result in reduced insight and denial of deficits. Ask the survivor for feedback about his or her performance. Together discuss other ways to approach the activity. It is very important for the survivor to feel that he or she is a part of the treatment team and that his or her opinion counts. It is important to remember that denial, inattention to detail, loss of motivation, emotional lability, and inflexibility are caused by biological dysfunction in many survivors and are not behaviors they engage in by intent (Prigatano, 1990).

Many programs offering services to persons with brain injuries are accredited by bodies such as the Commission on Accreditation of Rehabilitation Facilities (CARF) and the Joint Commission on Accreditation of Healthcare Organizations (JCAHO). These bodies require certain documentation standards, and so it is necessary to be a good record keeper; know what it is you are tracking before a session begins. A progress note about the survivor's goals and objectives written during the last five to ten minutes of a session is a good way to include the survivor in the evaluation process. Elements to consider in documentation include the following:

- State the type and amount of assistance required: Was it necessary to provide maximum physical assist, minimum verbal cuing, etc.?
- Record the amount of time required to complete a task. It is helpful to have "norms" for how long it takes a noninjured person to complete tasks you are timing.
- Record the frequency of a behavior you are encouraging (asking for help) or trying to extinguish (excessive talking).
- Record the length of engagement time in a behavior (attention to task).

• State the type of compensatory strategies used.

Notes and data collected via checklists each session can be totaled and summarized for monthly progress reports and qualitative assurance purposes.

CASE STUDIES

The experiences of three TBI survivors, Tina, Ed and Jim, demonstrate the benefits of horticultural therapy for this particular disability group at dramatically different levels during a lifelong recovery process. Tina uses horticultural therapy in a rehabilitation hospital as she struggles to overcome the physical and cognitive disabilities she sustained from a recent automobile accident. Horticulture therapy enables her to undertake rehabilitation through an activity she has always enjoyed. Ed uses horticultural therapy many years postinjury, as he tries to improve his independent work and living skills and to transfer these skills to a new vocation. Jim uses horticultural therapy as a leisure and vocational activity that can help boost his self-esteem and help him relate to others. The experiences of each individual demonstrate the benefits of this nonthreatening, nurturing therapy.

Tina

Before her automobile accident in April 1995, Tina enjoyed life as a wife, mother of two young daughters, and a hospital secretary. Her traumatic brain injury took away her ability to walk, to talk, to eat regular foods, to hold her children. Two-and-a-half months after her accident, she was transferred from an acute-care hospital to a 141-bed specialty hospital for physical medicine and rehabilitation. Both of her legs and one arm were in casts; she couldn't stand alone. She was depressed and she resisted making an effort. She cried during her regular physical therapy due to physical pain and her mental frustration.

Like many TBI survivors, Tina remembered what her life was like before her brain injury; the accident affected only her short-term memory. She was trapped with a brain that could no longer make her body work. Several weeks after arriving at the rehabilitation hospital, Tina cried bitterly and spelled out a heart-wrenching message on her communication letter board, "How did this happen? Did anyone else get hurt?"

Along with speech, physical and cognitive therapy sessions, Tina began visiting the rehabilitation hospital's horticultural center. The center has a state-of-the-art greenhouse, enabling garden and therapy work space for TBI and stroke survivors. Immediately, the staff could see Tina begin to relax. Tina had always been an active gardener. In this warm, sunny setting, filled with the smell of earth and plants, Tina could enjoy a piece of her life that seemed normal.

While Tina produced plant cuttings and placed them in soil, she worked to improve her mobility, balance, endurance, memory, and sequencing skills. When her therapist told her to stand, she didn't resist as she often did in physical therapy sessions. She stood from her wheelchair in order to reach plants she wanted to cut. When her therapist gave her scissors to hold, she didn't concentrate on improving her muscle strength and coordination; she concentrated on the plants that needed her attention. In this environment, Tina began to communicate more freely with her psychologist, who often visited the horticultural center with her. She began to come to terms not only with her accident and injury but her need to work hard to make improvements, despite the pain of therapy. Working with plants was more than therapy for Tina. Coming to the center became a goal in itself, as Tina was promised more horticultural therapy time as a reward for a good physical or speech therapy session. As her muscles began to strengthen, Tina's morale began to improve. She cried less and laughed more. She stood longer. She worked with more confidence.

When asked if horticultural therapy really helped, Tina emphatically and rapidly spelled out the following answer on the letter board she cradled in her lap: "THIS IS THE ONLY JOY FOR ME."

Ed

Ed was always described as the all-American kid, a college freshman who not only enjoyed sports but worked with children to coach them in football, baseball, and wrestling. Ed was in an automobile accident in 1977, which left him with a traumatic brain injury. His physical disabilities include decreased vision in his left eye, dysarthria, right hemiparesis, and a right-sided tremor involving his arm and leg. When walking, Ed leans towards his right side and often uses toe-heel steps. This once gentle person is prone to fights and aggressive outbursts, which are compounded by his disinhibition and inappropriate behavior around females. Ed continues to be quite verbal and has been able to talk about his problems.

Since 1985, Ed has participated in various residential programs for brain-injured adults, including one that utilizes horticultural therapy in four main areas of treatment: vocational assessment, work readiness, leisure skill development, and supported employment. As Ed became more capable of monitoring his behavior, he became involved in this horticultural therapy program. The reason was not necessarily Ed's love of plants, though he did enjoy them. Ed's work with plants provided training for future supported employment and work readiness.

Ed began working two-and-a-half days per week as an interior landscaper at a rehabilitation center. His primary tasks are to water plants, clean leaves, and prune plants that are placed throughout the hospital and in staff offices. To prepare for his rounds, Ed is required to check his work cart for water, supplies, and his plant checklist (which helps him remember proper care for the plants). Ed receives maximum support and supervision for his tasks; a job coach is always at his side. While Ed is learning a lot about plants, other work-readiness goals are:

- to learn how to communicate with those around him by making conversation that is appropriate in content, volume, and length;
- to knock on doors before entering offices and to introduce himself;
- to be quiet if a doctor or nurse is on the phone and to continue working;
- to say "thank-you" and "good-bye" at appropriate times and to move on;
- to focus and stay on task and to improve his ability to follow increasingly complex directions and sequences;
- to dress appropriately for work;
- to get to work on time;
- to judge how much time is needed to complete his tasks; and
- to monitor his break time independently.

Watching Ed push his cart through the hospital hallways makes it clear that he takes his job seriously. He is able to set up his watering cart and maneuver it to each plant location. He carefully uses a device to check a plant's moisture level to determine if the plant should be watered. Occasionally, Ed requires a cue to stay focused or needs a simple reminder of the next step in the plant-care sequence. Recently, Ed chose not to use a hand-written chart of watering steps to track each individual plant activity. He decided he was familiar with the routine and no longer needed the written format.

His job coach reports that through his work, Ed's attitude and behavior have matured. He is no longer an irresponsible, confrontational adolescent but a person with growing self-awareness and a renewed determination to assume responsibility for his life. The job coach confirmed that his communication skills have improved and that he demonstrates the ability to accept feedback about his work performance. He has become increasingly motivated to get to work on time, dress well, and improve his grooming skills.

Ed is now thirty-five years old. His coach observed, "He seems to understand the beginnings of what 'being professional' means."

Jim

Jim has always enjoyed gardening and taking care of plants. He still does, fifteen years after a work-related accident left him with a traumatic brain injury.

Since his accident, Jim, who is now fifty-eight years old, has participated in several residential programs and has progressed to living independently in an assisted-living community apartment program. Several years ago, Jim used his free time to get involved with a rehabilitation facility's horticultural therapy program.

Plants and vegetables have always been a hobby for Jim. He enjoys buying plants and putting them in his apartment. He takes care of them independently, but needs some assistance in making cuttings or starting rootings. This past summer, he suggested planting an outdoor garden. The garden produced enormous tomatoes that Jim delighted in sharing with the rehabilitation facility's staff.

Jim says he enjoys seeing things grow. It is an emotionally beneficial activity. He works in a supported employment program four days a week, and the plants and summer gardening give him a recreational activity that he can share with the community where he lives and works.

Jim's wife and son visit him regularly. His wife encourages his gardening activity and recently sent him a fence to put around his tomato garden. Jim takes great pride and pleasure in his accomplishments in his garden or tending his plants.

FUTURE

It is ironic to suggest that one of society's oldest healing arts could be in financial jeopardy. However, in today's reality of managed health care, limited resources, daily-rate guidelines, and insurance restrictions, finding money to reimburse horticultural therapy services is a major challenge.

As a nontraditional therapy, horticultural therapy is rarely if ever reimbursed as a line item; providers must fold horticultural therapy into a package of reimbursable services that address each client's needs. In Ed's case, for instance, horticultural therapy applied to work-readiness rehabilitation can be reimbursed. For Tina, horticultural therapy was part of a total therapy package geared to meet her acute rehabilitation needs, ranging from physical and cognitive to behavioral and emotional recovery.

Another irony, the funding crunch for horticultural therapy, comes at a time when the very nature of such therapy fits in particularly well with a movement called "real-life rehabilitation" for people with brain injuries. Real-life therapies are based on the belief that how people live, what they do vocationally-educationally, and how they spend their leisure time should be the focus of TBI's post-acute, comprehensive rehabilitation services. This model targets TBI survivors like Jim and resources like horticultural therapy to help teach TBI clients the skills they need to progress in real life—to overcome the chronic problems of social isolation, unemployment, divorce and family upheavals, lack of leisure activities, and diminished independent living skills.

Without a doubt, horticultural therapy integrates the skills developed in formal therapies into every aspect of daily life. Horticultural therapy is not something designed solely for a hospital setting. Rather, it teaches skills that are carried over into a person's community, home, work, and leisure activities. As everyone agrees, TBI survivors need life-long supported care. Their progress and length of recovery can vary so dramatically that some survivors are ready for rehabilitation after five years, some ten years after the injury.

As a lifelong therapy, horticultural therapy is very cost-effective. Dollars spent on horticultural therapy at any stage of a person's rehabilitation contribute to a real-life activity that survivors who need lifelong support can continue to build on. As seen through the case studies, horticultural therapy can even fulfill different needs at different times in a survivor's recovery process. Initially, horticultural therapy may be strictly therapeutic but later fill a void that ranges from occupational to social to leisure. It may be something new in a person's life or reminiscent of their life before brain injury.

A key challenge for proponents of horticultural therapy is to make sure the benefits of this particular therapy are recognized by case managers, who must fight for the funds to continue its usage. Horticultural therapy must not be viewed as the dispensable "fun with plants" therapy; rather, its myriad of therapeutic and real-life benefits must be repeatedly emphasized and documented in order to be taken seriously.

The experts agree that the limitations of horticultural therapy for TBI survivors must be reviewed honestly. If the skills learned in horticultural therapy cannot be transferred out of therapy into occupational settings for the majority of people, this particular aspect of its therapeutic value may need further evaluation.

Though the debate continues over how to pay for horticultural therapy, those familiar with its merits agree that on many levels one of the oldest healing arts still works—a real-life therapy that can play a vital role not only in the recovery but in the life of a brain-injury survivor.

KEY ORGANIZATIONS

American Academy of Physical Medicine and Rehabilitation: 127 South Michigan, Suite 1300, Chicago, IL. 60603. Call (312) 922-9366.

American Horticultural Therapy Association: 362A Christopher Avenue, Gaithersburg, MD. 20879. Call (301) 948-3010.

Brain Injury Association, Inc. (formerly National Head Injury Foundation, Inc.) 1776 Massachusetts Avenue NW, Suite 100, Washington, DC 20036. Call (202) 296-6443; Family Helpline (800) 444-6443. Chapters and state associations all over the country.

GLOSSARY

ataxia: Muscular incoordination characterized by tremors and unsteady gait

ADL: Activities of daily living, which are tasks performed daily, such as dressing, bathing, grooming, and feeding

aphasia: Loss of the ability to understand and/or communicate through speech, reading, and writing, featuring two types: **expressive aphasia**, in which a client knows what heor she wants to say, but cannot say it; **receptive aphasia**, in which a client cannot understand the meaning of the spoken or written word

brain injury: A traumatic insult to the brain resulting in temporary or permanent cognitive, physical, and behavioral/emotional changes

confabulation: To make up a story or response when one does not know or remember what occurred and to be frequently unaware that one is confabulating

cue: A verbal or nonverbal reminder to elicit or modify a response (e.g., What's next on your schedule?)

disinhibition: Lack of control over impulses, in thoughts as well as actions

dysarthria: Neurologically based speech difficulty resulting from the inability to coordinate musculature for speech production; slurred speech due to paralysis or weakness of the mouth and facial muscles

feedback: Sharing perceptions or evaluations to increase awareness of behavior, for both staff and clients

field cut: Visual disturbance in which part of the visual field is not seen

figure ground: The ability to distinguish the foreground (main object) from the background

frontal lobe: Front area of brain responsible for motor functions and executive functions that include abstract thinking, judgment, impulse control, and problem solving

hemiplegia: Paralysis of one side of the body

paresis: Diminished motion due to muscle weakness

perseveration: Repeated response or thought, in which the client is unable to move to a different statement or thought

pre-set: A set of instructions, strategies, or reminders that are reviewed with the client before the client begins an activity

pro-social behaviors: Behaviors typically desired and accepted by society (e.g., rules of common courtesy, resolving conflicts with verbal resolutions rather than physical aggression)

shearing: Type of brain lesion that results from traumatic brain injury and refers to tears in nerve fibers

strategy: A specific technique developed to help the client accomplish a task or exhibit the desired behavior (e.g., a checklist to help the client remember morning hygiene tasks such as brush teeth, wash face, and comb hair) or an exercise to control behavior (e.g., "Count to ten and then tell me what is upsetting you")

TBI: Traumatic brain injury of two types: a **closed brain injury,** in which there is injury to the brain but the skull is not fractured; and an **open brain injury,** in which there is an injury to the brain and the skull is fractured

treatment plan: Goals developed by the team and the client which target the skills and behaviors the client needs to develop in the areas of residential, vocational, and recreational

visual scanning: The ability to move the eyes smoothly in vertical and horizontal planes to track an object

ADDITIONAL RESOURCES

Archives

Archives of Physical Medicine and Rehabilitation. Suite 1310, 78 East Adams Street, Chicago, IL. 60603. Call (312) 922-9371.

Books

Educating Families of the Head Injured: A Guide to Medical, Cognitive, and Social Issues. Dana S. DeBoskey, Jeffrey S. Hecht, and Connie J. Calub. Gaithersburg, MD: Aspen Publishers, Inc., 1991, p. 30.

Journals

Brain Injury. Taylor & Francis, Ltd., Publishers. 4 John Street, London WC1N 2ET, UK.
The Journal of Head Trauma Rehabilitation. Aspen Publications, Inc.
NeuroRehabilitation. Andover Medical Publishers, Inc. 125 Main Street, Reading, MA. 01867.
"A Transformation to Real-Life Rehabilitation." David L. Strauss and Vicki L. Schaffer, ReMed. *Continuing Care Magazine*, July/August, 1995, pp. 31-35.

REFERENCES

Brain Injury Association, Inc. (1995). Fact Sheet, Traumatic Brain Injury. Washington, DC: Brain Injury Association, Inc., p. 1.
Bray, L., Carlson, F., Humphrey, R., Mastrilli, J., Valko, A. (1987) Physical Rehabilitation. In *Community Re-Entry for Head-Injured Adults.* Edited by M. Ylvisaker and E.M. Gobble. Boston: Little, Brown & Co., pp. 25-86.
Dow, P. W. (1989) Traumatic Brain Injuries. In *Occupational Therapy for Physical Dysfunction.* Edited by Catherine Trombley. Baltimore: Williams and Wilkins, pp. 484-509.
O'Shanick, G.J., and Zasler, N.D. (1990) Neuropsychopharmacological Approaches to Traumatic Brain Injury. In *Community Integration Following Traumatic Brain Injury.* Edited by J.S. Kreutzer and P. Wehman. Baltimore: Paul Brookes Publishing, pp. 15-27.
Prigatano, G.P. (1990) Neuropsychological Deficits, Personality Variables, and Outcome. In *Community Integration Following Traumatic Brain Injury.* Edited by J.S. Kreutzer and P. Wehman. Baltimore: Paul Brookes Publishing, p. 14.

ReMed Recovery Care Centers. Statistical Research (1996) and Staff Reports (1992). 625 Ridge Pike, Conshohocken, PA.

Sbordone, R.J. (1990) Psychotherapeutic Treatment. In *Community Integration Following Traumatic Brain Injury.* Edited by J.S. Kreutzer and P. Wehman. Baltimore: Paul Brookes Publishing, p. 140.

Straus, Martha C. and DiDonato, Barbara A. (1994) *Horticulture Handbook.* ReMed. Conshohocken, PA, p. 1.

Chapter 6

Developmental Disabilities and Horticultural Therapy Practice

Pamela Catlin

INTRODUCTION

A developmental disability is defined as an impairment which originates before the age of eighteen, which may be expected to continue indefinitely, and which constitutes a substantial disability. Such conditions include pervasive developmental disorders, cerebral palsy, and mental retardation (Powers, 1989). There are a number of issues to consider when establishing programs for people with developmental disabilities. This population has both large numbers and widely varying degrees of functioning within it. Programs must be developed to meet the many different needs and ability levels of the individuals being served.

Life expectancy for some people with developmental disabilities, such as people with Down's syndrome, is increasing, as is the frequency of Alzheimer's disease in a large portion of people with Down's syndrome over the age of thirty. Because of these factors, attention to the recreational and continued therapeutic needs of the older individual with developmental disabilities should be addressed.

More and more individuals are living independently or semi-independently. They need vocational skills, age-appropriate leisure skills, community activities, and an increased sense of responsibility. The majority of these people have the ability and desire to be productive members of their community. Statistics, however, show that only 7 to 23 percent of adults with mental retardation are employed full-time, and only 9 to 20 percent are employed part-time (ARC, 1993). Sixty-three percent of individuals out of school for three to five years are unemployed (ARC, 1994).

The purpose of this chapter is to present the many considerations encountered in developing and implementing a horticultural therapy (HT) program for people with developmental disabilities. The learning objectives for this chapter are the following:

1. Identify and describe the following characteristics and needs of the individuals:
   - physical
   - mental
   - social
2. Identify steps to establishing an appropriate treatment plan and use of HT
3. Identify considerations of the future of HT with this population
4. Identify resources available in this area

SIGNIFICANCE OF HORTICULTURAL THERAPY PROGRAMMING

Physiological Benefits

Individuals have many opportunities to move physically in a horticultural therapy program. Projects can be constructed to enhance fine and/or gross motor skills. Up and down physical exercise is automatically a part of an outdoor garden program. Walking to obtain supplies or water plants is a natural part of the routine. Even for those who take a more passive approach to the outdoor garden, just being in the fresh air and sunshine is a benefit.

One of the biggest challenges that many people with developmental disabilities face is adequate physical fitness. The horticultural therapy program does not meet all those needs but it is a step toward that end goal of more physical activity. Exercise tends to be more fun when it is purposeful, and having more fun results in more exercise.

Psychological Benefits

The primary psychological benefit from HT is an increase in self-esteem. Properly designed HT programs can meet the needs of individuals with varying levels of ability. The span of horticultural tasks is wide enough to provide activity that is satisfying and functional to almost anyone. Working with plants, either in the job setting or recreationally, is something that these individuals can have in common with the general population in their community. It is one more way of "fitting in." Handling living plants, watching them grow, and being able to create something that can be given as a gift, are all facets of building a healthy sense of self.

Activities can be designed to enhance creativity and self-expression. Projects such as making a garden collage and flower arranging are stepping stones to understanding oneself and one's surroundings better. The physical involvement in projects, such as turning the garden soil, raking leaves, or weeding can be an excellent avenue for redirecting aggressive behaviors, a common occurrence within this population.

Recreational Benefits

Horticultural activities can provide a broad range of age-appropriate leisure skills for individuals who have tended to use television as their primary recreational outlet. Gardening is listed as one of the favorite leisure activities in the United States. The National Gardening Association has estimated that about three-fourths of the households in the United States are involved in some form of gardening. A horticultural therapist can develop a program in which people with developmental disabilities can discover the joys of gardening as well.

Socialization Benefits

Improving socialization skills is a goal for individuals with developmental disabilities. The HT program offers many opportunities for social interaction. Activities often are done in group settings where learning to work together, sharing, and communicating effectively are natural parts of the process. In a vocational program the participant learns social skills appropriate for the workplace. Many opportunities for socializing with the general public arise naturally by visiting public gardens, nurseries, greenhouses, and parks. Participating in community fairs and flower shows allows the individual with a disability to share something in common with others.

The plant program provides a means for participants to create something they can give to others, opening them to another facet of social interaction. Learning age-appropriate skills in an HT program, the adult client is placed on the same level as a home gardener and can share this interest with other adults.

Cognitive Benefits

A horticulture program can stimulate increased cognitive development. Daubert and Rothert (1981) state that HT activities can be designed to provide a basis for teaching number concepts through projects such as repotting cuttings and sowing seeds. Vocabulary can be expanded through learning plant terminology. More complex skills can be taught using the plant program, such as geography, plant requirements, plant reproduction, plant parts, man's use of plants, ecosystems, and seasonal changes.

Vocational Benefits

Providing employment skills is an increasing focus of programs for adults with developmental disabilities. The field of horticulture offers many entry-level jobs that these individuals can be trained to do. Working in the nursery or greenhouse setting is also a safe, nonthreatening environment to learn basic job skills such as staying on task, following complex directions, being on time, wearing appropriate dress, and accepting feedback. These skills may be transferred into a nonhorticultural job setting.

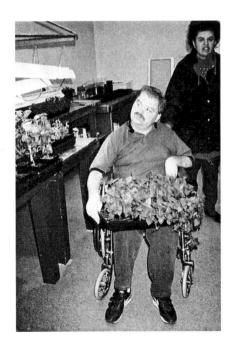

Harvesting the Bounty: Basil and other herbs are successfully grown and sold in vocational settings.

UNDERSTANDING THE CLIENT POPULATION

Definition of Developmental Disability

A key element in identifying a developmental disability is that it originated before the age of eighteen and may be expected to continue indefinitely. A common misperception is that any individual with a developmental disability has mental retardation. As a therapist, it is important to know the extent of a person's disability and to understand the various characteristics and issues that are a part of it. Following are descriptions of many of the conditions that fall into this category and the special considerations that people with these disabilities need in the HT program.

Autism

Autism is a physical disorder of the brain, characterized by impaired communication, extreme self-absorption, detachment from reality, and abnormal responses to sensory stimulation. According to Powers (1989), approximately 360,000 people in the United States have autism. It is the fourth most common developmental disability and occurs in four to five out of every 10,000 births. Aggressive behavior is typically displayed, often toward oneself, with severe actions such as head banging. One in four individuals with autism has seizures. Individuals may also have a diagnosis of mental retardation (70 percent), blindness, or deafness.

Some of the common medications used to control aggression with this population are thioridazine (Mellaril), chlorpromazine (Thorazine), and haloperidol (Haldol). These medications have side effects of sleepiness, problems with movement, blurred vision, and phytosensitivity. Phytosensitivity is an exaggerated sensitivity of the skin to the sun, resulting in severe sunburns or rashes. When involving an autistic individual in HT programming, all of these drug side effects and the possibility of seizures must be taken into consideration (Powers, 1989).

These individuals need programs that provide consistency and present little negative stimulation such as loud noises or overactivity in the immediate surroundings. Safety measures must be strictly observed at all times to avoid self-injury.

Cerebral Palsy

Approximately two out of every 1,000 people have some type of cerebral palsy (CP), a condition of muscular weakness and difficulty in coordinating voluntary movement owing to developmental or congenital damage to the brain. Typically, 50 percent of these individuals are prone to seizures. Other physical symptoms that may be present are respiratory problems, hearing loss (5 to 15 percent of the population), visual impairment, and contractures (shortening of the muscles and other tissues). Scoliosis, an abnormal lateral curvature of the spine, can be present, though this can be surgically repaired.

The location of the brain injury will determine other problems such as mental retardation (which occurs in 25 percent of individuals with CP), language disorders, and learning disabilities. Attention deficit hyperactive disorder (ADHD) occurs in approximately 20 percent of the children with CP. Individuals with CP may also experience sensory impairments. Movement is especially challenging for people who have both motor and sensory impairments. Not only is it difficult to make a specific motion, but it is also difficult to determine how much pressure is being applied (Gersh, 1991). For example, in a repotting activity, the gardener would have difficulty getting his or her hands to move to the proper position around the plant and he or she could also damage the plant stem and roots by pressing too hard.

Being aware of these impairments is important when designing the HT program. A high percentage of people with CP have average to above average intelligence. This is often not recognized due to difficulty with speech and movement. It is important to establish an accurate baseline for the individuals when developing an HT program for this population. The amount of mobility people with CP have varies greatly. Areas should be free of obstacles when designing a work space for people with CP.

Individuals may have different needs in terms of tools, and they are often the best ones to assist in determining what those needs are (see Table 6.1). The gardener working indoors and using mouth-held tools or a head wand will need an elevated work surface to ensure comfort in movement. The individual using toe-held tools will very likely need the floor surface raised somewhat for greater ease. Outdoors, a garden made from a bag of potting soil laid flat would be the most accessible for this person.

The cases of two avid female gardeners who are close in age illustrate how people with cerebral palsy garden in completely different manners. Lavon did everything with tools held in her mouth. She instructed me to make a long-handled, miniature tool set. We padded the handles of the tools with layers of Styrofoam for less wear on her teeth. With these tools, she did everything from mixing soil to transplanting tomato seedlings.

TABLE 6.1. Tools to Aid Gardeners with Cerebral Palsy

Challenge	Tool
Minimal or no use of arms	• head wand
	• mouth-held tools
	• universal cuff holding tool
	• toe-held tools
	• raised work surface
Difficulty with verbal communication	• communicaton board

Jeanne did everything with her toes. She requested that a small table, about three inches in height, be built. This allowed her to do indoor work holding simple tools, such as serving spoons, in her toes. One year at the fair, she received a purple ribbon for best hanging plant in a macramé hanger. She had macraméd the hanger, made the pot in ceramics, and planted the plant all with her toes. In the outdoor garden, Jeanne would garden in bags of potting mix, laid flat on the patio surface. Many of my gardeners with CP have chosen to work with their hands. Devices such as universal cuffs to hold a large spoon help relieve some of the frustration that comes with the lack of muscle control.

Down's Syndrome

Approximately one in every 800 to 1,100 live births is a baby with Down's syndrome (DS), a genetic disorder associated with the presence of an extra chromosome 21. Approximately one-quarter million families in the United States have a child with Down's syndrome (National Down's Syndrome Congress, 1994). For the affected child, this usually results in a delay in physical, intellectual, and language development. There is a wide variation in mental abilities and behavioral and physical development. For educational and vocational purposes, individuals are placed into one of three categories based on the level of retardation they experience. Those categories are mild, moderate, or severe mental retardation. Together with mental and behavioral disabilities that are a part of Down's syndrome, there are often physical disabilities that need to be addressed. Between 30 and 50 percent of people with DS have heart defects, and 8 to 12 percent have gastrointestinal tract abnormalities (National Down's Syndrome Congress, 1994).

Hearing loss and vision problems are often associated with DS, and 10 to 20 percent experience Atlantoaxial instability, a physical condition in which the ligaments between the first two neck bones are loose, resulting in reduced muscle strength and tone.

A concern has developed regarding older adults with Down's syndrome. The life expectancy of someone with DS has increased and it is now estimated that 80 percent will reach the age of 50. Studies are showing that all persons with DS past the age of 30 apparently have the plaques and tangles in their brains indicative of Alzheimer's disease. It is estimated that only 10 to 40 percent of these individuals actually develop the symptoms of Alzheimer's disease (National Down's Syndrome Congress, 1993). Appropriate leisure-time activities and socialization skills need to be developed for this group.

To address behavior issues, a successful vocational HT program involves training in appropriate communication and socialization skills as well as work sills. Goals that address increased fine and gross motor control are often incorporated in the HT sessions (see Photos 6.2 and 6.3).

Planting the garden, a favorite part of any gardening program.

Mixing soil in preparation for a planting project.

Mental Retardation

Mental retardation is a developmental disorder based on the following three criteria: intellectual functioning level (IQ) is below 70 to 75; significant limitations exist in two or more adaptive skill areas; and the condition is present from childhood from the age of eighteen or under. Between 2½ to 3 percent of the general population has some level of mental retardation. This translates to six to seven and one-half million people. The Association of Retarded Citizens (ARC) (1994) documents that one out of ten American families will be directly affected. ARC estimates that 87 percent of those individuals with mental retardation will be only mildly affected, slightly slower than normal in learning information and skills. The remaining 13 percent will have IQs below fifty and more serious limitations.

The *Digest of Data on Persons with Disabilities* (1992) notes that 15 percent of the population with mental retardation live in some type of institution and the remaining 85 percent live with families or in independent or semi-independent living situations. As states continue to deinstitutionalize, the latter percentage will increase.

The HT program should provide chronologically age-appropriate activity that also matches ability level. Behavior issues as well as fine and gross motor coordination are often challenges to be addressed in the HT setting. Repetitive tasks that do not require abstract thinking tend to be the job skills most readily learned. Many basic horticulture jobs such as watering, potting, and seeding do require making important judgments. In cases such as this, close supervision is necessary until the worker has consistently demonstrated the ability to correctly do the job.

Other Developmental Disabilities

People with spina bifida, fetal alcohol syndrome, and learning disorders fit into the category of having developmental disabilities. Spina bifida is a congenital defect in which part of the spinal cord protrudes through the spinal column, often resulting in neurological impairment. Fetal alcohol syndrome, caused by the mother's consumption of alcohol during pregnancy, is a cluster of birth defects that can include mental retardation and

Watering new seedlings.

physical impairments. Learning disorders is a large category made up of conditions characterized by difficulty in accomplishing certain tasks. This is most often diagnosed in school-age children. They have varying degrees of issues depending on the severity of the disorder.

Most individuals with developmental disabilities need individualized planning in terms of their work, and social and home life. These plans should include the development of skills in communication, socialization, work, daily living, and leisure time. It has been demonstrated that an individual's success in a job is not as closely related to IQ as it is to levels of interest and ability. This finding demonstrates the importance of dealing with the individual and his or her hidden potential.

HORTICULTURAL THERAPY AND TREATMENT

Identification of Treatment Issues

In a horticultural therapy program, treatment issues for people with developmental disabilities are based on the following areas:

a. health and physical development
b. cognitive development
c. communication skills
d. psychosocial skills
e. self-help skills

The initial identification of what treatment issues need to be addressed is achieved through testing and determining a baseline. A baseline is a standard guideline showing an individual's level of functioning at the time of testing. Two tests to determine baselines are the Vocational Attitudes and Behavioral Rating Scale and the San Francisco System.

Some treatment issues appropriate to an HT program are

a. attention to task,
b. age-appropriate leisure activity,
c. ability to follow simple and complex instructions,
d. appropriate interpersonal communication skills, and
e. improved or maintained fine and gross motor skills.

Objectives to Meet Treatment Goals

Objectives are the steps taken to achieve a goal. The time frames and methods of measurement can vary greatly from one individual to another, depending on their functioning levels. Objectives can start small and build up incrementally to meet the goal.

Identification of Treatment Issues and Goals

The following are five examples of plans presenting treatment issues, goals, HT goals, and objectives:

1. Treatment Issue: Attention to task

Goal: Individual will consistently stay on task for fifteen-minute time segments.

HT Goal: Individual will demonstrate five consecutive times the ability to stay on task while potting up rooted cuttings, for a minimum of fifteen minutes each time.

Objectives:

- Individual will maintain attention to task of potting for a minimum of five minutes, in a heavily trafficked area, for five consecutive sessions.
- Individual will maintain attention to task of potting for a minimum of ten minutes, in a heavily trafficked area, for five consecutive sessions.
- Individual will maintain attention to task of potting for a minimum of fifteen minutes, in a heavily trafficked area, for five consecutive sessions.

2. Treatment Issue: Appropriate leisure activities

Goal: Individual will participate in age-appropriate leisure activities other than television for one month.

HT Goal: Individual will care for the outdoor garden at living quarters on a daily basis for the month of July.

Objectives:

- Individual will groom the plants in the garden for a minimum of thirty minutes on Tuesdays and Saturdays during the month of July.

- Individual will water garden for twenty minutes on Mondays, Wednesdays, and Fridays during the month of July.
- Individual will harvest vegetables and flowers from the garden on Thursdays and Sundays during the month of July.

3. Treatment Issue: Ability to follow complex directions

Goal: Individual will follow multistep instructions on a consistent basis.

HT Goal: Individual will be able to carry out five-step seedling transplanting process, unassisted, three consecutive times.

Objectives:

- After receiving full instruction, individual will demonstrate how to fill trays with soil correctly three consecutive times.
- After receiving full instruction, individual will demonstrate how to correctly use dibble on soil-filled tray three consecutive times.
- Individual will correctly demonstrate first two steps in sequence three consecutive times.
- After receiving full instruction, individual will demonstrate how to remove seedlings from seedbed correctly three consecutive times.
- After receiving full instruction, individual will correctly demonstrate transplanting seedlings into soil-filled tray.
- Individual will demonstrate correctly all steps learned, in sequence, three consecutive times.
- After receiving full instruction, individual will demonstrate watering seedlings correctly three consecutive times.
- Individual will correctly demonstrate all steps in sequence three consecutive times.

4. Treatment Issue: Appropriate interpersonal communication skills

Goal: Individual will display appropriate social skills in a group setting a minimum of one hour each day for a minimum of one month.

HT Goal: Individual will communicate daily in an appropriate manner during the hour-long plant class for the month of July.

Objectives:

- Individual will refrain from:*
  — swearing
  — hitting
  — name calling
  — shouting
  for the first fifteen minutes of plant class.
- * This objective can then be expanded to thirty minutes, forty-five minutes, and then an hour.

5. Treatment Issue: Improved fine motor control

Goal: Individual will demonstrate improved fine motor skills measured by time required for completion and quality of work.

HT Goal: Individual will create a minimum of one corsage daily for five consecutive days.

Objective:

- Individual will create one or more corsages in a one-hour time span, following verbal and physical instruction, receiving physical assistance as needed for five consecutive days.

ADAPTATION OF HORTICULTURAL ACTIVITIES TO MEET TREATMENT OBJECTIVES

It is important to create the setting most likely to achieve the results desired in developing the horticultural activities. Building a sense of self-esteem is important. A project such as transplanting annual seedlings can be complex, involving all the steps from the beginning to the final product. The same project can be broken down into steps to simplify and allow for greater success for others. This approach is especially helpful in a group setting where participants have varying levels of ability. Lower-functioning individuals might carry out the mixing of the soil and filling of the cell packs, while higher-functioning individuals would carry out the actual transplanting. Creating a program environment that is nonthreatening, safe, and structured is advised since many individuals with developmental disabilities also have behavior issues. Individuals with severe cognitive deficits often lack the ability to think abstractly and will take what they hear quite literally. A horticultural therapist comments:

I learned this in a humorous manner at a time when I had been working in the field for only a few years. I had placed a bouquet of our garden flowers on the table and provided all the participants with paper, paints, and brushes, saying "I would like you to paint these flowers." I turned my attention away for just a moment and upon returning to the group, saw one of the individuals actually painting the flowers, instead of painting a picture of them! I could only laugh as she was doing exactly what I had instructed her to do. It was a great lesson in choosing words wisely.

In planning the horticultural activity, remember the needed space, tools, materials, and staffing. The following examples show how to set up activities to meet goals and objectives.

1. Activity: Potting Up Rooted Cuttings

Purpose:	To increase individual's ability to stay on task
Time:	Build up from five minutes to fifteen minutes
Space Required:	Workbench in greenhouse area
Materials Needed:	• one full flat of rooted cuttings
	• twenty-five, four-inch pots
	• prepared soil mix

Procedure:

Week 1

1. Review with individual the process for potting up rooted cuttings.
2. Discuss with individual the goal of staying on task for fifteen minutes and the process of starting out with five minutes and building to the full time.
3. When individual is ready, have them start potting and stop them at five minutes.
4. If individual removes attention from task, stop, bring their awareness to the action, and start step three again.
5. Continue steps three and four until individual stays on task for the full five minutes.

Week 2

1. Follow steps one through five from week 1, changing from five minutes to ten minutes.

Week 3

1. Follow the same steps as in weeks 1 and 2, increasing time to fifteen minutes.

2. *Activity: To Care for the Outdoor Garden for the Month of July*

Purpose:	To provide age-appropriate leisure activity on a daily basis
Time:	fifteen to thirty minutes
Facilities Required:	Outdoor garden at group home

Materials Needed:	• Tuesday and Saturday
	  - hand clippers
	  - hand cultivator
	  - bags for cleanup
	
	• Thursday and Sunday
	  - large basket(s)
	  - hand clippers
	  - bucket of warm water
	
	• Monday, Wednesday, and Friday
	  - hose
	  - water wand
	  - egg timer
	  - rain gauge
	
	• Every Day
	  - sun hat
	  - gloves
	  - sunscreen
	  - drinking water

Procedure:

Monday, Wednesday, Friday

1. Check rain gauge. Follow further steps only if less than one inch of water has accumulated.
2. Get water wand and hose from garage.
3. Connect hose to faucet at side of house.

4. Connect water wand to hose.
5. Turn water on and adjust pressure to proper level.
6. Set timer for twenty minutes.
7. Water garden by directing water to soil surface and walking along garden, covering all of soil surface, repeating until timer goes off.
8. Turn off water.
9. Coil hose into circle.
10. Take water wand off hose and store both hose and wand in garage.
11. Empty rain gauge and put back in its place.

Tuesday, Saturday

1. Cut off all dead blossoms, cutting just above a leaf, and place in bag.
2. Remove any brown or dead leaves on plant or on soil surface and place in bag.
3. Using hand cultivator, gently pull out weeds, being sure to get roots. Place in bag.
4. Check for insects or diseases and notify staff if any problems.
5. After at least thirty minutes have passed, throw bags of weeds, etc., in garbage or composter and put away tools.

Thursday, Sunday

1. Cut bouquet of fresh flowers, cutting stems just above a leaf.
2. Place immediately into bucket of warm water and set inside.
3. Using clippers, cut any vegetables that are ready to harvest and place in basket(s).
4. Bring vegetables into kitchen and wash off with cold water in sink.
5. Ask staff for further instruction.

Observation, Documentation, and Client Feedback

It is important to be the observer as well as the facilitator when working with an HT program, noting

1. how the activity is flowing for each individual and any behavior issues.
2. how the physical set up of the program area is working.
3. who is in attendance.
4. outside influences (anything distracting that doesn't occur within the HT activity itself such as fire drills or loud arguments in other work areas).

Documenting these observations is vital in doing client evaluations, supplying data to support changes in the work area and looking at how to improve an HT activity and the program as a whole. They are also important from a safety standpoint. Noting that John ate the soil placed in front of him or that Sally did not handle tools safely will help you to avoid such situations in the future. Charting the progress of an individual toward his or her goal is the only sure way of determining whether he or she has met that goal. For example, in the activity of making the corsage, charting the number of corsages made in the hour and the amount of physical assistance needed on a daily basis will show if there was improvement in fine motor skills over the week. Communicating with those individuals involved is important in helping them to have a realistic view of their skills. Discussing one's achievements and areas needing attention at the end of a session assist both the therapist and the client.

A number of methods of evaluation appear in the chapter on documentation. Each facility serving people with developmental disabilities will usually have a method of documentation that they want all programs to follow. Horticultural therapists working as private contractors may not be required to follow the facility's format and thus would need to select a method of evaluation that would best suit their particular program.

CASE ILLUSTRATIONS AND IMPLICATIONS FOR HT PRACTICE

Case Study One

Joe, a twenty-year-old male with Down's syndrome, entered a work training program with the goal of working in the community. The primary obstacle to Joe's employability is his inability to follow multiple-step directions. He has tested out at two steps and must test out at four steps to be accepted into the supported employment program.

Individualized Goal: Joe will be eligible to be considered for supported employment in the community.

HT Goal: Joe will demonstrate, unassisted, the four-step process of preparing trays with soil-filled pots, consistently for five consecutive days.

HT Objective: Joe will learn the four-step process of preparing trays with soil-filled pots.

The selection of the horticulture program as the site for working on this goal was based on the fact that this is Joe's favorite area of work at the center. He participates in plant classes as a recreational option and has few behavior problems when in this area. The staff felt he would achieve success more readily than he might in some of the other areas.

Horticultural Activity

Purpose:	To enable Joe to follow four-step directions with only one verbal cue
Facilities:	workbench in greenhouse, table, comfortable working temperature
Staffing:	one staff person
Materials:	• twenty-four, four-inch pots • three plastic nursery flats • prepared soil mix

Procedure:

1. Staff demonstrated twice the complete step-by-step procedure to Joe and then walked him through steps 2 through 7, giving cues only as needed. This was repeated until he could do the entire process with only the cue to prepare trays, and could do it correctly five days in a row. Charting was also done to keep a record of Joe's progress, and it was explained to him that not only would he qualify for the supported employment program upon completion of the goal but he would also be taken out for lunch when he achieved the five consecutive days of successfully following the instructions.
2. Bring pots from storage shelves to the left side of the workbench.
3. Place the tray to the right side of the workbench.
4. Fill one pot to the rim with soil mix.
5. Place the filled pot in the top tray.
6. Continue steps four and five until the tray is full.
7. Repeat the procedure a total of three times each day for seven days.

Evaluation:

1. Day One: Joe required a verbal cue for each step for the first and second time. On the third time, he carried out all steps but the pots were not filled to the instructed level.
2. Day Two: Joe required a verbal cue for each step the first time through. Pots were filled to correct level only with prompting. The second time through the steps, Joe carried out all steps without prompts, although half of pots were underfilled. On the third try, he carried out all the steps with no prompts and the soil level was correct in all pots.
3. Day Three: Joe came into the greenhouse expressing confidence in his ability to follow the steps. He did carry out each step correctly

and this time had all but three pots in the flat filled to the proper level. After being shown those pots and asked to start over, Joe did so two times with no mistakes.

4. Day Four: Staff gave Joe the instructions and he carried out the steps successfully all three times. He made sure to note to staff that he had filled the pots to the correct level each time.

5. Days Five to Eight: Joe continued to demonstrate correctly the process three times each day. The one day that he needed assistance was on Monday following the two days off for the weekend. Joe was taken out to lunch on day eight.

6. Due to the length of time this training took, staff felt that before he was ready for the community supported employment program, Joe should continue with training in at least two other projects in the horticulture area requiring multistep directions.

Case Study Two

Sue, a twenty-five-year-old female diagnosed with mild mental retardation and emotional disabilities lives in a group-home setting. She has been exhibiting inappropriate behavior toward housemates, including hitting, yelling, and breaking belongings. In order to remain in the current living situation, Sue must show improvement in her interactions with others in the home.

Individualized Goal: Sue will remain living in current residence.

Objective: Sue will work on projects cooperatively with housemate(s) a minimum of three times a week.

HT Objectives: Sue will water and/or groom patio plants in cooperation with a housemate for a minimum of twenty minutes on Monday, Wednesday, and Friday, displaying acceptable behavior for a period of one month.

Caring for the patio plants was selected because going out to the garden is one of Sue's loves and it was gardening season. When Sue displayed aggressive and inappropriate behavior during the session, she was sent inside and not allowed to watch television for two hours. When Sue displayed appropriate behavior during a twenty-minute session, she was rewarded with a treat of iced tea and cookies out on the patio immediately following the gardening time.

Horticultural Activity

Purpose:	To develop acceptable socialization skills
Facilities:	patio with planters, accessible water supply, patio chairs or picnic table

Staffing: one staff person

Materials:
- sunscreen
- sun hats
- watering cans
- hand clippers
- trash bags
- pitcher of iced tea
- glasses
- plate of cookies

Procedure:

1. A staff member demonstrated to Sue and her partner steps 2 through 6 in grooming and watering the plants correctly, emphasizing how to share tools and how to work as a team. She shared with Sue that every time she worked for twenty minutes without getting into an argument, there would be cookies and iced tea waiting for her, and that if she was uncooperative she would go inside with no television. The staff then remained on the patio while the two women worked together in order to observe and to be available to answer questions regarding the garden.
2. Remove old flowers ("dead head") from all flowering plants.
3. Remove all dead leaves from plants and soil surface.
4. Gently loosen soil around edge of pots using hand cultivator.
5. Fill water cans at faucet, filling so that you can carry them easily.
6. Water containers until water comes out drainage holes in bottom.

Evaluation:

1. Week One: Sue's behaviors were documented during each gardening session. She was sent in each day during the first week's sessions for very aggressive behavior. Negative behaviors displayed were swearing at a staff person and trying to hit the housemate with a cultivator. Staff made the decision to withdraw the use of the hand cultivator and sharp cutters until her behavior improved consistently.
2. Week Two: Sue showed an improvement in behavior with only minor arguing over sharing the water jugs, which did not result in her going inside. Staff initiated discussion about Sue's behavior and the importance of getting along while sharing in refreshments at the end of the sessions. Sue mentioned her desire to go out to a movie, and it was decided that if she continued to work cooperatively during the gardening sessions through the end of the month, staff would take her to a movie.

3. Week Three: This was a week of conflict for Sue. On the first two days of the week, she fought with housemates all day, including in the gardening sessions that lasted for only ten minutes. Sue was reminded of the movie and told that the movie would be postponed for a week. If behavior were to be positive, however, she could still plan on going. On day three, Sue spoke of looking forward to gardening, and when it was time, she gardened a full thirty minutes with no negative behavior.

4. Week Four: This week was much like week two. She was encouraged to make bouquets from the flowers in the garden to place in her room and the room of the housemate helping her, which she did. Sue was able to stay out for the full length of gardening each session. The good behavior carried beyond the gardening sessions and into the house with only minor negative behavior in terms of language and without outbursts or attacks on others. Sue received praise from others in the house for the bouquets she had made.

5. Because Sue's behavior had not consistently improved at the end of the month, the staff made a decision to continue the process for another month. By the end of the second week of August, Sue had completed three weeks of successful gardening sessions and was rewarded with a trip to the movies. One of the keys to the improvement was the continuation of offering Sue and the partner a garden "project" such as the flower arranging or pressing flowers. Staff noted that without the creative project, Sue seemed more agitated and less cooperative. By the end of the second month, Sue's behavior was not 100 percent improved, but the change was significant and resulted in staff agreeing that Sue could remain living at the group home with the understanding that she would continue to work on the same overall goal.

FUTURE PRACTICE

Future Challenges

HT has its share of challenges in securing its future in the area of programs for people with developmental disabilities. Funding for HT programs or programs serving those with developmental disabilities is an ongoing challenge. Organizations must always be on the lookout for financial resources. Some of the ways in which support may be obtained are through grants, sponsorship by benevolent organizations, donations from individuals, profit from program sales, and third-party payments. As more and more social service agencies experience cuts in their public funding, the competition for private moneys will be greater and greater. An HT program that incorporates a product to sell, such as nursery stock or herbs, will very likely be the one to succeed over the purely recreational program in the future, though they both serve a definite purpose.

Job availability is an area that has plenty of room for growth, no pun intended. The "green" industry is expanding and with that come entry-level positions that could be filled with very capable individuals who have developmental disabilities. Lack of knowledge about the facts and the positive benefits of hiring people with disabilities is now and will continue to be one of the primary blocks in obtaining gainful employment.

Many agencies that serve people with developmental disabilities are not reaping the benefits of HT because they are unaware of what horticultural therapy is and how it can serve the needs of the facility and its clients. A continuing challenge is to make HT a household word when it comes to agencies serving people with special needs. Currently, knowledge of HT and having a trained horticultural therapist in the area go hand in hand and there are far too many cities and towns where there are no horticultural therapists to provide this awareness.

Proposed Plans to Address Changes

One way in which to establish a greater level of acceptance is to develop increased standards of professionalism in the field. This can be done through documentation and evaluation of current programs, through carrying out research, and through developing standards that can then be shown to prospective funding sources or employers. Educating the horticultural industry and the general public about HT and people with developmental disabilities is also a primary step toward opening up more job opportunities. Providing information regarding HT on a national basis to all agencies serving people with special needs is a step toward developing more program sites.

KEY ORGANIZATIONS

Administration on Developmental Disabilities: Office of Human Development Services, Wilbur J. Cohen Federal Building, Room 20201, 300 Independence Avenue, SW, Washington, DC 20201. (202) 245-7719.

Association for Children and Adults with Learning Disabilities: 4156 Library Road, Pittsburgh, PA 15234. (412) 341-1515 or (412) 341-8077.

Association for Persons in Supported Employment (APSE): 5001 West Broad Street, Suite 34, Richmond, VA 23230. (804) 282-3655.

Association for Retarded Citizens (ARC): 500 E. Border Street, Suite 300, Arlington, TX 76010. (817) 261-6003 or (817) 277-0553 (TDD).

Autism Society of America: 1234 Massachusetts Avenue, NW, Washington, DC 20005. (202) 783-0125.

Fetal Alcohol Syndrome Effects: 800-462-5254.

National Association of State Mental Retardation Program Directions, Inc.: 113 Oronoco Street, Alexandria, VA 22314. (703) 683-4202.

National Council on the Handicapped: Suite 814, 800 Independence Avenue, SW, Washington, DC 20591. (202) 267-3846.

National Down's Syndrome Congress: 1605 Chantilly Drive, Suite 250, Atlanta, GA 30324. (404) 633-1555 or 800-232-6372.

National Down's Syndrome Society: 800-221-4602.

President's Committee on Mental Retardation: Wilbur J. Cohen Building, Room 4723, 300 Independence Avenue, SW, Washington, DC 20201. (202) 245-7634.

Spina Bifida Association of America: 800-621-3141.

United Cerebral Palsy Associations, Inc.: 800-872-1827.

ACRONYMS (Kenison, 1992)

A	Autism
AAMR	American Association on Mental Retardation
ADA	Americans with Disabilities Act
ARC	Association for Retarded Citizens
BD	Behavioral Disorder
BHS	Behavioral Health Services
CP	Cerebral Palsy
DD	Developmental Disability
DDD	Division of Developmental Disabilities
DES	Department of Economic Security, U.S.

DHS   Department of Health Services, U.S.

DOL   Department of Labor, U.S.

DVR   Department of Vocational Rehabilitation

ED   Emotional Disability

HI   Hearing Impairment

IPP   Individual Program Plan

IQ   Intelligence Quotient

LD   Learning Disability

MD   Multiple Disabilities

MDSSI   Multiple Disabilities-Severe Sensory Impairment

MIMR   Mild Mental Retardation

MOMR   Moderate Mental Retardation

MR   Mental Retardation

O&M   Orientation and Mobility

OT   Occupational Therapy

PT   Physical Therapy

PWI   Project with Industry

RT   Recreation Therapy

SLI   Severe Language Impairment

SMR   Severe Mental Retardation

SSI   Supplemental Security Income

VI   Visual Impairment

RESOURCES

Books

Daubert, James R. and Rothert, Eugene A. (1981). *Horticultural Therapy for the Mentally Handicapped*. Glencoe, IL: Chicago Horticultural Society.

Hewson, Mitchell L. (1994). *Horticulture as Therapy*. Homewood Health Centre, 150 Delhi Street, Guelph, Ontario N1E 6k9, Canada.

Hudak, J. and Mallory, D. (1980). *The Melwood Manual.* Upper Marlboro, MD: Melwood Publishing.

Moore, Bibby (1989). *Growing with Gardening: A Twelve-Month Guide for Therapy, Recreation, and Education.* University of North Carolina Press, P.O. Box 2288, Chapel Hill, NC 27515-2288.

University of Georgia/Chicago Horticultural Society. (19__). *The Growing Connection in Therapy* (videotapes). Personal Adult Learning Services, Georgia Center for Continuing Education, Department FG, University of Georgia, Athens, GA 30602.

Journals

Journals of Therapeutic Horticulture. American Horticulture Therapy Association, 362 A Christopher Avenue, Gaithersburg, MD 20879.
   -vol. 1, 1986
   -vol. 2, 1987
   -vol. 3, 1988
   -vol. 5, 1990
   -vol. 6, 1991

REFERENCES

ARC (September 1993). *Introduction to Mental Retardation.* Arlington, TX: ARC.

ARC (January 1994). *Employment of People with Mental Retardation.* Arlington, TX: ARC.

Daubert, James R. and Rothert, Eugene A. (1981). *Horticultural Therapy for the Mentally Handicapped.* Glencoe, IL: Chicago Horticultural Society, pp. 3-6.

Digest of Data on Persons with Disabilities (1992). Washington, DC: National Institute on Disability and Rehabilitation Research. pp. 28-29, 78-79, 152, 170-171.

Gersh, Elliot (1991). *Children with Cerebral Palsy, A Parent's Guide.* Rockville, MD: Woodbine House, pp. 2-24, 57-77, 118.

Kenison, Rita (1992). *Arizona CHILD FIND Project Advocate Training Manual.* Phoenix, AZ: Arizona Department of Education.

National Down's Syndrome Congress (1993). *Facts About Down Syndrome.* Atlanta, GA: National Down's Syndrome Congress.

National Down's Syndrome Congress (1994). *Facts About Down Syndrome.* Atlanta, GA: National Down's Syndrome Congress.

Powers, Michel D. and Volkmar, F. (1989). *Children with Autism, A Parent's Guide.* Rockville, MD: Woodbine House, pp. 3-10, 57-61.

Chapter 7

Mental Illness and Horticultural Therapy Practice

Barbara A. Shapiro
Maxine Jewel Kaplan

INTRODUCTION

The process of combining the life elements of plants and people establishes the central core of horticulture therapy. The intrinsic nature, the environmental conditions and the physical makeup of both people and plants combine to form a system of growth for each. As a treatment modality used in psychiatric settings, horticulture therapy programs strive to create environments that promote and allow for growth. Because the horticulture world has changed little over the years, the legends and fables about trees and flowers are especially illuminating of the symbolism adopted from the plant cycle. Plant life cycles and ecosystems coincide and interweave with our human life cycles. Each shapes the other's present existence and determines its future. This interrelatedness, with its striking parallels in life processes, connects plant and animal worlds to the human experience in revealing ways. The shared biological processes of reproduction, food requirements, proper environmental conditions, and death readily draw emotional responses. These elements in plants form metaphors for the psychological processes of human growth. Such connections are the backdrop of daily experiences using horticulture as a therapeutic medium.

Generations have been dependent on the natural environment for food, shelter, and clothing. This dependency has established a basic relationship between each person and the environment. Gradually, Western people have built a cushion of insulation between themselves and their natural environment. While this cushion serves them well in many ways, it also

The Menninger Clinic horticulture therapy herb gardens (1967).

has deepened their alienation from the earth on which they live. Reestablishing these ties can happen in a variety of forms, which then can be used for positive gains for the individual and for the larger society.

Patients who are admitted to psychiatric hospitals are no longer able to function in society. They are considered either in danger of harming themselves or harming others. Some come voluntarily while others are brought in through an emergency status. The stresses of life have become too great for them to handle—they need the secured protection of a hospital setting with professional staff to guide and assist them back on the road to recovery.

Every patient population has specific problems and specific needs. With some familiarity of mental illness and behavioral disorders, horticultural therapists can program their activities to meet the needs of the mentally ill population they serve. Most horticultural therapy programs for the mentally ill are inpatient programs.

Demographics

Psychiatric patients come from varied racial, religious, or ethnic backgrounds and can be of all ages. They come from all over the country and all over the world, and are either visiting or living in the hospital locale

upon the time of admission. Many of the adult patients are married; many others are single adults living at home with their parents for emotional and financial support; while still many others have no family and support system whatsoever and are oftentimes homeless. The majority of the children and adolescents are from foster care or residential facilities/treatment centers, yet some children do come from home directly. Demographics may vary in public versus private institutions. Most private institutions admit patients who "private pay," and therefore would probably not serve the indigent and homeless population nor the foster care and residential children. They can be selective; public institutions cannot.

Physical Health

The physical health issues of patients are usually of secondary significance in admission to a psychiatric hospital. Often physical health is exacerbated by the patient's mental health problems. The physical problems are addressed and treated, when possible, during their period of hospitalization.

Mental Health Issues

Upon admission, the patient is interviewed and examined by a physician. A complete physical and history is taken and an assessment is made. A tentative diagnosis or diagnoses is determined by using the *Diagnostic and Statistical Manual of Mental Disorders* (DSM-IV) criteria, the "Bible" of psychiatry, and treatment is initiated.

Social Issues

The social problems and needs of patients will often vary because of the socioeconomic status. A public or private facility will often be an indicator of a patient's socioeconomic status. The majority of adult patients in publicly supported facilities has been on public assistance most of their adults lives and has had little or no employment history and minimal educational background with no or few employable skills. Many patients have drug and/ or alcohol-related problems secondary to their psychiatric history, some are HIV positive, and some have criminal records. There is a high rate of recidivism with this population mostly due to noncompliancy with medication after discharge and failure to continue in outpatient programs.

Learning Objectives

The learning objectives for this chapter are the following:

1. This chapter focuses on horticulture therapy as it applies to psychiatric settings. The purpose is to give an overview of psychiatric diagnosis in order to provide clear descriptions of diagnostic categories in order to enable clinicians and horticulture therapists to communicate about, study, and treat people with various mental disorders. In order to understand and communicate how clinicians can apply horticultural therapy techniques, it is important to understand how people are diagnosed and classified by the diagnostic criteria from the DSM-IV from the American Psychiatric Association. The DSM-IV is the diagnostic criteria that improves the reliability of diagnostic judgments.
2. The chapter will describe treatment teams and how horticulture therapy is incorporated into the overall psychiatric treatment plan.

HISTORY OF HORTICULTURE THERAPY

Dating back to the 1790s, when patients paid for psychiatric care by working in gardens, improvement was observed in such individuals. In 1817, Friends Hospital in Philadelphia began the first gardening program for mentally disturbed people. Emphasizing humane treatment and the value of productivity helped provide a public atmosphere supportive of better treatment. With this change in attitudes, the public soon began pushing for improved conditions for those hospitalized. Around the turn of the century, therapeutic work programs began to appear. References to garden therapy began to be seen in occupational therapy literature and gradually, horticulture was recognized for its therapeutic value.

Horticulture Therapy at the Menninger Clinic

As Dr. C. F. Menninger, Founder of the Menninger Clinic in Topeka, Kansas, noted in his article in the *Bulletin of the Menninger Clinic,* May 1942,

Peonies are very healthy flowers, they have no aches and pains, they make no outcry and there are no anxious and troubled faces to

Dr. Charles Menninger, founder of the Menninger Clinic, established the tradition of using nature and gardening in hospital care.

comfort. They just grow and bloom. That is why I fall more and more in love with them. They have helped me to keep my emotional and intellectual equilibrium. Growing peonies has helped me to satisfy an inborn curiosity to watch things grow. There is a gratification of the sense of sight in color and color combinations, of the sense of smell in perfumes and odors, and to that inner aesthetic sense of beauty and charm that has, I believe, made a better physician of me. My whole nature improved, my horizons widened, and my appreciation increased in a way that aided me in my vocation. Hope never dies in a real gardener's heart. Thousands of scientists have chosen horticulture as a hobby and are better for it too.

This belief in the value of the plant world for human health was the inspiration for horticultural therapy at the Menninger Clinic. Dr. Menninger's love of plants established a tradition of using nature and gardening in hospital care. Soon vegetable gardens, flower beds, and a greenhouse contributed to the ways people worked with plants to promote human growth. Vocational training programs were added to assist patients in developing work skills and to provide work experience in the field of horticulture.

Rhea McCandliss of the Menninger Foundation understood the value of people/plant interactions for mental health. She investigated the needs and interests of health institutions in horticulture programs. The results of her survey indicated that trained professionals were needed in the field of horticulture and mental health. This finding spurred action. As a result, in 1971, the Menninger Foundation, together with Kansas State University, began the first student training program for horticulture therapists. Active today, the internship assists students in combining horticultural and thera-peutic skills. The two institutions continue to work together to assess the needs of the profession in order to respond more effectively to people/ plant interactions and to expand and refine the program.

Rhea McCandliss was the first horticulture therapist on staff at the Menninger Clinic.

Rhea McCandliss developed a program utilizing vegetable gardening, flower beds, and later a glass greenhouse to grow houseplants and various bedding plants for the grounds.

Table 7.1 shows the prevalence of various mental disorders during a one month period (current prevalence) and during a year, as determined by a multisite epidemiological and health services research survey conducted between 1980 and 1984. The survey, called the Epidemiologic Catchment Area (EAC) Study, was supported by the National Institute of Mental Health (NIMH) and improves the diagnosis, treatment, and prevention of mental disorders and brings hope to millions of people who suffer from mental illness and to their families and friends.

The number of affected adults is based on estimates of the U.S. resident population from the 1990 census of 184 million persons aged eighteen and over. Some people have more than one mental disorder. Therefore, the numbers for each type of disorder, if added together, will be more than the total number for all individuals diagnosed with mental disorders.

Mental illness is characterized by symptoms and/or impairment in functioning (see Table 7.2). Although the numbers continue to vary, approxi-

TABLE 7.1. Number of U.S. Adults (in millions) with Mental Disorders, 1990 (based on five epidemiologic catchment areas [ECA] sites)

Disorder	One-Month Number (in millions)	Percent	One-Year Number (in millions)	Percent
Any mental disorder and substance-use disorder covered in survey	28.90%	15.70%	51.70	0.28
Any mental disorder except substance use disorders	23.90%	13.00%	40.70%	0.22
Schizophrenia/Schizophreniform disorders	1.3	0.70%	2.0	1.10%
Depressive (affective) disorders	9.6	5.20%	17.5	9.50%
Manic-depressive illness (Bipolar disorder)	1.1	0.6%	2.2	1.2%
Major Depression	3.3	1.8%	9.2	5.0%
Dysthymia	6.1	3.3%	9.9	5.4%
Anxiety Disorders	13.4	7.3%	23.2	12.6%
Phobia	11.6	6.3%	20.1	10.9%
Panic disorder	0.9	0.5%	2.4	1.3%
Obsessive-Compulsive disorder	2.4	1.3%	3.9	2.1%
Somatization disorder	0.2	0.1%	0.4	0.2%
Antisocial Personality Disorder	0.9	0.5%	2.8	1.5%
Severe Cognitive Impairment	3.1	1.7%	5.0	2.7%
Substance-use disorders	7.0	3.8%	17.5	9.5%
Alcohol abuse/dependence	5.2	2.8%	13.6	7.4%
Drug abuse/dependence	2.4	1.3%	5.7	3.1%

Note: Somatization disorder is a chronic psychiatric condition characterized by multiple physical complaints for which there are no apparent physical causes.

Source: Prepared by the Office of Scientific Information, National Institute of Mental Health, a component of the National Institutes of Health, Public Health Service, U.S. Department of Health and Human Services. Adapted from Reiger, D.A. et al., "The De Facto US Mental and Addictive Disorders Service System." Archives of General Psychiatry, February 1993.

TABLE 7.2. Major Forms of Psychiatric Disorders

COGNITIVE DISORDERS

Cognitive disorders involve impairment of a person's ability to think or recall information. Such impairments may be of short duration or chronic in nature. Cognition can be affected by substance abuse, stroke, injury to the brain, or degenerative diseases such as Alzheimer's, which produce multiple cognitive deficiencies and affect memory. With more severe forms of cognitive disorders, confusion and disorientation to time and place may occur, and impulse control may be compromised. In addition, the ability to speak, understand others, or move may be affected.

PSYCHOTIC DISORDERS

Psychotic disorders are typified by hallucinations and/or beliefs not supported in reality. Hallucinations involve seeing or hearing things others do not hear or see, such as hearing voices or seeing others who are not present, or feeling bodily sensations with no apparent cause. Delusions or beliefs that are persistent but not substantiated in reality may also be present, or may be the primary symptoms of a psychotic disorder. Psychotic disorders such as schizophrenia may include severe impairments to social functioning and self-care deficits. Psychiatric medications are the treatment of choice for psychotic disorders.

MOOD DISORDERS

Mood disorders are characterized by disturbance in affect. Included in this category are depression and bipolar disorder, among others. Depressions are typified by anhedonia (the inability to feel pleasure), sleep alterations, appetite loss or an increase in appetite, diminished energy, crying spells, and difficulty concentrating. When depression alternates with extreme elevations in mood, referred to as mania, bipolar disorder may be present. Manic episodes may include persistent sleeplessness, euphoria, grandiose beliefs, pressured speech, impulsive or disinhibited behavior, hyperactivity, and racing thoughts. Mood disorders may be effectively treated with antidepressant or mood-stabilizing medications.

ANXIETY DISORDERS

Anxiety disorders include frequent and undue nervousness or anxiety, fears, or phobic reactions that may interfere with daily tasks, or panic-attack symptoms of racing heart, shortness of breath, sudden intense fright, chest pains, and a feeling of doom. Persons having panic attacks frequently believe they are having a heart attack and may initially present in an emergency room. Anxiety reactions may have no clear stressor or may be related to specific situations, such as social situations or the work environment. In addition, anxiety may be related to a specific trauma or series of traumatic events. Anxiety disorders may respond to cognitive therapies, relaxation training, desensitization, or antianxiety medications.

Adapted from DSM-IV by Martha C. Straus and J. Scott Soud.

mately 20 percent of the population experiences a mental disorder. Some common indicators of mental illness are depression, feelings of anxiety that are not proportionate to a possible cause, physical complaints having no organic cause, any sudden change of behavior or mood, unreasonable and unrealistic expectations of self or others, and failure to achieve potential. Schizophrenia and depression commonly cause people to seek mental health services, although phobias, alcohol abuse or dependence, and dysthymia are also seen as frequently. Many homeless people in America account for a large percentage of the mentally ill. Many are migrants, refugees, drug abusers, individuals with severe personality disorders, and people who are displaced, unemployed, and severely or chronically mentally ill.

There is a lack of definitive causal factors in mental illness. Several factors have been shown to have a relationship to the occurrence of mental illness. Physiological factors include defective genes, disturbance in neurotransmission, activity of endorphins, disturbance in the immune system, hormone imbalance, abnormal blood factors, malnutrition, vitamin deficiencies, low blood sugar, sensory deprivation, and sleep and dream deprivation. Psychological factors include mental attachment and deprivation, sibling position, parental behavior and child-rearing practices, conflict, stress, and coping styles. Sociocultural and spiritual factors include age, sex, race, marital status, social class, religious beliefs and values, migration, roles, ethnic mores, lack of participation in the community, lack of social support system, overcrowding, rapid social change, and availability of and impediments within health care systems. Biological factors that explain mental disorders are being accepted increasingly due to the complex interchanges occurring in the nervous system.

DIAGNOSTIC CODES

The specified diagnostic criteria for each mental disorder are used as guidelines for making diagnoses, so that clinicians from all fields can be in agreement. Proper use of these criteria requires specialized clinical training that provides both a body of knowledge and clinical skills. In the United States, the official coding system in use is the International Classification of Diseases. Most DSM-IV disorders have a numerical code that precedes the name of the disorder in the classification. The DSM-IV diagnosis is usually applied to the individual's current presentation. The following specifiers indicating severity and course may be listed after the diagnosis: mild, moderate, severe, in partial remission, in full remission, and prior history. After a period of time in which the full criteria for the disease are no longer met, individuals may develop symptoms that suggest a recurrence of their original disorder.

Principal Diagnosis

The principal diagnosis is chiefly used as the reason for the visit and/or for admission into a hospital. This will be the main focus of attention in treatment. Dual diagnosis is also a possibility; for example, a substance-related diagnosis may be accompanied by another diagnosis, such as schizophrenia. Sometimes it is unclear which diagnosis should be considered principal. Each condition may have contributed equally to the need for admission and treatment. A principal diagnosis of an Axis I disorder is indicated by listing it first. The remaining disorders are listed in order of focus of attention and treatment. When a person has both an Axis I and an Axis II diagnosis, the principal diagnosis will be assumed to be on Axis I unless the Axis II diagnosis is followed by a qualifying phrase, such as "(Principal Diagnosis)" or "(Reason for Visit)."

The specifier "provisional" can be used when there is a strong presumption that the full criteria will ultimately be met for a disorder, but not enough information is available to make a firm diagnosis. This can happen when (1) an individual is unable to give an adequate history to establish the diagnosis, and (2) certain diagnoses require a length of time for certain symptoms to manifest.

If criteria are currently met, one of the following severity specifiers may be noted after the diagnosis: mild, moderate, or severe. If criteria has not been met, one of the following specifiers may be noted: in partial remission, in full remission, or prior history.

Disorders Usually Diagnosed in Infancy, Childhood, or Adolescence

1. Mental Retardation
2. Learning Disorders
3. Motor Skills Disorder
4. Communication Disorders
5. Pervasive Developmental Disorders
6. Attention Deficit and Disruptive Behavior Disorders
7. Feeding and Eating Disorders of Infancy or Early Childhood
8. Tic Disorders
9. Elimination Disorders
10. Other Disorders of Infancy, Childhood, or Adolescence
11. Delirium

12. Dementia
13. Amnestic Disorders

Mental Disorders Due to a General Medical Condition

1. Catatonia
2. Personality Change
3. Mental Disorder

Substance-Related Disorders

1. Alcohol-Related
2. Amphetamine-Related
3. Caffeine-Related
4. Cannabis-Related
5. Cocaine-Related
6. Hallucinogen-Related
7. Inhalant-Related
8. Nicotine-Related
9. Opioid-Related
10. Phencyclidine-Related
11. Sedative-, Hypnotic-, or Anxiolytic-Related
12. Polysubstance-Related
13. Other (or Unknown) Substance-Related

Schizophrenia and Other Psychotic Disorders

1. Schizophrenia

    a. Paranoid Type
    b. Disorganized Type
    c. Catatonic Type
    d. Undifferentiated Type
    e. Residual Type

2. Schizophreniform Disorder
3. Schizoaffective Disorder
4. Brief Psychotic Disorder
5. Shared Psychotic Disorder
6. Psychotic Disorder Due to General Medical Condition

    a. With Delusions
    b. With hallucinations
    c. Substance-Related

Mood Disorders

1. Depressive Disorders
2. Bipolar Disorders

   a. Cyclothymic Disorder
   b. Mood Disorder Due to General Medical Condition
   c. Substance-Related

Anxiety Disorders

1. Panic Disorder with or without Agoraphobia
2. Phobia—specific, social
3. Obsessive-Compulsive Disorder
4. Posttraumatic Stress Disorder
5. Acute Stress
6. Generalized
7. Due to General Medical Condition
8. Substance-Related

Somatoform Disorders

1. Conversion Disorder
2. Pain Disorder
3. Hypochondriasis
4. Body Dysmorphia

Factitious Disorders

1. Predominantly Psychological Signs/Symptoms
2. Predominantly Physical Signs/Symptoms
3. Combined Psychological/Physical Signs/Symptoms

Dissociative Disorders

1. Amnesia
2. Fugue
3. Identity Disorder
4. Depersonalization

Sexual and Gender Identity Disorders

1. Sexual Dysfunctions
2. Sexual Desire Disorders

3. Sexual Arousal Disorders
4. Orgasmic Disorders
5. Sexual Pain Disorders
6. Sexual Dysfunction Due to a Generalized Medical Condition
7. Paraphilias (Exhibitionism, Fetishism, Frotteurism, Pedophilia, Sexual Masochism/Sadism, Transvestic Fetishism, Voyeurism)

Gender Identity Disorders

1. Children
2. Adolescents/Adults

Eating Disorders

1. Anorexia Nervosa (Restricting, Binge-Purge)
2. Bulimia (Purging, Nonpurging)

Sleep Disorders

1. Primary Sleep Disorders
2. Dyssomnias (Insomnia, Hypersomnia, Narcolepsy, Breathing-Related, Circadian Rhythm)
3. Parasomnias (Nightmare, Sleep Terror, Sleep Walking)
4. Sleep Disorders Related to Another Mental Disorder
5. Other Sleep Disorders

Impulse-Control Disorders not Elsewhere Classified

1. Intermittent Explosive Disorder
2. Kleptomania
3. Pyromania
4. Pathological Gambling
5. Trichotillomania

Adjustment Disorders

1. With Depressed Mood
2. With Anxiety
3. With Mixed Anxiety/Depressed Mood

4. With Disturbance of Conduct
5. With Mixed Disturbance of Emotions/Conduct

Personality Disorders

- These are coded on Axis II: Paranoid, Schizoid, Schizotypal, Antisocial, Borderline, Histrionic, Narcissistic, Avoidant, Dependent, Obsessive-Compulsive.

Multiaxial System

A multiaxial system involves an assessment on several axes, each of which refers to a different domain of information that may help clinicians plan treatment and predict outcome.

- Axis I   Clinical Disorders
- Axis II   Personality Disorders
              Mental Retardation
- Axis III   General Medical Conditions
- Axis IV   Psychosocial/Environmental Problems
- Axis V   Global Assessment of Functioning

HOSPITALIZATION

Upon admission, the presenting problems are noted. First, the patient is identified by age, race, and marital status to understand background information. The family history is taken to denote the patient's siblings and illnesses in other family members, either psychiatric or medical. The patient's personal history, including education, and work history are useful in determining strengths and assets of the patient.

The symptoms are then reviewed, such as anhedonia, agitation, anxiety, change in sleeping/eating pattern, hallucinations, depression, mania (presence/absence of manic episode), substance abuse, and suicidal/homicidal ideation. The duration of the illness is important, and if there is a past history of prior hospitalizations, this would also be denoted. In order to justify hospitalization, certain criteria have to be present. A patient must present signs and symptoms consistent with a mental disorder or illness in one of the following:

1. Disorientation or memory impairment
2. Delusions or hallucinations

3. Dependence on drugs or alcohol
4. Impaired reality testing
5. Paranoid thinking
6. Depression
7. Bizarre behavior
8. Agitation
9. Mania
10. Poor impulse control

The patient demonstrates a need for a structured therapeutic milieu if one of the following is present:

1. Potential danger to self, others, or property
2. Need for continuous skilled observation
3. Failure or inaccessibility of outpatient treatment
4. Inadequate social supports
5. Impaired social, family, or occupational functioning
6. Need for evaluation and regulation of medication
7. Legally mandated admission

TREATMENT APPROACHES

The nine treatment approaches described in this chapter are:

1. neurological,
2. stress adaption,
3. psychodynamic,
4. interpersonal,
5. ego development,
6. behaviorist,
7. humanistic,
8. family, and the
9. Bowen System Theory

Neurobiological Approach

This approach emphasizes a scientific study of the nervous system in order to explain and treat mental disorders. "Illness" is defined as a disturbance in the neurobiological system. As an example, children of schizophrenic parents have an increased risk of the illness, regardless of whether they are raised by the parent with schizophrenia or by another person. Therapy in this approach includes diagnostic aids such as computer tomography (CT) scans, electroencephalograms (EEGs), laboratory studies, radiographs, history of present illness, history of familial incidences of disorders, physical examination, and behavior observations to determine areas of dysfunction. Drugs that effect change in the neurobiological system are prescribed.

Stress-Adaptation Approach

This approach emphasizes the role of stress in the increased incidence of illness. Illness is viewed as a human reaction pattern to stress or maladaptation to it. Various risk factors may include prematurity, poor diet, chromosomal disorders, accidents, racial discrimination, or life events such as death of spouse, divorce, or marital separation; crisis-like maturation, such as transition into retirement; situational stress, such as loss of a job; or adventitious stress, such as earthquakes.

Therapy focuses on establishing a working relationship with the client, problem identification and steps in resolution, support of coping strategies, enhancement of self-esteem, anticipatory guidance, and prevention interventions.

Psychodynamic Approach

With this approach, influences of intrapsychic forces on observable behavior are emphasized. "Illness" is defined in terms of behavior disorders that originate in conflicts occurring before six years of age among the id, ego, superego, and/or environment. Anxiety is then experienced as a result of these conflicts. Excessive use of mental defense mechanisms leads to serious behavioral disturbances. Mental processes used to reduce anxiety and conflict by modifying, distorting, or rejecting reality frequently include a number of defense mechanisms:

1. *Repression*—major response keeps painful thoughts, feelings, and impulses from consciousness.
2. *Denial*—response acknowledges no awareness of a painful event.
3. *Reaction Formation*—response expresses feelings opposite to those being experienced.
4. *Projection*—response ascribes to another person or object the unacceptable thoughts and feelings.
5. *Rationalization*—response justifies behavior by an attempt to explain it logically.

6. *Undoing*—response cancels the effect of another response just made.

7. *Displacement*—response is misdirected from original person or object to an alternate or safer object.

8. *Sublimation*—response partially substitutes socially acceptable activities for unacceptable impulses.

9. *Regression*—response deals with anxiety by behaving at a level more appropriate to an earlier age.

10. *Identification*—response likens actions and feelings to those of a significant other.

11. *Introjection*—response takes the behavior or thought of another into the ego structure.

12. *Isolation*—response blocks the feeling associated with an unpleasant, threatening situation or thought.

13. *Suppression*—response is conscious and which deliberately forces certain ideas away from thought and action.

Freud is recognized as the founder of the psychoanalytic school of thought. Therapy may involve psychoanalysis, in which an intense relationship with a psychoanalyst for a period of time is cultivated for the purpose of helping the person establish conscious control of affect and behavior. Through dream analysis, free association, interpretation, analysis of resistance and "transference" (ascribing to the psychiatrist the thoughts and feelings associated with parents or other important people), and neutrality, the therapist assists the patient in modifying his or her conflicts and behaviors. Supportive/expressive psychotherapy approaches are often the norm in treatment centers today, and are sometimes of a short-term nature as well.

Interpersonal Approach

The importance of interpersonal relationships and communication on behavior is emphasized. Sullivan is noted for developing an interpersonal theory of psychiatry where stages of growth and development are emphasized:

1. *Infancy*—lasts until the appearance of speech, which enables infants to change environment.

2. *Childhood*—lasts until the emergence of the need for peers.

3. *Juvenile*—lasts until the need for close relationships.

4. *Preadolescence*—lasts until puberty and the beginning interest in the opposite sex.

5. *Early Adolescence*—lasts until the development of relationships with the opposite sex.
6. *Late Adolescence*—lasts until the establishment of a stable love relationship with another.

Anxiety is seen to arise from interpersonal situations involving tension, disapproval, rejections, etc. Therapy uses a principle called "elucidation," which states that a behavior change can occur when one can identify, conceptualize, and evaluate one's behavior. The therapist is a participant observer and not a neutral object. The focus of interview is on exploring (1) anxiety experiences, (2) avoidance behaviors, and (3) the interpersonal context in which these occur.

Ego Development Approach

The development of identity throughout the life span was developed by Erikson. "Illness" is characterized by problems with identity formation, relationships, or society, may give rise to stunted or frozen development. Behavioral disorders result from unresolved conflicts and/or failed accomplishments during each stage of the life cycle. Erikson's eight developmental stages represent critical periods for the emergence of specific emotional capacities.

In infancy (birth to eighteen months), trust is learned if needs are met in a consistent and satisfying manner. Confidence, realistic trust, hope, optimism, and the ability to form relationships later in life stem from trust. Due to mistreatment, an infant may develop "mistrust," which later reflects hostility, suspiciousness, and a general feeling of dissatisfaction.

In early childhood (eighteen months to three years), autonomy develops from reassuring experiences in which the child is allowed to exercise self-control of his or her behavior without being subjected to experiences beyond his or her capabilities. Socially acceptable behaviors of holding and letting go, on which toilet training focuses during this stage, become generalized to other aspects of living. The development of autonomy leads to self-control without loss of self-esteem, a sense of pride and good will, the ability to initiate yet to be cooperative, and appropriate generosity and withholding. Imposition of overcontrol can lead to "shame and doubt."

In late childhood (three to five years), the child develops "initiative"; that is, the ability to undertake and plan tasks, the pleasure of being active, and the experience of a sense of purpose. Pleasure in attack and conquest aid in developing conscience. The person grows to develop and strives to utilize his or her potential in a socially appropriate manner. "Guilt," accompanied by self-restriction and denial, can result from an unsuccessful negotiation of this stage. The person fails to develop his or her potential.

In the school age (six to twelve years), the major task is "industry," characterized by involvement in the world, construction and planning, development of relationships with peers, development of specific skills, and identification with admired others. A sense of competence and the pleasure of diligence develops "inferiority," which is the feeling that one is unworthy and inadequate, which can result from hindrances.

Adolescence (twelve to twenty years) is where the development task is "identity," characterized by a confident sense of self, commitment to a career, and finding one's place in society. Successful resolution leads to the ability to work toward long-term goals, self-esteem, and emotional stability. The danger is "role" confusion, characterized by feelings of confusion, lack of confidence, indecision, alienation, and possibly acting-out behavior. Unsuccessful resolution may require the adult to spend life-long energies attempting to resolve remaining conflicts.

In young adulthood (eighteen to twenty-five years), "intimacy" is the major development task. The person develops the ability to love, to develop commitments to other persons, and to enter true mutual relationships. "Isolation" is the danger when the person remains distant from others, withdraws, enters into superficial relationships, or develops prejudices.

In adulthood (twenty-eight to sixty-five years), "generativity" is the task. The adult becomes responsible for guiding children in the creation and development of productive and constructive tasks. Failure leads to "stagnation," personal impoverishment, and self-indulgence.

At old-age (sixty-five years to death), the last feeling is characterized by feelings of acceptance, importance, and self-worth about the value of one's life, that is, his or her "integrity." "Despair," the negative outcome of this stage, is the sense of loss, a feeling of life's meaninglessness, and the feeling that life's goals have not been achieved and that it is too late to start over.

The therapy in this approach is on establishing trust not obtained early in life and helping the patient to (1) gain insight into unconscious motivations, (2) reduce anxiety, and (3) reignite development where it may have become derailed.

Behaviorist Approach

This approach emphasizes observable and measurable behavioral processes. "Behavior" can be classified as behavior excess, behavioral deficit, distortion of reinforcing stimuli, distortion of discrimination stimuli, and aversive behavior. Pavlov and Skinner contributed to the development of the behaviorist school of thought. Two schools of thought have developed.

Behavioralism

All behavior follows learning principles; therefore, behavior may become maladaptive but is not considered abnormal. (Respondent conditioning, operant conditioning, and reinforcement are concepts used.)

Cognitive Behaviorism

Behavior is influenced by cognition independent of the stimulus. Important variables determining behavior include plans, beliefs expectancies, encodings, and competencies. Feelings are believed to follow thoughts. Therapy involves a "functional analysis," which is an analysis of the manifest behavior. Techniques frequently used in behavioral therapy are systematic desensitization, flooding, implosion, positive reinforcement, programs such as assertiveness training, relaxation exercises, token economy, and sex therapy. Cognitive therapy focuses on changing the internal contingencies, such as one's expectations, distortions, self-injunctions, self-reproaches, and sequence of thoughts, to effect a behavior change. Techniques in cognitive therapy include verbal probing, reality testing, thought substitution, role playing, self-monitoring, assignment of tasks, use of humor, and reflection.

Humanistic-Existential Approach

This approach emphasizes the holistic view of humans, their individuality and intrinsic worth, the importance of experiencing the present, and the personal meaning of experience. According to Rogers, Maslow, and Frank, abnormal behavior is a consequence of the following:

1. Rogers—An incongruence exists between one's self-image and experience.
2. Maslow—The spectrum of basic needs are at some point not satisfied (air, food, water, safety, love, belonging, self-actualization).
3. Frank—Lack of meaning of life may result in illness.

"Self" is a central concept because one's evaluation of life is related to views of the self: Who am I? What can I do? What am I able to do? What do I want to be? Human beings strive for self-actualization.

Incongruence can develop between the ideal self and the real self and/or reality. This causes dissatisfaction, anxiety, and activation of a self-defense mechanism. Continuous feedback about behavior is given to children and others. These impressions are integrated, denied, or accepted as truth, thus affecting one's experience of oneself. The importance of accepting one's feelings and not denying them and of recognizing one's own values and beliefs and not generally accepting the values of others is stressed.

Therapy emphasizes the demonstration, by therapists, unconditional positive self-regard (genuine acceptance), empathic understanding (ability to perceive another's world), and attunement with the patient to facilitate the experience of his or her own worth, uniqueness, and authenticity.

Family Approach

Differing conceptual views of family theory and family therapy have developed. There is no accepted typology or diagnostic categorization of families.

Minuchin views the family as a structured and organized social system within which the individual lives and to which he or she responds. Transactional patterns develop that control the interaction and behavior of family members. Maladaptation is noted in the transactional patterns (i.e., disengagement with no or minimal contact or enmeshment and overinvolvement between family members). Therapy is directed toward initiating change in the family structure by modifying and/or clarifying boundaries, rules, and expectations.

Satir and Haley note the interactional framework in families. The double-bind theory offers a way to view the development of dysfunction in a family system. Its characteristics are as follows:

- The individual is in an important relationship that requires being able to understand communication.
- The other person in the relationship is communicating two orders or levels of messages that are contradictory and mutually exclusive.
- The individual is unable to react to or make a comment about either of the messages and therefore is in a double bind.

Dysfunctional communication is produced by denying, rejecting, or disqualifying the relationship, by differing punctuation in the interactions between two persons, and by having symmetrical or rigid interaction patterns.

Therapy is directed toward change in the interaction patterns of the family system (change in the structure or transaction pattern, goal setting, giving tasks, symptom prescription, advertising symptoms, and reframing behavior).

Bowen System Theory

The Bowen System theory (BST) presents a conceptualization of the emotional system of families over several generations. "Illness" is viewed as a level of undifferentiation transmitted from past generations. Theoretical bases include sibling position, triangles, and family projection process.

Triangles are the basic unit of the emotional system. When tension is experienced, each person will attempt to obtain the outside position. If the tension increases, one of the persons will triangle another, and a larger and larger interlocking system is thus formed.

Family projection process is when anxiety is experienced by the mother, who may respond by becoming sensitive to the child and overconcerned. Mother's overattachment to the child is supported by the father. The child may become anxious, demanding, and unable to function autonomously. The child may as an adult relate in turn to its own child as similarly enmeshed, fused, and undifferentiated manner.

A multigenerational transmission process occurs when the family projection process involves multiple generations, with one child in each generation becoming less differentiated and less able to function.

The emotional cutoff concept describes the process of separation from parents to resolve the binds of these emotional attachments. The more intense the cutoff from parents, the more likely that a person and his or her children will reenact these problems in life.

Differentiation is a concept related to a state of being and becoming more responsible for self at emotional and intellectual levels.

The nuclear-family emotional system identifies relational patterns between mother, father, and children.

Societal regression occurs when a society is exposed to chronic anxiety. This can promote regression. An example of regression response is overuse of drugs in society.

Therapy from a Bowenian perspective focuses on reducing reactivity and increasing differentiation. Expression of feelings is not encouraged or interpreted, but members are assisted in thinking about the process of family dynamics and history and reestablishing contact with family members. The therapist uses his or her skill to remain out of the interlocking triangles, thereby increasing therapeutic flexibility and effectiveness.

EFFICACY

The need to have available data on the efficacy of psychiatric treatment has never been more urgent. Currently being debated are the priorities for coverage under the potential new systems of national health care. For psychiatry, however, the challenges are complicated by an additional hurdle that holds the potential for significantly influencing the outcome of the debate. Perhaps no other field of medicine experiences the general stigma and often disparagement as does psychiatry, fueled in part by a general lack of understanding of the status of scientific progress in the field. Mental illness fails to capture the public interest unless it strikes individuals or their family members. No other health problem is as shrouded by public indifference and ignorance.

Of all mental illnesses, none suffers as much from indifference and ignorance as schizophrenia. For many years, people afflicted with schizophrenia, along with their families who support them, have borne the burden of this lack of public concern. It has led to isolation and further heartache.

Children's mental disorders are frequent, often chronic, and precursors of adult dysfunction. Most of the nine million children with mental disorders go untreated. The successful treatment of childhood disorders represents a major public health concern. There is an array of therapeutic interventions that can bring meaningful relief to children with serious, chronic mental disorders. Their successful application is a wise investment, given the pain, long-term disadvantages, and financial costs associated with untreated childhood behavioral and emotional disorders.

Depression in late life occurs in the context of numerous social and physical problems that often obscure or complicate diagnosis and impede management of the illness. There is no specific diagnostic test for depression; therefore, an attentive and focused clinical assessment is essential for diagnosis. The prevalence of major depression in nursing home populations is high and is generally unrecognized and untreated. About 800,000 persons are widowed each year, and most of them are old and experience varying degrees of depressive symptomatology. Most do not need formal treatment, but those who are moderately or severely dysphoric appear to benefit from treatment.

For reasons that are difficult to understand, the value of psychiatric treatments for all mental disorders has been disparaged by the general public and by those responsible for the reimbursement for the care of the mentally ill. The critical determinants for value in health care involve the relationship of efficacy to cost in an equation in which value equals efficacy/

cost. While practicing in an era of cost-containment, every horticultural therapist should have a goal of providing the highest quality of care.

The first part of this chapter contained the concepts and principles essential to caring for clients who have psychiatric problems. The content was presented in as practical a format as possible so that horticultural therapists can be guided in caring for the client who is experiencing behavioral or mental health problems in whatever setting care is being rendered.

Health care is changing in many ways. Meeting the challenge can be an enormous undertaking.

- Psychiatric care is shifting away from the hospital to day treatment and outpatient care.
- Payment for psychiatric care is increasingly based on *competitive pricing*.
- Payment will be determined, at least in part, on the basis of demonstrated *outcome*.
- Funding for health care is changing from fee for service to *capitation*. Care providers in capitated environments *must* know and *control their costs* in order to price their services appropriately or risk financial disaster.
- Power and *control* are shifting from individual practitioners and the institutions that provide care *to the payer*.
- Health care is moving toward comprehensive, regional, vertically *integrated systems of care*. Payers are looking for a "one-stop shop" for their health care needs.
- Patients can no longer choose any doctor in any setting at any time. Patients are *limited to providers* in defined networks.
- The practice of psychiatry will *almost always* involve a third party overseeing and managing the care.
- *Information systems technology* will play an increasingly important role in psychiatric care, including facilitating the creation and retrieval of *medical records*, speeding up the process of *billing and collections*, coordinating different facets of treatment, and promoting *menu-driven* practice guidelines for diagnosis and treatment.

Most long-term psychiatric hospitals are closing down because of dwindling government health care dollars. The thrust is directed more toward acute care, short-term treatment programs, and community-based outpatient support systems. Working in an acute-care hospital is a challenge. Patients are in for short stays of one week to one month. Horticultural therapy programs must provide short-range activities that can be fulfilling and meaningful before discharge. Currently, the political climate is cutting back on all health care funding. Many outpatient clinics have been closed along with the long-term psychiatric hospitals. There is little money available to create new mental health support systems.

HORTICULTURAL THERAPY AND TREATMENT

Identification of Treatment Issues

When working with patients who have mental illness, it is essential that the horticultural therapist have a broad understanding of the various psychiatric disorders, the terminology used, and a deep sensitivity toward the needs of their patients. The therapist is but one member of a multidisciplinary team that consists of doctors, medical students, social workers, nurses, psychiatric aides, other recreational staff, and sometimes teachers when children are involved. Each member of the team has specific goals and objectives. All goals are directed toward achieving the wellness and ultimate discharge of the patient.

The horticultural therapist's treatment focus should be the following:

1. Recreational—Horticultural therapy activities should be diversional and relaxing. When patients are involved in potting plants, they are not thinking about their problems; they are thinking about the task. Horticulture can also provide good leisure skills upon discharge. It can be an inexpensive and accessible hobby.
2. Educational/Vocational—Horticultural therapy is also instructional. It imparts horticultural knowledge and develops skills with some degree of expertise. This acquisition of knowledge and skills may and can ultimately lead to employment in the horticultural field.
3. Therapeutic—Most important of all, horticultural therapy activities can be used therapeutically. This is what sets horticultural therapists apart from gardeners and horticulturists. The end goal doesn't necessarily have to be a beautiful potted plant or a manicured garden, but should focus on how the activity helps the patient. If the patient feels a sense of accomplishment in propagating or potting a plant, and if the patient can sustain life in that plant, then the patient gains a sense of control in a small part of his or her life. Plants are a symbol of life. The health of the patient is the primary goal for which the plant is one tool used to achieve that goal.

After an initial interviewing process and history of the patient and before the multidisciplinary treatment plan is determined, specific patient issues are identified. The horticultural therapist will then set goals to meet these issues. As professionals, horticultural therapists function to help treat or relieve the patient's symptoms, as well as to give them insight into the needs of living things by using plants and nature as examples or symbols. An informal assessment of the patient's functioning and interest levels is made by the horticultural therapist before the first session. Treatment should be appropriate and directed to the diagnosis and symptoms.

Identification of Treatment Goals and Objectives

Some general goals are listed in Table 7.3. Once the needs of the patient are ascertained, the horticultural therapist must decide a plan of action to achieve success for the patient. It is important for the therapist to communicate to the patient the objective of the treatment and his or her goals for the patient, and it is essential that the patient be a willing participant to achieve a measure of success.

Example: An isolated patient who does not participate in any programs, keeps to him or herself, and spends most of his or her time in the bedroom.
Goal: To increase socialization skills.
Objective: Patient will agree to go with the therapist to the horticulture therapy site and remain there for thirty minutes.

Example: Agitated patient who paces.
Goal: To reduce agitation and hyperactivity.
Objective: Patient will participate in an outdoor nature walk for sixty minutes without incident.

Example: Patient with dependent personality disorder.
Goal: To have patient assume responsibility.
Objective Patient will water and/or check plants on the unit once a day for one week.

See Figures 7.1 and 7.2 for sample goals.

ADAPTATION OF HORTICULTURAL THERAPY ACTIVITIES TO MEET TREATMENT OBJECTIVES

When working with psychiatric patients as inpatients, the therapist must be aware that some materials can be potentially dangerous. Depending

TABLE 7.3. Goals for a Program with Psychiatric Patients

1. Horticultural work develops a sense of responsibility when caring for plants. Plants are living things that require a commitment and in turn a sense of responsibility. This is accomplished in a less threatening environment.

2. Horticultural activities increase socialization and appropriate interaction. Shared interest in plants is expressed and patients talk more freely about their past horticulture experiences. While working, defenses are often lessened and patients talk more openly about their feelings.

3. Horticulture provides an opportunity for self-expression and creativity. Flower arranging, crafts, dish gardens, and garden designs allow individuals to express themselves.

4. Horticultural work can help lessen feelings of depression. Looking at a beautiful flower or working with your hands in the dirt can provide a feeling of inspiration and renewal for life. Planting and finishing a project increases feelings of accomplishment and self-esteem.

5. Horticulture increases leisure skill development and recreational activities. Gardening is a skill that patients can take home with them. Continuing education programs, master gardeners courses, and visiting garden centers and local gardens are all opportunities for community involvement. Yard and garden work are activities that are valued by the general public.

6. Increasing functional goals such as attention span and concentration, improving problem-solving skills, and increasing independence are life and vocational skills needed to work in the real world. Patients who like what they are doing show more motivation and willingness to change.

7. Prevocational and vocational skills are developed in a real work situation. Plants are started, grown, and sold. Each step develops and reinforces many skill-building opportunities that can help patients get and keep jobs.

8. For patients with organic impairments, horticulture provides sensory-stimulating activities. Feeling velvety leaves or crushing dried mint both stimulate different senses and stimulate the patient to be in contact with the environment. Sitting in a garden or wandering around a circular path reduces agitation and has a calming effect on many patients.

Created by Martha C. Straus.

Figure 7.1. Sample Goals and Objectives as Related to Diagnoses

Diagnoses	Symptoms	Goals	Objectives
Mood Disorder Depression	Isolative	Pt. will agree to go with therapist to A.T. program site and remain for thirty minutes	Increase socialization
Mood Disorder Mania	Agitated with Pacing	Pt. will participate in outdoor nature walk	To reduce agitation and hyperactivity
Anxiety Disorder, Obsessive/Compulsive	Constantly Washing Hands and Cleaning Room	Pt. will mix potting soil with gloves for N.T. activity	To reduce stress and anxiety associated with dirt and germs
Schizophrenia	Auditory Hallucinations	Pt. will concentrate on repotting plants for 30 minutes	To direct attention from voices to N.T. activity
Dependent Personality Disorder	Unable to Make Decisions About Dress, Menu, Activity, etc.	Pt. will water (6) plants on unit windowsill two times per week	To assume responsibility for something

Source: Westchester County Medical Center Psychiatric Institute Department of Recreational and Expressive Activities.

Figure 7.2. Sample Goals and Objectives as Related to Diagnoses

Focus	Sample Goals
Cog./Phys.	Patient will focus attention to interact with leisure modality with his or her hands once during HT intervention
Cog./Soc.	Patient will attend to HT for a two-minute conversation after one week of HT
Cognitive	Patient will respond to HT's verbal stimuli by looking at HT when addressed
Cognitive	Patient will make eye contact with HT for thirty seconds
Cognitive	Patient will name place and time of his or her environment consistently during interventions
Cognitive	Patient will maintain eye contact and respond to three questions coherently and relevantly during HT intervention within one week
Cognitive	Patient will focus attention to complete an HT task for fifteen minutes daily within one week
Cognitive	Patient will focus attention to maintain discussion for ten minutes within one week
Cognitive	Patient will focus attention to HT task for ten minutes within one week
Cognitive	Patient will indicate awareness of HT by maintaining eye contact for thirty seconds during HT intervention
Cognitive	Patient will focus attention to complete leisure task cooperatively for ten minutes within one week
Cognitive	Patient will respond to visual stimuli with direct eye contact twice during intervention
Cognitive	Patient will respond verbally or otherwise to HT within one week
Leisure Ed.	Patient will identify three leisure resources
Leisure Ed.	Patient will identify two ways he or she can change unhealthy attitudes or adopt healthy ones
Mot./Cog.	Patient will participate during one activity daily for three minutes for one week
Mot./Cog.	Patient will participate for five continuous minutes for one activity
Mot./Social	Patient will initiate interaction with HT one time during each activity within one week
Motivation	Patient will attend one HT program per day for forty-five minutes
Motivation	Patient will participate in one HT activity per day without demonstrating unusual behavior
Motivation	Patient will initiate one HT activity per day for one week

187

Figure 7.2 (continued)

Focus	Sample Goals
Motivation	Patient will participate during three activities on/off the unit for one week
Motivation	Patient will participate during one activity per day for five minutes continuously with encouragement from HT
Motivation	Patient will participate during one scheduled activity per day with the assistance of others for one week
Physical	Patient will use his or her hands to hold an HT implement (will focus eye contact) for ten minutes within one week
Physical	Patient will use his or her hands to grasp an object once during HT intervention
Soc./Cog.	Patient will verbally respond to HT one time per week
Soc./Cog.	Patient will socialize with HT using sentence and staying on topic for five minutes
Soc./Cog.	Patient will maintain a conversation with HT for three minutes within one week
Social	Patient will maintain social conversation with HT for five minutes during daily intervention
Social	Patient will socialize with HT for five minutes without interrupting
Social	Patient will maintain social interaction with one person for five minutes during leisure time
Social	Patient will initiate interaction with one peer on unit per day within one week
Social	Patient will respond verbally to HT during 1:1 visits and leisure activity within one week
Social	Patient will respond verbally or otherwise to HT within one week
Social	Patient will interact on 1:1 basis or in small group in an appropriate manner for five minutes
Social	Patient will initiate conversation with one peer one time per day for five minutes
Social	Patient will interact verbally with HT for three minutes two times daily
Social	Patient will interact spontaneously during leisure activity
Social	Patient will respond appropriately to HT during 1:1 visit

upon the severity of their mental illness, the danger levels can vary from patient population to patient population and institution to institution. For example, an acute-care facility may be different from a rehabilitation facility. In an acute-care facility, patients are in less control of their actions and less responsible for them; whereas in a rehabilitation center patients would have moderate control. Each psychiatric facility sets its own policies. In addition to its restrictive policies on horticultural materials, there is usually a patient/staff ratio requirement.

Generally speaking, the horticultural therapist should keep in mind that materials such as those listed below and the like should be judiciously used.

1. Clay, ceramic, glass containers—can be broken and used as a weapon.
2. Rope, string, twine, ribbon, wire, shoelaces—can be used to choke others or hang themselves or others.
3. Balloons, plastic bags—can be used for suffocation.
4. Spray paints, rubber glue, model glue—toxic inhalants can give "highs."
5. Pointed scissors, clippers, loppers, knives, sticks, stakes—can be used to cut, stab, or mutilate themselves or others.
6. Cactus, roses, or other plants with thorns—can be used for self-mutilation.
7. Known toxic or allergenic plants—can be ingested by patient.
8. Craft items such as shells or rocks—can be used as weapons.

Patient care units are kept locked at all times. Patients must be stabilized, with written orders by their doctor, for permission to leave the unit and go to horticultural therapy (HT). This decision is determined by the multidisciplinary team.

On locked units, patients are often restricted from having personal possessions because of the potential dangers involved. When the items listed above are used in HT sessions, it is essential that these materials to be retrieved before patients return to the unit. In facilities that treat the less acute mental illnesses, such as rehabilitation institutes, residential facilities, halfway houses, or group homes, this may or may not be a problem. However, it is always important for the therapist to be vigilant.

Propagation, potting, and transplanting are core tasks for horticultural therapy activities. Fast-growing plants such as coleus, Swedish ivy, and wandering Jews propagate in a week or less, which can provide short-term gratification. Taking the plant for themselves provides the patient with an opportunity to nurture and accept responsibility for another living thing in a nonthreatening way. Plants and nature can be wonderful metaphors for "life." The needs of the plants for nourishment and care and the environment in which they grow can be directly related to human life.

Sample Horticultural Therapy Activity

Dish Garden: analogous to societal living and compatibility

1. Materials: container, soil, compatible plants
2. Procedure: Group discussion on dish garden project, and discussion about choice of container, soil, and selection of plants
3. Goals: Getting along in society/family/roommate/etc., listening to directions, and following directions
4. Task: Patient to complete dish garden
5. Closure: Cleanup, patient given plant-care instruction for follow-up care at home

Nature Walks

1. Dormancy of winter
2. Rebirth of spring
3. Splendor and maturity of summer
4. Senescence of fall

Activities for horticultural therapists are endless. There are as many activities as the imagination is willing to create. Nature encompasses almost everything. One could do outdoor gardening (which might result in cooking, baking, making potpourri), nature photography, drawing from nature, writing poetry about nature, birdwatching, butterfly watching, or building a scarecrow. Indoor gardening could include innumerable propagation of cuttings, seed propagation, nature crafts, terrariums, sand art, rock tumbling, or wreath making. It is important to remember that your horticultural activities need not to be restricted to plants but can encompass the entire "natural" world.

OBSERVATION, DOCUMENTATION, AND CLIENT FEEDBACK

Horticultural therapy sessions are usually time limited, anywhere from forty-five to sixty minutes or whatever duration your schedule provides. It is important for the therapist to make and note observations of the patient's conduct and behavior during the session:

1. Did the patient meet the goals and objectives set for this HT session?
2. How did the patient relate to the therapist?
3. How did the therapist relate to the patient?
4. How did the patient relate to the HT group?
5. Was the patient passive, aggressive, alert, distracted?
6. What was the patient's attention span?
7. Did the patient understand the activity?
8. Could the patient participate independently?
9. Did the patient enjoy the activity?

The sample Group Effectiveness Evaluation Form (Figure 7.3) can be used as one way of measuring patient satisfaction. This is simply another modality other than direct observation to get patient feedback. If the patient expresses an interest in returning to your HT program, this is yet another form of client feedback, particularly of patient satisfaction.

Documentation is required for all patients. Each institution has its own policies, procedures, and style. Samples of descriptive language (Figure 7.4), as well as a list of abbreviations (Figure 7.5), are useful when charting progress notes or incident reports.

Case Illustration

Mr. B was a thirty-two-year-old, single, white male, diagnosed with bipolar disorder. Mr. B was refusing medications because he said he was a "naturalist"; that is, he didn't want to put any unnatural chemicals into his body. His behavior was manic. The horticultural therapist working with him related this story: "I approached Mr. B. and said, 'Allow me to tell you about my vegetable garden. I tilled it, fertilized it, planted it, weeded it, and watered it, and nothing happened. The plants barely grew. They didn't die, but they certainly didn't flourish. I wondered what could be wrong. I finally decided to test the soil and discovered the pH to be very, very low. What could I do to raise it to normal? Lime it; and so I did. It worked! It neutralized the soil and brought it into balance.' I went on to say, 'There should be a balance in nature. Sometimes our chemistries and nature's chemistries are out of balance. Taking the medication the doctor prescribed, like adding lime to the soil, put nature back into balance.' This was something the patient could comprehend. He said I gave him some 'food' for thought."

FIGURE 7.3. Sample Questionnaire to Evaluate Group Effectiveness

Group Effectiveness Evaluation—Patient

Therapist: _____
Group Title: _____
Date: _____

the whole time 5
a lot 4
an even amount 3
a little bit 2
not at all 1

	1	2	3	4	5
Did you talk with other members of the group?	1	2	3	4	5
Did you talk with the therapists?	1	2	3	4	5
Did you talk because of the activity that was done in this group?	1	2	3	4	5
Did you feel trust for the other members of the group?	1	2	3	4	5
Did you feel trust for the therapist?	1	2	3	4	5
Did you discuss your problems?	1	2	3	4	5
Did you discuss how to solve your problems?	1	2	3	4	5
Did you discuss your feelings?	1	2	3	4	5
Did the group activity help you talk about your problems or feelings?	1	2	3	4	5
Did other members of the group help you talk about your problems or feelings?	1	2	3	4	5
Was the activity in this group of interest to you?	1	2	3	4	5

Group Effectiveness Evaluation—Therapist

High 4
Moderate 3
Minimal 2
Not Observed 1

	1	2	3	4
Did patient interact verbally with the therapist?	1	2	3	4
Did patient interact verbally with the other group members?	1	2	3	4
Did patient interact verbally with others through this structured modality?	1	2	3	4
Did patient display trust of others?	1	2	3	4
Was patient guarded/suspicious?	1	2	3	4
Did patient display psychotic thinking?	1	2	3	4
Did patient display concrete thinking?	1	2	3	4
Did patient display an increase in appropriate affect?	1	2	3	4
Did patient discuss problem solving?	1	2	3	4
Did patient discuss feelings?	1	2	3	4
Did patient discuss feelings related to the modality?	1	2	3	4
Did patient relate modality experience to life experience?	1	2	3	4

FIGURE 7.4. Descriptive Adjectives

Flexibility:
yielding
changeable
amenable
adaptable
conventional
persistent
habit-bound
stubborn
persevering
unbending
obstinate
Emotional Warmth:
overindulgent
doting
affectionate
sentimental
kindly
considerate
cool
unresponsive
detached
unfeeling
hardened
rejecting
frigid
Sociability:
intrusive
meddlesome
gregarious
convivial
intimate
comradely
companionable
agreeable
accessible
hesitant
reserved
reticent
shrinking
secluded
withdrawn
solitary
isolated
Dominance:
derisive
derogatory
scornful
sarcastic
argumentative
overcritical
nagging
outspoken
frank
tactful

soft-spoken
complimentary
praising
flattering
mealy-mouthed
apple-polishing
eulogistic
Physical Hostility:
malicious
embittered
quarrelsome
surly
provocative
resentful
irritable
grouchy
petulant
grudging
civil
inoffensive
unresentful
agreeable
gentle
gracious
conciliatory
ingratiating
oily
fawning
Anxiety:
shameless
conscienceless
unscrupulous
incorrigible
Self-Esteem:
self-exalting
pompous
conceited
boastful
vain
cocky
confident
self-respecting
modest
unassuming
humble
self-doubting
self-deprecatory
forlorn
self-abasing
Ideation:
delusional
ruminative
daydreaming
fanciful
musing

contemplative
thoughtful
matter-of-fact
literal
unreflective
unimaginative
stolid
vacuous
impulsive
incontinent
reckless
rash
impetuous
excitable
hasty
abrupt
restless
mobile
self-possessed
cool-headed
deliberate
controlled
restrained
staid
terrified
panicky
agitated
tremulous
apprehensive
tense
fretful
uneasy
composed
calm
nonchalant
unconcerned
cool
bland
stolid
imperturbable
phlegmatic
Mood:
euphoric
elated
frivolous
buoyant
gay
jovial
light-hearted
cheerful
placid
sober
serious
solemn
mirthless

grave
gloomy
brooding
dejected
disconsolate
despondent
hopeless
Guilt:
self-condemning
self-reproachful
remorseful
ashamed
chagrined
regretful
concerned
indifferent
unfeeling
unreformed
Cynical:
unrepentant
hardened
dictatorial
autocratic
high-handed
masterful
forceful
assertive
decisive
cooperative
assenting
conforming
compliant
acquiescent
imitative
deferent
timid
meek
servile
Ambition:
grandiose
pretentious
aspiring
enterprising
persistent
eager
self-satisfied
complacent
lackadaisical
indifferent
listless
indolent
apathetic
lethargic

FIGURE 7.5. Westchester County Medical Center Psychiatric and Correctional Services Approved Psychiatric, Alcoholism, and Correctional Health Abbreviations

Ψ	Psychiatry, psychiatric, psychiatrist	COA	Children of Alcoholics
↑	Increase, Increased	COSA	Children of Substance Abusers
↓	Decrease, Decreased	CPEP	Comprehensive Psychiatric Emergency Program
Δ	Change	CPP	Center for Preventive Psychiatry
+	Positive		
−	Negative	CPS	Children's Protective Services
A	Assessment/Absent	CPZ	Chloropromazine
A/VH	Auditory/Visual hallucinations	Cs	Conscious
AA	Alcoholics Anonymous	CS	Children's Services
Abs	Abstinent	CSC	Community Services Center
AC	Alcoholism Counselor	CSS	Community Services Support
ACOA	Adult Children of Alcoholics	CSW	Certified Social Worker
ADHD	Attention Deficit Hyperactivity Disorder	CT	Crisis Team
ADL	Activities of Daily Living	CV	Children's Village
ADM	Admission	CS	Close Watch
AGPN	Attending group progress note	CXR	Chest X Ray
AIMS	Abnormal involuntary movement scale	d	Day
		DARE	Domestic Abuse Rehabilitation Encounters
ALANON	12-step program for significant others		
		D/C	Discontinue
ALATEEN	12-step program for teenagers	DCMH	Department of Community Mental Health
ALD	Alcoholic Liver Disease	DD	Developmental Disability
AOB	Alcoholic Odor to Breath	DDG	Dual Diagnosis Group
AOS	Attending on Site	DDP	Drinking Driver Program
APN	Attending Progress Note	DMV	Department of Motor Vehicles
ARCS	AIDS-Related Community Services	DNKA	Did not keep appointment
		d/o	Disorder
ASIS	Alcoholism Services Information System	DOC	Department of Corrections
		DOLD	Date of Last Drink
ATS	Alcoholism Treatment Services	DOLSA	Date of Last Substance Abuse
Att	Attendance		
BAC	Blood Alcohol Concentration	DSS	Department of Social Services
BB	Big Book (used in AA)	DT	Delirium Tremens
BF	Black Female	DTATI	Direct Treatment Alternatives to Incarceration
BM	Black Male		
BIB	Brought in By	DTP	Day Treatment Program
BIBA	Brought in by Ambulance	Dtr	Daughter
C/L	Consultation/Liaison	DWAI	Driving While Impaired
CAC	Credential Alcoholism Counselor	DWI	Driving While Intoxicated
		E/A	Excused Absence
Clt.	Client	EA	Employability Assessment
CM	Case Management	EFTs	Endocrine Function Tests
CMHC	Community Mental Health Center	EOC	Educational Opportunity Center
CND	Clean and Dry	EPA	Employment Performance Appraisal
C/O	Complaint		

EPRA	Employment Program for Recovered Alcoholics	MR	Mental Retardation
EPS	Extra Pyramidal Symptoms	MSE	Mental Status Examinations
FA	Frequent Awareness	MSPN	Medical Student Progress Note
FOI	Flight of Ideas	Mtgs.	Meetings
FTD	Formal Thought Disorder	N/A	Not applicable
FU	Follow-Up	NA	Narcotics Anonymous
G	Group	NARANON	12-step program for significant others
GATS	Greenburgh Alcoholism Treatment Services	NARATEEN	12-step program for teens
GCI	Guidance Center Industries	NCA	National Council on Alcoholism
GED	General Educational Diploma Group	NCAC	National Certified Alcoholism Counselor
HMO	Health Maintenance Organization	NFA	Not for Admission
HP	Higher Power	NOS	Not Otherwise Specified
HPI	History of Present Illness	NPN	Nursing Progressive Note
HTP	House Tree Person Test	O	Objective
HS	Hour of Sleep	OASAS	Office of Alcoholism and Substance Abuse Services
ICD	International Center for the Disabled	OCD	Obsessive-Compulsive Disorder
ICM	Intensive Case Management	OD	Officer of the Day
ID	Identify	ODAT	One Day at a Time
I/J	Insight/Judgment	OMH	Office of Mental Health
IOR	Ideas of Reference	OMS	Organic Mental Syndrome
IP	Identified Patient	OOC	Out of Control
IPA	Intermediate Psychiatric Aide	OP	Order of Protection
IQ	Intelligence Quotient	Orientedx3	Oriented for time, place, and person
L	Long		
LD	Learning Disability	OT	Occupational Therapy
LOA	Loosening of Associations	OTC	Over the Counter
LTG	Long-Term Goal	p	after
LWOC	Leave Without Consent	P	Plan
M-D	Manic-Depressive	PC	Personal Contact
MDDS	Multidisciplinary Diagnostic Summary	PCS	Pleasantville Cottage School
MI	Mental Illness	PD	Personality Disorder
MAOI	Monoamine Oxidase Inhibitor	P/D	Per Day
MAST	Michigan Alcoholism Screening Test	PE	Physical Examination
MGM	Maternal Grandmother	PGF	Paternal Grandfather
MGP	Maternal Grandparent(s)	PGM	Paternal Grandmother
MHA	Mental Health Association	PGPs	Paternal Grandparents
MHATI	Mental Health Alternatives to Incarceration	PhD	Psychologist
		PIC	Private Industry Council
MHC	Mental Health Center	PIER	Psychiatry Institute Emergency Room
MICA	Mentally Ill Chemical Abusers	PMA	Psychomotor Activity
Min	Minimum	PMD	Personal MD
MMPI	Minnesota Multiphasic Inventory	p.o.	Protective Observation
		PO	Probation Officer
MPS	Mandated Protective Services	ppd	Packs per day

FIGURE 7.5 (continued)

PRBA	Physician Risk Benefit Note	TC	Therapeutic Community
PRN	As needed	TCA	Tri-Cyclic Antidepressant
PROCN	Psychiatric Resident on Call Note	TID	Three Times a Day
		TOP	Temporary Order of Protection
PROST	Protestant	Uncon	Unconscious
Psa	Psychoanalytic, Psychoanalyst	VA	Veterans Administration
Psol	Psychologist, psychology	Vac	Vacation
P/W	Per week	VESID	NYS Vocational, Educational Sciences for Individuals with Disabilities
q	every		
Q/A	Question/Answer	VOA	Volunteers of America
QD	Every day	WAIS	Wechsler Adult Intelligence Scale
QID	Four times a day		
QOD	Every other day	WAIS-R	Wechsler Adult Intelligence Scale Revised
SO	Significant Other		
SPA	Senior Psychiatric Aide	WARC	Westchester Association for Retarded Citizens
SSD	Social Security Disability		
SSI	Supplemental Security Income	WF	White Female
STG	Short-Term Goal	WM	White Male
SW	Social Worker	WISC-III	Wechsler Intelligence Scale Children (3rd Edition)
SWM	Single White Male	WESTHAB	Shelter
SWPN	Social Work Progress Note	WESTHELP	Shelter
Sx	Symptoms	wk	Week
TASC	Treatment Alternatives to Street Crime	W/O	Without
TAT	Thematic Apperception Test	Wu	Work-up
TB	Tuberculosis	X	Time(s)
TBD	To Be Determined		

BIBLIOGRAPHY

Bowen, Murray. (1978). *Family Therapy in Clinical Practice*. New York: J. Aronson.

Diagnostic and Statistical Manual of Mental Disorders, Fourth Edition. (1994). Washington, DC: The American Psychiatric Association.

Fishman, H.C. and Rosman, B.L. (Eds.). (1986). *Evolving Models for Family Change: A Volume in Honor of Salvador Minuchin*. New York: Guilford Press.

Freud, S. and Oppenheim, D.E. (1958). *Dreams in Folklore*. New York: International Universities Press.

Gedo, J.E. and Pollack, G.H. (Eds.). Freud: The Fusion of Science and Humanism: The Intellectual History of Psychoanalysis. *Psychological Issues*, 34/35, 474 pp.

Haley, J. (1971). *Changing Families: A Family Therapy Reader*. New York: Grune & Stratton.

Hoffman, E. (Eds.). (1996). *Future Visions: The Unpublished Papers of Abraham Maslow*. Thousand Oaks, CA: Sage Publications.

Lowry, R.J. (1973). (Ed.). *Dominance, Self-Esteem, Self-Actualization: Germinal Papers of A.H. Maslow*. Monterey, CA: Brooks/Cole Pub. Co.

Maslow, A.H. (1968). *Toward a Psychology of Being*. Princeton, NJ: Van Nostrand.

_____. (1970). *Motivation and Personality*. New York: Harper & Brothers.

McFarland, Gertrude K. and Wasli, Evelyn L. *Nursing Diagnoses and Process in Psychiatric Mental Health Nursing*. (1986). Philadelphia: J.B. Lippincott-Ravin Publishers.

Minuchin, S. (1974). *Families and Family Therapy*. Cambridge, MA: Harvard University Press.

Minuchin, S. and Fishman, H.C. (1981). *Family Therapy Techniques*. Cambridge, MA: Harvard Press.

National Institute of Mental Health. The value of psychiatric treatment: its efficacy in severe mental disorder. (1993). vol. 29, No. 4 National Institute of Mental Health.

Nye, R.D. (1996). *Three Psychologies: Perspectives From Freud, Skinner, and Rogers*. Pacific Grove, CA: Brooks/Cole Pub. Co.

Perry, H.S. and Gawel, M.L. (Eds.). (1953). *The Interpersonal Theory of Psychiatry*. New York: Norton.

Piers, M.W. (Eds.). (1972). *Play and Development: A Symposium, with Contributions by Jean Piaget and Others*. New York: Norton.

Popov, Y. and Rokhlin, L. (Eds.). (196_). *Psychopathology and Psychiatry: Selected Works*. Moscow: Foreign Languages Pub. House.

Rothgeb, C.L. (Ed.). (1971). Abstracts of the Standard Edition of The Complete Psychological Works of Sigmund Freud. Rockville, MD: National Institute of Mental Health.

Satir, V. (1983). *Conjoint Family Therapy*. Palo Alto, CA: Science and Behavior Books.

Skinner, B.F. (1974). *About Behaviorism*. New York: Knopf; Random House.

Sullivan, H.S. (1964). *The Fusion of Psychiatry and Social Science*. New York: Norton.

_____. (1965). *Collected Works*. New York: Norton.

Chapter 8

Children and Youth
and Horticultural Therapy Practice

Thom Pentz
Martha C. Straus

INTRODUCTION

Statement of the Challenge and Issues

Children represent society's most valuable resource. In particular, the mental and emotional health of children, as a resource, is one to be carefully fostered and nurtured. As such, children represent a focal point for the application of society's many other resources to ensure their healthy growth and development.

Purpose and Learning Objectives of Chapter

This chapter will undertake an exploration of child development, the principles that characterize development, and the forces that impact upon it. Within this framework, therapy with children will be discussed as an undertaking designed to facilitate and reinforce the normal developmental course or to remediate misdirections and blockages in the developmental passage. Horticultural therapy, in particular, will be presented as a viable and effective means for enhancing the healthy development of children through their engagement in gardening and related activities.

Significance

Horticultural therapy provides an opportunity for children to work cooperatively with other children while learning new skills and new information. Horticulture is also an activity where children can participate and grow plants equal to those grown by adults. Working with helping adults in the context of specifically designed horticultural activities promotes healthy interaction and developmental growth in the child. Skills learned in gardening activities are transferable to other settings and offer the experience of success, which reinforces a sense of mastery and heightened self-esteem. Horticultural therapy also provides an engaging setting in which to implement a broad range of therapeutic techniques that can be applied to the healthy management of behavior, feelings, and relationships. In working with nature, children learn about their environment, the cycles of development, and about themselves.

Client Population

Of the 63 million children in the United States today, approximately 15 percent suffer from emotional and behavioral problems that warrant mental health services (President's Commission on Mental Health, 1978). Of these, 3 to 8 percent, or approximately ten million children, are seriously emotionally disturbed (Knitzer, 1982). Untold numbers of other children are psychologically at risk and would benefit from preventive services.

Some of the environmental risk factors associated with higher rates of mental health problems in children include poverty, parental psychopathology, physical or sexual abuse, parental divorce, and serious childhood illness.

In the United States, approximately fourteen million children are living in poverty (Horowitz and O'Brien, 1989). Approximately 40 to 50 percent of children will experience firsthand their parents' divorce and will spend an average of five years in a single-parent home (Glick and Lin, 1986). Even conservative estimates of the incidence of child abuse are disturbing. Approximately one million official reports of child abuse or neglect are made in the United States each year (American Humane Association, 1984). Other surveys estimate three million incidents of physical abuse and one-and-a-third million incidents of sexual abuse involving children in the United States yearly (Gallup Poll, 1995).

Utilization data show that about 5 percent of children in the United States receive outpatient mental health treatment and 1 percent receive hospital or residential treatment center treatment yearly. (See Figure 8.1.) Such data would indicate that about 60 percent of children in need are not receiving appropriate mental health treatment (Taube and Barrett, 1985).

Despite recent significant changes in the health care system in the United States, children's mental health needs remain underserved. Development of viable and effective modalities for delivering mental health treatment to children remains a crucial societal challenge.

FIGURE 8.1. Childhood Psychiatric Diagnoses

The Diagnostic and Statistical Manual of Mental Disorders, Fourth Edition (1994), uses a diagnostic system to categorize clinical problems based primarily upon descriptive information and symptoms. Diagnoses for childhood problems are divided into five general categories based upon the aspect of functioning most disturbed: intellectual, developmental, behavioral, emotional, and physical. Children's problems are complex, and more than one diagnosis may be required to describe disruptions to a child's functioning. The table shows diagnoses commonly used with children, related symptoms, and prevalence rates.

Diagnosis	Prevalence	Symptoms
Childhood Depression	0.4% to 8.8% (McCracken, 1992)	Depressed mood, somatic complaints, irritability, social withdrawal, low self-esteem, loss of interest in activities
Anxiety Disorder	3% to 5%	Anxious mood, excessive worry especially about performance, unrealistic fears, insecurity, perfectionism, neediness for approval and reassurance
Attention Deficit Hyperactivity Disorder	3% to 5%	Pattern of persistent and excessive inattention, hyperactivity, distractibility, impulsivity
Oppositional Defiant Disorder	2% to 16%	Recurrent pattern of negativistic, defiant, disobedient, hostile behavior

DEVELOPMENTAL PSYCHOLOGY AND HORTICULTURAL THERAPY

The course of childhood is characterized by persistent and progressive change as the child traverses a path leading from a relatively helpless and dependent infant to a competent and independent adult. The understanding of children and their needs is prefaced by an understanding of this developmental course and the factors that guide and influence it.

Principles That Characterize Child Development

Psychologists who study child development have identified a number of principles that tend to characterize the process by which children grow from infancy through childhood and into adulthood.

1. Continuity and Progression

Development is a continuous process marked by the unfolding of genetically determined plans and potentials in interaction with the environment to which the child is exposed.

2. Dynamism and Interaction

Each child is born with a biological foundation that guides and influences subsequent development. Environmental experiences interact with biological givens to stimulate and reinforce or to alter the pattern and direction of growth. The child and his or her environment have a dynamic and reciprocal relationship. Children play an active role in affecting their environment and its response to them.

Development is marked by a balance between complementary processes of assimilation and accommodation. In assimilation, the child incorporates new experiences into his or her conceptual and characterological framework in such a way as to preserve a sense of identity, integrity, and continuity. In accommodation, the child makes changes in cognitive and behavioral patterns in order to cope with and adapt to the demands and realities of the environment (Piaget and Inhelder, 1969).

3. Variation

There is variation in the rate of development between individuals and within individuals over time.

4. Normative Sequence

A normative pattern for development can be identified particularly when the environment promotes and permits the unfolding of the genetic map.

5. Epigenesis

There is a cumulative effect to development. Developmental growth proceeds and builds upon the foundation of previous developmental achievements.

6. Optimal Tendency

The developing organism strives to achieve its maximum potential.

Developmental Stages

Many theorists have sought to explain the development of the child from a tiny package of DNA to a complex, functioning adult based upon a series of stages through which children pass. Stage theories incorporate the developmental principles of continuity and epigenesis and generally represent development as progressive mastery of a series of challenges or tasks. Erik Erikson has focused on the social and emotional development of the child (see Figure 8.2), while Jean Piaget has focused on the cognitive development of the child, which involves the growth of the child's intellectual and perceptual abilities.

Developmental Lines

The developmental growth of the child can be traced and understood in terms of progressive changes that occur across various functional areas called "developmental lines."

One developmental line involves the cognitive growth of the child, represented by changes in the way that the child perceives and understands the world around him or her. Piaget has described the child's progressive growth in understanding the world in terms of sensorimotor, then concrete, and then abstract concepts (Piaget and Inhelder, 1969). Appreciating how the child views, understands, and communicates with his or her world is essential in understanding, interacting with, and helping children.

Another developmental line involves interpersonal relationships and is represented by growth in the establishment and constructive utilization of stable attachment relationships. This same line would involve the learning of pragmatic social skills.

A third example of a developmental line would trace the growth of skills for differentiating, identifying, expressing, controlling, and managing feelings. Other developmental lines include self-image and self-esteem, motor skills, and language development. Obviously, the developmental lines do not proceed independently of one another. There is a dynamic interaction that occurs between the developmental lines. Cognitive and communication capacities certainly impact upon the negotiation of interpersonal relations as well as the understanding and management of affective experience. On the other hand, the quality of significant attachment

FIGURE 8.2. Erik Erikson: The Eight Stages of Man

Erikson's psychosocial theory of development focuses on the psychosocial motivations and needs that drive human development. He conceived of human development in terms of eight stages. Each stage is characterized by a psychosocial task the individual needs to master. If the task is mastered, the individual incorporates a positive attribute into the developing personality and moves on to the next stage. Erikson also described negative outcomes that result when an individual is not able to successfully master the developmental task.

1. *Trust vs. Distrust* (zero to one year). Infants develop the capacity to trust caregivers and people in general based upon their experiences of sensitive and timely nurturance and affection from parental figures. Distrust develops when needs are not met.
2. *Autonomy vs. Shame and Doubt* (one to two years). Children learn to have control over body functions. They are allowed to venture away from parents and explore the world with some degree of independence. They develop a sense of separateness and mastery. Overrestrictiveness and criticism can produce a sense of shame and self-doubt.
3. *Initiative vs. Guilt* (three to five years). Based upon growing motor and intellectual abilities and expanded experience, children display more initiative and assume more responsibility in exploring and affecting their environment. Caregivers who may not tolerate the child's growing independence and initiative may instill a sense of guilt over misbehavior.
4. *Industry vs. Inferiority* (six to eleven years). Children encounter the expanded environment and challenges of school. They develop a sense of self-worth through academic accomplishment, gratifying interactions with adults and peers, and successful participation in activities. A sense of inferiority may develop when children are not able to successfully master the developmental challenges of the school-age years.
5. *Identity vs. Role Confusion* (twelve to nineteen years). Adolescents face physical changes and increasing demands for independent functioning as they negotiate this transitional stage to adulthood. They progressively achieve a unique sense of independent identity and worth. Some adolescents are left with a sense of confusion about their identity and role in life.
6. *Intimacy vs. Isolation* (young adulthood: twenties and thirties). Young adults develop the capacity to establish close relationships with others or remain isolated from meaningful attachments.
7. *Generativity vs. Stagnation* (middle adulthood: forties and fifties). Middle adults master their assumption of responsible and productive roles in the family, workplace, and community. Conversely, the failure to achieve successful productivity can result in a sense of stagnation.
8. *Integrity vs. Despair* (late adulthood: sixty years and over). Older adults look back upon and evaluate their lives and achieve a sense of satisfaction and acceptance, or they may despair over a sense of not having achieved a sense of meaning and fulfillment for their life.

Horticultural Therapy provides an excellent opportunity to practice prevocational work skills, such as signing in and out on a time card.

relationships has been shown to influence the cognitive development of the child (Bell, 1970).

Biopsychosocial Model of Child Development

The forces that guide and influence child development can be conceptualized in terms of a biopsychosocial model.

1. Biological Forces

Developmental sequences and potentials are genetically encoded and unfold when provided with sufficient environmental stimulation. Interpersonal attachment (Bowlby, 1969) and language development (Chomsky, 1980) have been shown to be essentially programmed into the human organism. Constitutional differences in children's temperament, activity levels, and impulsivity have also been demonstrated (Chess and Thomas,

1986). Capacities for attention, concentration, and impulse control are affected by the child's neurological makeup. Individual differences in cognitive capacity impact upon children's learning, their understanding of the world, and their management of the educational experience of school. Biochemical factors have been shown to influence how children experience their emotions (Izard, 1982).

2. Psychosocial Forces

Psychosocial factors influencing the child's development include the quality of attachment relationships with parents, delivery of nurturance and cognitive stimulation, exposure to opportunities for peer interaction, and experience of mastery and success.

3. Cultural Forces

Cultural forces significantly impact how children are raised and grow. On their way to adulthood, children must negotiate the many cultural values incorporated into child-rearing practices, such as gender roles, as well as many cultural challenges, such as sexually transmitted diseases and violence.

DEVELOPMENTAL MODEL OF CHILDHOOD PSYCHOPATHOLOGY

From a developmental standpoint, psychopathology occurs when the normal developmental course is altered or impeded and the child's adaptive functioning is disrupted. To achieve better understanding of how such developmental disruptions may occur, this chapter will focus on three specific developmental lines: interpersonal relations, self-image and self-esteem, and affect management. Within that context, this chapter will consider how biopsychosocial factors may serve to reinforce or disrupt progress along the developmental lines.

Interpersonal Relations

The human infant appears to be biologically programmed or predisposed to function as a social creature. The quality of the attachment relationship that develops between the infant and its parents depends upon the

When choosing seeds to sow, a student must use reading, problem-solving, and decision-making skills.

nature of the parenting (Ainsworth, 1985). Parental characteristics of warmth, sensitivity, availability, and consistency promote positive attachments. A child who receives reliable nurturance develops a sense of trust in parents and in the world as well. Children's perceptions of their parents and their feelings about the relationships they have with their parents tend to generally influence their views of and responses to people in general. Children exposed to psychological or physical abuse, rejection, neglect, or disruption in caretaking often experience difficulty developing trusting and secure attachment relationships.

During the preschool and elementary school years, the development of interpersonal relationships focuses progressively more on peer interactions and relationships. Children who are not provided with opportunities to play with peers may fail to develop skills for interacting and getting along with other children. Children who are challenged by communication or processing deficits may experience particular difficulty understanding and managing social situations. Children who are hyperactive, impulsive, and distractible may have a difficult time sustaining healthy peer interactions.

Sowing seeds provides an opportunity for cooperative activity.

Self-Image and Self-Esteem

As the child traverses the developmental line of self-image and self-esteem, the goals are to develop a healthy sense of self as a separate and unique individual and to feel good about oneself. A child's sense of self-esteem is at first founded largely upon what is reflected from significant adults in his or her life. When children feel cared about by important grown-ups in their life, they are more likely, in turn, to care about themselves. When children experience a sense of rejection, they tend to question their self-worth. The egocentric nature of young children makes them prone to view the world as revolving around them. Consequently, they may tend to misinterpret developmental experiences and to view themselves as responsible for interpersonal events beyond their control. Children who face separation or loss, such as occurs when parents divorce, may blame themselves for causing the situation and thereby experience a disruption in self-esteem.

During middle childhood, a child's self-esteem is based largely upon the sense of mastery and competence derived from various accomplishments. School becomes a central arena for mastering challenges, both educational and interpersonal. Children who experience significant academic frustration in school, perhaps as the result of a learning disability or attention deficit, may have a difficult time feeling good about themselves. Interpersonal conflict or isolation may also undermine the growth of self-esteem.

Affect Management

Affect management represents a significant developmental line as the child grows in his or her ability to recognize, differentiate, express, manage, and control various feelings. How a child learns to identify, express, and manage a broad range of feelings depends largely on what is modeled and taught by significant adults in his or her life. In some families, feelings may be expressed in an out-of-control and perhaps frightening manner. Other families may have a difficult time openly expressing various feelings of anger, sadness, or affection. The quality of a child's attachment relationships, in addition to the teaching and modeling of affective communication, will influence the child's comfort and skills with the expression of feelings. Children who are exposed to traumatic experiences at an early age may feel overwhelmed by feelings and have a difficult time understanding and managing them. The manner in which children experience their feelings is also determined by the neurochemical substrates that mediate the experience of all human emotion. Certain individuals are predisposed to experiencing depression or affective lability based upon their biochemical makeup. Such children face greater challenges in learning to understand and modulate their affective experience.

ASSESSMENT AND TREATMENT OF CHILDREN WITH DEVELOPMENTAL DISTURBANCES

An appreciation of normal developmental principles and processes, a corresponding understanding of developmental disruption, and a knowledge of the factors that foster or impede healthy development will guide therapeutic efforts to remediate developmental problems and to facilitate healthy developmental progress.

Effective therapeutic intervention begins with a thorough evaluation of the child. Assessment of the child's status and progress along various developmental lines can be compared with normative expectations for children of that age. Corresponding strengths and deficits form a profile, which enhances the understanding of the whole child and guides the development of therapeutic interventions.

Before embarking upon a course of assessment and therapy with an elementary school child, it will be important to have an appreciation of the normative developmental profile that characterizes children during the middle childhood years.

Normative Profile of Middle Childhood

Most cultures identify the age of six or seven years, when formal education typically begins, as an age when the child enters a new stage of life. In the child's broadening world, educational, interpersonal, and social forces build upon and interact with all that has gone before to create new and more complex psychological and behavioral patterns.

Physical Growth

Physical growth during the middle childhood years is typically slow, relative to other stages of development. This period of physical quiescence seems to allow energies to be focused on educational and social development in preparation for the significant physical and psychological changes ushered in by puberty and adolescence. Physical size and skill are important insofar as they impact upon the child's adjustment to the school setting. His or her ability to master the academic, interpersonal, and social challenges of the school environment are essential to the child's development of a sense of industry and competence. Physical size and skill may impact upon children's sense of self as they attune to the reactions and relative abilities of peers in classroom activities and athletic games. The average six-year-old has mastered the motor skills to participate in school-age play and activities. A progressive refinement in motor skills finds the nine- or ten-year-old capable of and interested in participating in a broad variety of games and activities. Assessments of the school-age child in terms of physical development focus upon attributes of size, strength, speed, precision, and flexibility in both gross and fine motor areas.

Cognitive Development

A central task marking the school-age years is the development of intellectual skills and the ability to apply them to the solution of intellectual problems. During the middle childhood years, the child develops increased freedom, flexibility, and control in thinking, including a more sophisticated understanding of relationships not only between objects and events but also between symbols. Piaget named this stage "concrete operations" to describe how a child learns to use symbols rather than concrete objects to perform acts of cognition (Piaget and Inhelder, 1969). In this stage, the child's thinking becomes less egocentric and he or she can entertain in thought a broader range of possibilities, including alternative courses of action. Development of the capacity for cognitive reversibility—to think an action and then to think it undone—is what allows the child to try out different courses of action mentally.

One of the cognitive operations that develops during middle childhood is classification, the ordering of objects into common groups based upon common qualities which the child is able to identify. Piaget has described how children at different stages enjoy activities that "feed" the growth of cognitive skills developing during that stage; children during middle childhood typically enjoy collecting., an activity that "feeds" the development of classification.

One outgrowth of the child's expanding cognitive abilities is reflected in the enrichment of the child's capacity to integrate information from various sensory and perceptual modalities to develop a better understanding of the environment and more effective ways of interacting with it. There is also growth in a child's imaginative life and ability to use imagination for mastering and dealing with experience.

Cognitive development is reflected in a child's building of a more complex and accurate internal (cognitive) representation of the world, its objects and people, and their interrelationships.

Language Development

The school-age child also reflects significant growth in the understanding and effective use of language. This development in language is both a product of and contributor to the child's cognitive development. During the middle childhood years, the child masters the mechanics of language, articulates more clearly, and speaks in longer and more complex sentences. During the course of elementary school, the size of a child's expressive and receptive vocabulary is typically doubled. The child becomes more proficient in using language to mediate thought and to communicate ideas, which results in a broadening of horizons in understanding and interacting with the world. Growth in cognitive and language abilities during middle childhood opens the way for communicating verbally with children about their ideas, experiences, feelings, and relationships.

The Sense of Industry

Erikson has identified the developmental task or challenge of middle childhood as the achievement of a sense of industry (Erikson, 1963). The child seeks to establish a sense of competence as a learner, a doer, and a more social being. The child's increasing intellectual maturity paves the way for advances not only in learning but also in communication and in more socialized ways of thinking and acting. The child becomes better able to delay a response, consider various behavioral options, and evaluate potential social consequences. The child pursues a sense of achievement and recognition through mastery of academic, problem-solving, and interpersonal skills as well as through incorporation of cultural skills and values.

The converse of a sense of industry and competence is a sense of inferiority or inadequacy. A sense of inadequacy may result when the child is exposed to limited opportunities for success or when the child's resources and abilities are overtaxed by the demands of experience. The child who experiences a sense of failure in school, in competition, and in

Understanding the needs of seeds and plants helps children understand the life cycles.

interaction with peers, and who feels frustrated in efforts to master life's challenges, may develop feelings of anxiety and/or depression. The child may also begin to exhibit acting-out behaviors that seem to vent feelings of frustration or to avoid frustrating situations.

Social and Emotional Development

Children's entry into school expands their social horizons and exposure to extrafamilial influences. Classmates and teachers become important parts of a child's social world. The process of identification, whereby children emulate and internalize parental attributes, is extended to other influential adults and peers as well. Consequently, children's identities are broadened and enriched. Peer acceptance and peer friendships become highly valued by elementary school children. The peer group thus assumes a more significant role in the development of a child's behavior, values, and attitudes.

Children's self-concept and self-esteem are significantly influenced by their sense of how others respond to, value, and appreciate them. Children also measure themselves against their perceptions of the achievements and abilities of peers. A sense of inadequacy can develop when a child feels rejected or neglected by significant adults or by peers, or when the child feels inferior in comparison to the achievements of peers.

Advances in cognitive and social skills allow children to appreciate the thoughts and feelings of other people. They develop a capacity for empathy, with better understanding of the subjective viewpoints and feelings of others. Capacities for interpersonal understanding, communication, and cooperation are also enhanced. The capacity for empathy also opens the door to deeper and closer friendships.

Cognitive development also expands and enriches children's emotional lives. They can identify and differentiate a broader range of feelings. Also, they are beginning to develop improved skills for managing affective experience, and there is increased internalization of skills for controlling impulses. Cognitive advances allow the child to consider alternative courses of action and evaluate potential consequences with a resulting growth in judgment.

PRINCIPLES OF PSYCHOTHERAPEUTIC INTERVENTION

Therapeutic interventions with children can focus on changing identified forces that cause a developmental disruption, remediating a developmental deficit, or modifying specific behavioral and interactional symptoms that are reflections of the developmental disruptions. While such interventions can adopt a multiplicity of methods and are guided by a variety of theories, certain common principles tend to guide most therapeutic efforts with children.

All therapies begin with and are centered around a therapeutic relationship. Development of a supportive attachment relationship is a positive, growth-promoting experience. Children can learn to trust the therapist and to turn to the therapist for support, guidance, and approval. Many times, children develop a transference relationship with the therapist, in which the child's perception of and reaction to the therapist is colored and influenced by the child's prior experiences and relationships with adults. The therapeutic relationship then represents an opportunity to examine and to correct sensitized and maladaptive ways of perceiving and relating to adults.

At a verbal level, depending upon the cognitive and emotional availability of the child to such an approach, therapy can represent an opportunity to talk about feelings and problems. Both current as well as past experience can be a focus of discussion. What happens during the therapy session can be examined and talked about. Enhanced self-awareness and self-understanding can be valuable precursors to emotional and behavioral change.

Education is an important therapeutic tool, and the therapist may often assume the role of teacher. In addition to promoting self-awareness and insight, verbal interaction in therapy can be applied to problem-solving and teaching skills and strategies for managing feelings, relationships, and experiences. Cognitive approaches may also be focused on helping children to view their experience in different and more productive ways.

Developmentally, many children may not be ready to respond to the more verbal therapeutic approaches. Nonetheless, therapy can proceed at an experiential level. As discussed, the relationship with the therapist can provide a corrective and healing psychological experience. Behavior modification serves in an informal or more structured fashion to selectively offer positive attention or rewards to healthy, adaptive behaviors. Therapeutic activities can be designed to offer constructive outlets for channeling aggressive energies. Activities also stimulate the child's imagination and creative energies and can be designed to provide a sense of mastery and accomplishment. Behavior modification techniques and various activities can offer an intrinsic reward to capture the child who is guarded and mistrustful or who suffers from a fragile or damaged self-esteem. In the long run, however, therapy itself should prove to be intrinsically rewarding as relationships and experiences foster a satisfying sense of developmental growth and mastery.

Planning for spring.

Group therapy expands therapeutic opportunities through the presence of and interactions with other children in the therapeutic setting. Here again, inherently therapeutic relationships can develop. Peer interactions afford an arena for examining and understanding interactional problems while teaching constructive social and interactional skills. Children can engage in shared activities and learn to benefit from mutual cooperation and problem solving.

IDENTIFICATION OF TREATMENT ISSUES APPROPRIATE FOR HORTICULTURAL THERAPY

Horticultural therapy for children provides a challenging and rewarding opportunity because a wide range of physical and cognitive functioning can be found within the same age range. Children can show incredible creativity and can wear out a therapist, all in the same group period. There seems to be something magical about bringing together children, soil, and plants, and offering the children a time when they can get "down and dirty," explore the world, and feel good about their accomplishments. Horticultural therapy programs are found in school systems, hospitals, botanic gardens, after-school recreation sites, and special education facilities.

The program focus depends on the mission of the facility and the needs of the participants, but may include the social, therapeutic, educational, or pre-vocational use of horticulture. The benefits to children for their participation in Horticultural Therapy are seen in four areas: cognitive development, psychological growth, social skill learning, and in prevocational work skills development.

Cognitive development for children in horticultural therapy is promoted in a variety of ways. For children who have trouble focusing and staying on task, garden activities engage the children and allow them to focus and increase their attention span. Gardening is an action-oriented activity that is useful for children with excess energy or poor attention span who need to channel their energies in positive and constructive ways. Horticulture is a fun science where new skills and ideas are explored, and problem solving and independent functioning are reinforced. Children can investigate scientific information using the garden as their laboratory. Those who like what they are doing exhibit improved concentration, increased attention span, and learn new information.

Psychological growth in children is fostered in a number of ways. When working with plants, children learn the responsibility of taking care of something that is dependent on them and learn that living organisms, even if they are sick, can get better and thrive if they are cared for. Gardening teaches the life lessons that plants, like children, develop in stages, each at their own pace, and that there are causes and effects for actions. For example, plants that are not watered, die. Children discover nature to have its own rhythms and cycle of development that man can not manipulate, just as children have their own developmental cycles. Gardening is something which a child can share and participate in with a parent or adult, which helps to integrate the child into the family or community, and which helps the child feel useful and productive.

Emotional benefits of horticultural therapy include improved self-esteem and self-confidence and enhanced affects-anagement skills. Excess energy is channeled into constructive activities. Social skills are modeled by the therapist, and the participants are always expected to exhibit respect and politeness in their interactions. Children learn to participate in cooperative tasks such as sharing tools, mixing soil, or planting seeds. Interpersonal relationships improve with both adults and other group members when children enjoy what they are doing. Successful activities increase self-esteem. Youngsters often find active participation in gardening to be calming and relaxing (Airhart, Willis and Westrick, 1987). While the environment may seem stimulating, children often find the greenhouse a place to refocus.

Prevocational work skills can be practiced even by the youngest of children. Following directions, staying on task, accepting support, and processing feedback are skills that everyone needs for work. These abilities can begin to be developed in a horticultural therapy program. Lessons do not seem so difficult when learned in the course of an enjoyable activity that also offers a plant reward at the end. Children learn that they need to work to the best of their abilities, which allows them to feel good about what they have accomplished.

GOALS RELATING TO HORTICULTURAL THERAPY

The makeup of the treatment teams and goals for children depend on what type of facility is providing the horticultural therapy program. After-school programs have a social or educational focus and include teachers,

The reward of picking a plant at the end of a successful activity helps with motivation and encourages cooperation.

horticulturalists, and families. Children who participate in horticultural therapy as part of treatment in a hospital often have social workers, therapist (psychologist or psychiatrist), occupational therapist, and a special education teacher as part of the team.

Areas targeted for treatment must relate to the child's diagnosis, past history, educational goals, family needs, and future goals. What can reasonably be accomplished in the time available with each client must also be considered. Incorporating the child's interests, strengths, and demonstrated past areas of success will aid greatly in motivating and succeeding with the youngster. Goals focus on broad areas of development, but also identify specific desired outcomes. Goals must be behavioral and measurable. (See Table 8.1.)

Objectives are steps taken to achieve identified goals and must be patient-specific and include time frames and methods of assessment and measurement. Most facilities require the therapist to write a plan that describes how patients will be helped to meet these objectives. Objectives must start small and then be increased as they are met. Children should be aware of and understand their goals and hopefully agree to the need to attain these goals and objectives.

Treatment focus for children participating in horticultural therapy may include the following:

1. Development of social skills and interpersonal relationships
   Goal: To reduce demanding behavior
   Objective: Patient will ask for supplies politely one time per group for one week.
2. Increasing self-esteem and self-confidence
   Goal: To increase self-esteem
   Objective: Patient will successfully complete one task daily for one week.
3. Mastery of some aspect of the environment to enhance a sense of control
   Goal: To accept responsibility for a plant
   Objective: Patient will pick a plant and care for it daily for one month.
4. Development of prevocational skills such as following directions, staying on task, and accepting feedback
   Goal: Increase ability to stay on task
   Objective: Patient will stay on task for ten minutes with one cue.

TABLE 8.1. Horticultural Therapy Curriculum for Limerick Extended Day Program

This chart shows protocol or curriculum for an after-school program for elementary and middle school children. An appropriate goal will be chosen for each participant and corresponding objectives written.

I. *Format*

Time:	forty-five-minute sessions twice a month
Location:	Greenhouse and gardens
Size:	three to four students
Materials:	Plants, soil, pots, flowers, and some small craft activities

II. *Definition*

Horticultural therapy uses plants to stimulate interest, increase attention and concentration, and motivate participants. Horticulture is an excellent leisure skill that children can use all their lives. Greenhouse participation focuses on prevocational work skills such as following directions and cooperative participation.

III. *Goals*

Cognitive Improvement:

1. Improve problem-solving and independent decision-making skills.
2. Increase ability to attend and/or concentrate by gradation of performance activity.
3. Improve ability to follow verbal directions and visual cues.

Physical Functioning:

1. Improve eye-hand coordination.
2. Increase fine and/or gross motor dexterity.

Psychosocial Skills:

1. Increase socialization skills with peers and adults.
2. Increase awareness of other group members' needs and feelings through utilization of cooperative tasks.
3. Improve ability to cope with losses, failures, and fears through the care of plants.

Prevocational Work Skills

1. Increase ability to cooperate and share responsibilities, supplies, and tools with other group members.
2. Increase ability to follow multistep directions.
3. Increase attention span and concentration.
4. Increase ability to seek assistance from group leader and group members.
5. Increase ability to give and receive feedback.

TABLE 8.1 (continued)

IV. *Methodology*

1. The horticultural therapist will provide necessary instructions and supplies, and will educate the group in basic horticulture skills.

2. The horticultural therapist will provide gradation of activities and tasks from simple to more complex levels, depending on students' needs and level of functioning.

3. The group will be designed to encourage independence and/or cooperative decision making.

4. Students will be encouraged to organize and initiate projects when able.

5. The horticultural therapist will assess performance and give feedback in the wrap-up discussion.

V. *Precautions*

1. Sharp objects will be inventoried daily.

2. Students are required to dress appropriately for the weather, both hot and cold.

3. Students will be encouraged to wear hats and sunscreen when outside.

As objectives are met, more advanced goals may be established to increase the participant's success. (See Figure 8.3.)

ADAPTATION OF HORTICULTURAL ACTIVITIES TO MEET TREATMENT OBJECTIVES

Children respond to clear expectations and simple explanations that help them feel secure and safe in a new environment. Knowing the rules in advance and understanding the consequences for not adhering to them give the child a sense of personal control. Posting and reviewing the rules briefly at the beginning of each session remind the participants of what is expected. Evaluating with the group the necessities for the rules, such as safety concerns, helps them understand their importance and functions to solicit participants' agreement to adhere to the rules. Some general guidelines for participation are included in Table 8.2.

When planning activities for children, directions should be kept short and simple. Have materials readily available to ensure children's attention. If the children arrive to the activity already charged, it is useful to take a

FIGURE 8.3. Treatment Goals and Objectives

The following are some examples of problems, goals, and objectives appropriate for horticultural therapy.

Problem:	The client is unable to follow directions and needs to be able to follow four-step directions in order to be eligible for supportive employment.
HT Goal:	Client will follow four-step directions consistently and independently.
Objective 1:	Client will correctly repot geraniums 80% of the time, with staff support, daily for one week.
Objective 2:	Client will correctly repot geraniums 80% of the time, using a cue sheet, daily for one week.
Procedure:	1. **Staff will demonstrate the four-step procedure.** 2. Using a cue card, staff will explain each step. 3. Client will work with staff reminders. 4. Continue step 3 until client has met first objective. 5. Staff assistance will fade from client, with cue cards used only when mistakes are made. 6. Continue step 5 until objective 2 is met.
Problem:	Patient is depressed with poor self-esteem.
HT Goal:	Patient will increase self-esteem by completing a daily project.
Objective:	Patient will choose one small project per group and complete it daily.
Procedure:	1. Staff will help patient choose an appropriate project. 2. With staff support, patient will complete that project daily. 3. With staff support, patient will process feelings about being successful.
Problem:	Patient has had a stroke and has poor use of her right hand.
HT Goal:	Patient will increase right-hand range of motion.
Objective 1:	Patient will disbud chrysanthemums daily for five minutes without a break.
Objective 2:	Patient will disbud chrysanthemums daily for ten minutes without a break.
Procedure:	1. Staff will demonstrate correct method for disbudding and assist patient for ten minutes daily. 2. Staff will reduce the amount of assistance daily until patient is working on her own.

TABLE 8.2. Sample of Greenhouse Rules

---

1. Follow all safety and hygiene rules.

2. Tools will be put away at the end of each group; write date and time down when receiving sharps.

3. There will be no swearing, abusive language, and/or threats.

4. Aggressive behavior, either verbal or physical, may result in suspension from the program.

5. Handle all property with care.

6. You may not leave the area without asking permission.

7. Eating is not allowed in the greenhouse.

8. Listen carefully and ask if you do not understand.

9. Treat your supervisor and co-workers with respect.

10. Dress appropriately for work.

11. Throwing of any object is prohibited.

12. Use the restroom before going to the greenhouse.

13. Walk on the sidewalks.

14. Check in by the second bell.

15. Try your best, have fun, learn something new.

"mini time-out" to get everyone refocused and ready to listen. Children must first be able to listen to directions before they are ready to participate. Activities should be structured and broken down into steps that each child understands and designed so that each child can participate in every part. Long explanations should be avoided. Make science fun, explore, ask "what if" and "why," and experiment with plants, such as adding extra fertilizer, sun, or water. Children learn more by doing.

Building in rewards at the end of the session, such as a plant for participation or a craft to take home, is an excellent motivational tool for children. Young children tend to have short attention spans and need immediate gratification. Projects with elementary-age children should be short and completed within thirty minutes. Children learn more and are more easily engaged when the activity is "hands on." Distractions should be kept to a minimum for children who have difficulty concentrating. Children should be encouraged to take turns and to try on different roles during the group.

The therapist needs to be ready to change the focus or plan if the group gets excited about a certain area or topic. It is helpful to ask children what they want to do and to try to accommodate their wishes if possible.

Planning and planting a vegetable garden teaches children delayed gratification and appreciation for the natural rhythms of the seasons. It also provides an opportunity to address environmental issues and to help youngsters understand various concepts, such as the food chain. Gardening provides a great way to burn off and channel excess energy. We have witnessed children carrying radishes like prizes and being amazed or proudly showing others their vegetables and understanding where food actually grows.

Tables, chairs, and other work spaces should be comfortable and not overwhelming, especially for small children. Short sturdy stools allow participants to reach sinks. To encourage independence, pots and other supplies need to be placed where children can reach them. Children's gardening tools, gloves, and even chairs help the participants feel appreciated and can be a real motivating device. A variety of caps or hats help to protect participants from the sun and can add to the fun.

Verbally processing with the group at the end helps children evaluate their participation immediately and provides any necessary feedback. Self-evaluation forms and points systems can give children precise feedback on their behavior. Clients can be asked to describe their reactions to the activity, their thoughts about how well they did, what they liked about the group and what they would like changed, and their wishes regarding what they would like to do next time. Asking participants how they felt initially about going to horticultural therapy and what their expectations were, and asking them to discuss their feelings at the end of the group provides interesting insight into the group. Children are often surprised that they found the activity creative and calming.

Children can be asked what their goals were for the day and how horticultural therapy helped them to meet their goals. This type of processing reinforces the therapeutic value of the group and also offers the child some insight into how horticulture can be used at home. For young children, questions should be very simple, such as "What did you like?" In providing feedback and recognition, the therapist should reward the process, not the product.

Documentation requirements will differ from facility to facility. Social programs may not require documentation, while therapeutic programs may have daily requirements that must relate to the problems and goals defined for each participant. Documentation needs to address the functioning of the patient related to their objectives on a particular day. In order to record clearly, the therapist must know the patient's individual objectives and have a system for observing and recording observations. The report must be objective and limited to describing what happened, without any interpretation or guessing. Special problems, concerns, or changes need to be addressed immediately with the primary careperson for the child.

Since documentation is based on observations, and observations need to relate to treatment plans, a therapist must know a patient's general functioning and abilities in order to document progress. Although time-consuming, documentation's importance cannot be overstated. It supports why and how you are working with each patient and the progress that is made. Without this information, there is no reason for someone to be in the program.

SUMMARY

Working with children in horticultural therapy is an exciting, challenging, and rewarding opportunity. The benefits for this group include cognitive improvement, prevocational skills training, psychological improvements, and social skills development. Children are able to learn about plants, their environment, and themselves. Programs for children take place in schools, community gardens, recreation centers, and hospitals. Groups need to be structured, accessible, and short term to gain the child's attention and keep him or her focused. Creative and hands-on activities with rewards built in will ensure the success of your horticultural therapy program.

Future Practice

It had been said many times and in many different ways, but our children are our future. With all the intensities of life today—crime, urban decay, single parents, working families, and the decline of extended families—horticultural therapy can bring peace and the natural rhythm of life to a child. School gardening programs, after-school recreational horticulture activities, and even family gardens help children regain contact with nature. Gardening allows us to stop trying to control our environment and learn to work within its parameters. The more children we can involve in horticulture and horticultural therapy programs, the better for us all. The challenge is to educate facilities to the benefits and needs of gardening activities and to provide the funding and professional support to ensure their success.

CASE STUDIES

Lonnie

A therapy group for nine- and ten-year-old boys who were students at a special education school for children whose education had been disrupted by emotional disturbance, had operated using a combination of activities and talking to address the social and emotional difficulties of the students involved. Group members participated in the planning of activities, and the boys conceived of a project involving the growing of vegetable plants. The project entailed shopping for seeds and other supplies, arranging to use the school greenhouse, planting and caring for the plants, cooperatively allocating and sharing various responsibilities, and finally discovering an equitable way to distribute the plants for taking home.

Lonnie was a ten-year-old boy who had exhibited longstanding academic and behavioral difficulties in school, including low frustration tolerance, poor impulse control, temper outbursts, aggressive behavior, depression, anxiety, and low self-esteem. He was removed from the care of his mother at two years of age due to physical and emotional neglect. Siblings had been physically abused. His father had died of a drug overdose. Lonnie spent time in institutional and foster care and was placed in a permanent foster care home at five years of age.

Lonnie's disruptive and chaotic developmental experience had stimulated intense feelings of anger, sadness, fear, and hurt, but had failed to promote the growth of skills for managing these feelings or to provide nurturant caretakers to buffer his emotional experience. Consequently, Lonnie reverted to oppositional and aggressive behaviors to define and protect himself and to create a sense of power and control, while fending off underlying feelings of sadness, helplessness, and fear. Interpersonally, Lonnie tended to be needy but also guarded and mistrustful.

Lonnie had been chided by other group members for his "greediness," their term for his perennial concern over getting his fair share of snacks. Lonnie was instrumental in conceiving the horticultural project, which clearly appealed to his neediness and acquisitiveness, but also to his penchant for creativity, his interest in nature, and his enjoyment of working with his hands. Lonnie tended to rely heavily on material gifts and rewards, especially food, to assuage his inner feeling of hunger and underlying neediness. Eventually, Lonnie would learn that no amount of material gratification would compensate for his sense of unfulfilled nurturance needs. On the other hand, he could discover ways to feel good about himself and about his relationships with other people. The gardening project appealed to Lonnie because he liked the idea of having something to eat. What Lonnie discovered, however, over the course of his participation in the project was that working cooperatively with his peers had its own rewards. Lonnie liked working in the dirt and allocating and sharing responsibilities for work and cleanup. He eagerly awaited trips to the greenhouse to assess the progress of the plants as they emerged from the soil and grew. At the conclusion of the project, Lonnie was the mediator who made sure that all group members had a fair share of the plants to take home. Lonnie accrued emotional benefits from the project, which extended beyond its conclusion in school. He spent valuable time with his foster father planning, planting, cultivating, and harvesting a vegetable garden at home. He discovered a shared interest with his foster father that increased the amount of time they spent together. He was also gratified to be cast in the role of provider, sharing responsibility for feeding his family. Over the course of the summer, Lonnie issued reports to his therapist and the group about the vegetables that were being harvested and consumed at home. The gardening project began with an appeal to Lonnie's material neediness, and in the long run served to reinforce his relatedness with others in a spirit of reciprocity and sharing, while enhancing his sense of competence and self-esteem.

Randy

Randy was a nine-year-old boy who had also displayed longstanding academic and behavioral difficulties in school. He had been referred for special education as early as first grade, due to recurrent problems with hyperactivity, impulsivity, distractibility, difficulty following directions, and peer conflicts that escalated at times into aggressive outbursts. Randy had been removed from the care of his drug-abusing mother as a toddler, following the death of an infant sister due to physical injuries. Randy showed signs of physical abuse himself. He was cared for by his father's family until he was seven years of age. At that time, his father was incarcerated and a paternal aunt became Randy's legal guardian.

Randy presented as a depressed and highly anxious child whose developmental experience exposed him to abuse, disruptions in caretaking, and loss. These early experiences had a significant impact on Randy's interpersonal and emotional development. Interpersonally, he tended to be both needy and guarded. He had a difficult time trusting and developing supportive attachment relationships with adults. With peers, he tended to be alternately withdrawn, provocative, or competitive. He lacked skills for managing his emotional experience and tended to become anxious and easily overwhelmed, as well as to act out his feelings through angry outbursts. Randy's attentional deficits together with his emotional difficulties served to undermine his sense of mastery, control, and self-esteem. Randy displayed low frustration tolerance and a lack of confidence that often led to avoidance and disruptive behavior when facing academic, social, or other challenges.

For Randy, the gardening project represented a nonthreatening and largely nonverbal activity that did not stimulate his anxiety and avoidant and oppositional defenses. Instead, the gardening activities proved enjoyable and calming for Randy. A less anxious and more relaxed Randy became more accessible to experiencing success in the project as well as to engaging in more productive interactions with his peers. Randy had tended to vacillate between shyness and withdrawal or misguided attempts to establish peer contacts through provocative remarks. His developmentally based difficulty establishing a sense of belonging and fitting in or feeling competent had interfered with his engaging in activities and developing friendships. The gardening project, however, served to enhance Randy's sense of mastery and confidence. Consequently, Randy responded to the need for interpersonal interaction and cooperation and displayed more relatedness and assertiveness with his peers. By completion of the project, Randy had become a more active, animated, constructive participant in the group. As with Lonnie, Randy also carried benefits of the project home with him. He developed a real interest and skill at gardening and not only established a vegetable garden at home, but also assumed responsibility for landscaping his aunt's yard. Actively contributing to the appearance and caretaking of his home environment served to reinforce Randy's sense of belonging in his adoptive family. At a more symbolic level, Randy conveyed his gratitude and consolidated his position in the family by assuming the role of provider by feeding his family with the produce harvested from his vegetable garden.

GLOSSARY

accommodation: Process by which children during the course of development make changes in cognitive and behavioral patterns in order to cope with and adapt to the demands and realities of the environment (Piaget)

affect management: Developmental task involving the ability of the individual to differentiate, express, and control feelings in a healthy manner

assimilation: Process by which children incorporate new experiences into their conceptual and characterological framework in order to maintain a sense of identity, integrity, and continuity

behavior modification: Process by which an attempt is made to control and alter behavior by reinforcing (rewarding) positive behaviors and ignoring or punishing negative behaviors

biopsychosocial model: Model of human development focusing on the interactive effect of biological, psychological, social, and cultural forces on the development of the individual

cognitive development: Growth of the child's intellectual and perceptual abilities

developmental lines: Discrete but interrelated functional areas across which the course of human development can be traced (e.g., cognition, affect management, motor skills, interpersonal relations)

epigenesis: Tendency of human development to proceed in a cumulative fashion with progressive achievements built upon a foundation of prior accomplishments

identification: Process whereby children emulate and internalize attributes of other individuals

optimal tendency: Characteristic of development whereby the organism strives to achieve its maximum potential

psychopathology: Disruption to adaptive functioning of the individual resulting from interference in the normal course of human development

sense of industry: Developmental task involving the child's establishment of a sense of competence as a learner, doer, and social being (Erikson)

transference relationship: Quality of relationship between psychotherapist and patient wherein the perception of and reaction to the therapist is colored and influenced by the patient's significant developmental experiences and relationships

RESOURCES

Books

Berk, Laura (1989). *Child Development*. Needham Heights, MA: Allyn & Bacon.
Lewis, Melvin (Ed.) (1996). *Child and Adolescent Psychiatry: A Comprehensive Textbook*. Baltimore: Williams and Wilkins.
Rose, S.D. and Edelson, J.L. (1987). *Working with Children and Adolescents in Groups*. San Francisco: Jossey-Bass.
Slavson, S.R. and Schiffer, M. (1975). *Group Psychotherapies for Children*. New York: International Universities Press.
Wiener, J.M. (1991). *Textbook of Child and Adolescent Psychiatry*. Washington, DC: American Psychiatric Press.

Journals

Journal of the American Academy of Child and Adolescent Psychiatry
Journal of the Exceptional Child

Organizations

American Academy of Child and Adolescent Psychiatry, 3615 Wisconsin Ave., NW, Washington, DC 20016.
American Psychological Association, 1200 Seventeenth St., NW, Washington, DC 20036.

REFERENCES

Ainsworth, M.D.S. (1985). Patterns of infant-mother attachments: Antecedents and effects on development. *Bulletin of the New York Academy of Medicine,* 61, 771-791.
Airhart, D.L., Willis, T., and Westrick, P. (1987). Horticulture training for adolescent special education students. *Journal of Therapeutic Horticulture,* Volume II, 17-22.
American Humane Association. (1984). *Trends in child abuse and neglect: A national perspective*. Denver: American Humane Association.
American Psychiatric Association. (1994). *Diagnostic and Statistical Manual of Mental Disorders*, Fourth Edition. Washington DC: American Psychiatric Association.
Bell, S.M. (1970). The development of the concept of object as related to infant-mother attachment. *Child Development,* 43: 1171-1190.
Bowlby, J. (1969). *Attachment*. New York: Basic Books.
Chess, S. and Thomas, A. (1986). *Temperament in clinical practice*. New York: Guilford.

Chomsky, N. (1980). *Rules and Representations.* New York: Columbia University Press.

Erikson, E.H. (1963). *Childhood and society.* New York: Norton.

Gallup Poll. (1995). Disciplining Children in America.

Glick, P.C. and Lin, S. (1986) Recent changes in divorce and remarriage. *Journal of Marriage and the Family,* 48, 737-747.

Horowitz, F.D., and O'Brien, M. (1989). In the interest of the nation: A reflective essay on the state of our knowledge and the challenges before us. *American Psychologist,* 44, 441-442.

Izard, C.E. (1982). *Measuring emotion in infants and children.* New York: Cambridge University Press.

Knitzer, J. (1982). *Unclaimed Children.* Washington, DC: Children's Defense Fund.

McCracken, J. (1992). The epidemiology of child and adolescent mood disorders. *Child and Adolescent Psychiatric Clinics of North America,* 1, 53-72.

Piaget, J. and Inhelder, B. (1969). *The psychology of the child.* New York: Basic Books.

President's Commission on Mental Health Task Panel Reports (1978). Volumes 1 and 2. Washington, DC: U.S. Government Printing Office.

Taube, C.A. and Barrett, S.A. (Eds.). (1985). *Mental health, United States, 1985.* Washington, DC: U.S. Department of Health and Human Services, no. (ADM) 85-1378.

ADDITIONAL RESOURCE

Freud, A. (1965). *Normality and pathology in childhood: Assessments of development.* NewYork: International Universities Press.

Chapter 9

Older Persons
and Horticultural Therapy Practice

Karen Haas
Sharon P. Simson
Nancy C. Stevenson

INTRODUCTION

Horticulture is a leisure pursuit of hundreds of thousands of older adults who enjoy tending flower and vegetable gardens or indoor plant collections. Many others participate in group activities such as garden clubs, community garden associations, plant societies, arboretum and botanic garden events, and county extension programs. For an increasing number of older adults, gardening is an essential therapeutic activity that helps in maintaining health, aids in rehabilitation from chronic diseases and impairments, or slows the affects of dementia.

This chapter focuses on older adults and how horticultural therapy can be used to enhance the lives of this special age population. The purpose of this chapter is to demonstrate how horticultural therapists can utilize plants and horticultural activities to improve the social, educational, psychological, and physical adjustment of older persons while improving their bodies, minds, and spirits. More specifically, the learning objectives for this chapter are to

1. provide basic information about older persons that is relevant to the practice of horticultural therapy with this special population.
2. review the health conditions of older persons who can benefit from horticultural therapy.
3. identify and describe horticultural therapy treatment issues, goals, objectives, and adaptations that address the physical, psychological, and social problems of older persons.

4. present future challenges in horticultural therapy practice with older persons and propose plans for addressing these challenges.

THE CLIENT POPULATION

The elderly are the fastest growing segment of the U.S. population. Every eighth American is age sixty-five or older. There will be 35.3 million older adults by the year 2000. This number will steadily increase and nearly double by 2030 when the older population will swell to seventy million.

This section provides an overview of the older population as a whole. It presents basic facts and figures about older persons that are important to horticulture therapists practicing with this special age population. Next, it reviews the health conditions of three types of older persons who could benefit from horticultural therapy: (1) those striving to maintain their health; (2) older persons undergoing rehabilitation for chronic diseases and impairments; and (3) older persons struggling with dementia. This section concludes with a discussion of the impact of health conditions on the lives of older persons.

Information About Older Persons

According to data from the U.S. Bureau of the Census reported by the American Association of Retired Persons (1995), overall characteristics of older persons are the following:

1. *Gender.* Older women outnumber men 19.7 million to 13.5 million. There are 146 women for every one hundred men.
2. *Minorities.* 14 percent of the older population are Black, Asian, Pacific Islander, American Indian, or Native Alaskan.
3. *Marital status.* 43 percent of women and 77 percent of men are married. Half (47 percent) of all older women are widows.
4. *Living arrangements.* The majority (68 percent) of noninstitutionalized older persons live in a family setting. About 30 percent live alone.
5. *Children.* About 80 percent of older adults have living children; of these, 66 percent live within thirty minutes of a child, 62 percent have at least weekly visits with children, and 76 percent talk on the phone at least weekly with children.
6. *Geographic distribution.* About half of older persons live in nine states: California, Florida, New York, Pennsylvania, Texas, Ohio, Illinois, Michigan, and New Jersey.

7. *Income.* Households headed by persons sixty-five and over have a median income of $26,512. Median income for males is $15,250 and $8,950 for females. About 3.7 million older persons are below poverty level.

8. *Housing.* Of the 20.8 million households headed by older persons, 78 percent are owners and 22 percent are renters.

9. *Employment.* About 3.8 million older persons (12 percent) are working or actively seeking work. Of these, over half (55 percent) are employed part-time. They constitute 2.9 percent of the U.S. labor force.

10. *Education.* The educational level of older persons is increasing. About 62 percent have completed high school.

Health Conditions of Older Persons Striving to Maintain Their Health

The following list describes the overall health characteristics of older persons:

1. *General health.* Most older persons have at least one chronic condition and many have multiple conditions.

2. *Restriction of activity.* Usual activities are restricted thirty-four days per year because of illness or injury. This number increases with age.

3. *Activities of daily living (ADLs).* About 6.1 million (23 percent) older people living in the community in 1986 had health-related difficulties with one or more activities of daily living. ADLs include bathing, dressing, eating, transferring from a bed or chair, walking, getting outside, and using the toilet.

4. *Hospital stays.* In 1993, older people accounted for 36 percent of all hospital stays and 48 percent of all days of care in hospitals. The average length of stay for older persons is 7.8 days compared to 4.9 for persons under sixty-five.

5. *Long-term care.* Over 60 percent of the American public have some experience with long-term care that includes a range of health, personal care, and social services for individuals lacking certain functional capacities.

6. *Nursing homes.* A small number (1.6 million) and percentage (5 percent) of all persons sixty-five years and over lived in nursing homes in 1990. This percentage increased dramatically with age and reached 24 percent for persons eighty-five and over.

Health Conditions of Older Persons Undergoing Rehabilitation for Chronic Diseases and Impairments

Normal aging involves a general decline in the functioning of all the body's systems and the emergence of chronic conditions. According to Robert Atchley, chronic conditions are "long term, leave residual disability, require special training for rehabilitation, or may be expected to require a long period of supervision, observations or care" (1994, p. 102). The leading chronic health problems for older persons are presented in Figure 9.1 (AARP, 1995).

Although most older persons are affected by one or more of these conditions, they become more frequent and debilitating with advancing age. These conditions are described below using information from the National Institute on Aging's *Age Pages* (1993). Symptoms for these conditions are summarized in Figure 9.2 (Pfeiffer, 1989).

FIGURE 9.1. Most Frequently Occurring Conditions per One Hundred Elderly Living in the Community in 1992.

Condition	% Frequency
Arthritis	49
Hypertension	35
Heart	31
Hearing	31
Orthopedic	18
Cataracts	15
Sinusitis	15
Diabetes	10
Tinnitus	10
Visual	10

FIGURE 9.2. Symptoms of Chronic Diseases and Impairments

Disease/Impairment	Symptoms
Arthritis	• Painful and stiff joints (osteo, rheumatoid)
	• Knobby finger joints (osteo, rheumatoid)
	• Multiple inflamed joints (rheumatoid)
	• Sudden excruciating joint pain, often in big toe (gout)
Hypertension	• Usually none
Heart	• Chest pain, sometimes radiating to jaw and arms, or back pain
	• Loss of consciousness
Stroke	• Loss of muscle control on one side of the body
	• Loss of ability to speak or understand speech
	• Diminution of vision
Hearing	• Sounds are muffled or faint
	• Difficulty in understanding speech
	• Inability to hear certain sounds—violin notes, bird songs
	• Continuous hissing or ringing sounds
Cataracts	• Hazy, fuzzy, or blurred vision, developing gradually and painlessly in one or both eyes
Sinusitis	• Pain over the sinuses
	• Fever, chills, headache
Diabetes	• Frequent urination
	• Excessive thirst
	• Weight loss
	• Numbness or sores in legs and feet
	• Blurred vision

1. *Arthritis.* There are two types of arthritis: osteoarthritis, which occurs with advancing ages, and rheumatoid arthritis, which can develop at any age. Osteoarthritis is a noninflammatory degenerative disease that destroys the protective barrier of cartilage in joints. As the cartilage wears away, bone ends become exposed and rub together, restricting movement. Nearly all older persons have some

degree of osteoarthritis. Rheumatoid arthritis is an autoimmune disease involving chronic inflammation that begins in the synovial membrane of joints and spreads to other joint tissue. It affects about 40 percent of older persons.

2. *Hypertension.* Blood pressure is the force exerted by blood flowing against the walls of the blood vessels. High blood pressure occurs when the heart must pump harder to force blood through the arteries to the lungs or to other parts of the body. Hypertension is easily detected and usually controllable. A person is considered to have elevated blood pressure when the systollic pressure reading is above 140 mm or the diastolic pressure reading is above 90 mm.

3. *Heart disease and stroke.* Heart disease develops when the normal flow of blood through the heart and to the body is impeded, the heart's pacemaker functions improperly, or the blood that supplies nutrients to the heart is blocked. Stroke is a sudden loss of consciousness due a sudden interruption in blood flow to part of the brain.

4. *Hearing impairments.* A gradual loss of hearing occurs for most people throughout their lives; nearly 90 percent of older persons have a hearing impairment. Presbycusis is caused by changes in the delicate workings of the inner ear that lead to difficulties understanding speech and possibly an intolerance for loud sounds. Conduction deafness involves blockage or impairment of the mechanical movement in the outer or middle ear so that sound waves are not able to travel properly through the ear. Central deafness is caused by damage to the nerve centers within the brain and affects the understanding of language.

5. *Orthopedic impairments.* Prevention and correction of impairments involve locomotor structures of the body such as skeleton, joints, muscles, and fascia. Falls are the most common cause of orthopedic injuries in older persons.

6. *Cataracts.* A change in the chemical makeup of the lens causes cataracts to form. A cataract can cloud one or both of the usually clear and transparent lenses and can block the passage of light needed for vision. Its underlying cause has not yet been determined. Removal of the clouded lens through surgery, one of the safest operations a patient can undergo, is the only effective way to treat cataracts.

7. *Sinusitis.* Inflammation of a sinus can be caused by viruses, bacteria, allergies, or other agents. This conditions can affect balance and the ability to pay attention.

8. *Diabetes.* Lack of insulin secretion causes inappropriate elevation of blood glucose and associated alterations in lipid metabolism. As

a result, the body is unable to use insulin properly. Older persons contracting diabetes frequently have other health problems such as obesity, atherosclerosis, diabetic retinopathy, diabetic neuropathy, and kidney and foot problems.

9. *Tinnitus.* This condition is characterized by ringing, buzzing, or other sounds in the ear that accompany certain diseases of the ear.

10. *Visual impairments.* Changes include the development or worsening of nearsightedness or farsightedness, reduced ability to adjust to darkness and bright light, impaired side vision, and declining ability to distinguish colors.

Health Conditions of Older Persons Struggling with Dementia

Dementia describes a group of symptoms that are usually caused by changes in the normal activity of very sensitive brain cells. Two common types of dementia in older persons are Alzheimer's disease and multi-infarct dementia. These types of dementia are described below using information from the National Institute on Aging's Alzheimer's Disease, Education, and Referral Center.

Alzheimer's Disease

Definition. This disease is the most common type of dementia in older persons. It is a chronic, degenerative, irreversible disorder that develops when nerve cells in the brain die. The disease affects the parts of the brain that control thought, memory, and language. The exact causes are not known and a cure does not exist.

Number afflicted. Approximately four million older persons in the United States are diagnosed as having Alzheimer's disease. This number will increase to fourteen million by the year 2000. The disease usually occurs after age sixty-five and the risk increases steadily with age. It is not a normal part of aging.

Symptoms. Progressive, multifaceted changes in intellectual abilities occur with loss of memory; judgment; abstract thinking; orientation with regard to time, place, and person; language skills; abstract thinking; and higher cortical functioning.

Personality and behavioral changes. Changes may include confusion, inability to carry out routine tasks, anxiety, irritability, agitation, withdrawal, petulance, paranoid ideation, and wandering.

Progression. Onset is gradual. The disease may last from one to ten or more years. Patients undergo progressive debilitation and become unable to take care of themselves. Figure 9.3 summarizes Dr. Barry Reisberg's conceptualization of functional capacities of victims during the seven phases of this disease.

Mortality. The disease is the fourth leading cause of death in the United States and accounts for more than 100,000 deaths annually.

FIGURE 9.3. Functional Capacities of Alzheimer's Victims During Phases of Disease

Phase	Functional Capacities
1. Normal	No impairment
2. Forgetfulness	No impairment but subjective concern about memory loss
3. Early Confusional	Inability to perform in demanding employment and social interactions apparent to intimates and associates
4. Late Confusional	Decreased ability to handle finances and shopping
5. Early Dementia	No assistance required with toileting or eating, but may have some difficulty choosing the proper clothing to wear; may require coaxing to bathe
6. Middle Dementia	Personal or hygienic dysfunction, or both, with following progression: • Difficulty pulling on clothing properly • Requires assistance with bathing; may develop fear of bathing • Inability to handle mechanics of toileting • Urinary incontinence • Fecal incontinence
7. Late Dementia	Speech and motor dysfunction, with the following progression: • Ability to speak limited to a few words • All intelligible vocabulary lost • All motor abilities lost • Stupor • Coma

Multi-Infarct Dementia

Definition. Multi-infarct dementia is caused by a series of strokes that damage or destroy brain tissue. Untreated high blood pressure is the most important risk factor for multi-infarct dementia. Other main causes are high blood cholesterol, diabetes, and heart disease. Sometimes, it is difficult to distinguish multi-infarct dementia from Alzheimer's disease. Sometimes, victims can have both multi-infarct dementia and Alzheimer's disease.

Number afflicted. Most multi-infarct dementia usually affect people between the ages of sixty and seventy-five.

Symptoms. Confusion and problems with recent memory occur, as well as wandering or getting lost in familiar places, moving with rapid and shuffling steps, loss of bladder or bowel control, emotional problems such as laughing or crying inappropriately, difficulty following instructions, and problems handling money.

Personality and behavioral changes. Changes may include confusion, inability to carry out routine tasks, anxiety, irritability, agitation, withdrawal, petulance, paranoid ideation, and wandering.

Progression. While no treatment can reverse damage that has already been done, additional strokes can be prevented through medicines, good health habits, and nutrition.

Mortality. Heart disease including stroke is the leading cause of death for older persons.

THE IMPACT OF HEALTH CONDITIONS ON THE LIVES OF OLDER PERSONS

Health conditions affect an older person's ability to perform a range of common activities needed for personal self-maintenance and independent community residence. These activities of daily living or ADLs, are divided into two categories: physical ADLs and instrumental ADLs.

1. Physical ADLs refer to basic personal care tasks: bathing, dressing, eating, transferring from a bed or chair, walking, getting outside, and using the toilet.
2. Instrumental ADLs refer to more complex activities needed for independent living: handling personal finances, preparing meals, shopping, traveling, and doing housework.

About 6.1 million (23 percent) of older people living in the community have difficulty with one or more physical ADLs and 7.6 million (28 percent) had difficulty with one or more instrumental ADLs (AARP, 1995). These percentages increase sharply with age.

Declining health can diminish an older person's ability to perform physical and instrumental ADLs. These conditions are long term and may leave residual disability. Rehabilitation and supervised care that includes horticultural therapy can help reduce personal losses and restore an older person's level of functioning.

HORTICULTURAL THERAPY AND TREATMENT

Identification of Treatment Issues

The horticultural therapist must have an understanding of the individual's stage in the aging process and their chronic health issues in order to develop an effective program for older adults. A treatment team establishes the issues to be addressed and refers the individual to various program areas or services within the facility or community that relate to the treatment areas. Each of the individual program areas then identifies specific goals and objectives that relate to the overall treatment issues. The treatment team may consist of a physician, occupational therapist, physical therapist, recreational therapist, dietician, and social worker. The older adult and family members also are part of this process.

The treatment focus for older adults appropriate for horticultural therapy include

1. living and self-care independence,
2. physical health maintenance,
3. physical health improvement,
4. cognitive functional-level maintenance,
5. age-appropriate leisure-skill development,
6. emotional status improvement, and
7. social interaction with peers.

The treatment issues will relate to the older adult's reasons for seeking treatment, history, living arrangement, and functional level, as well as his or her motivation, willingness to participate in treatment, and personal interests.

Identification of Treatment Goals and Objectives

Once the treatment focus is established, the horticultural therapist identifies the area to be addressed and establishes the initial goal(s) and objectives. Methods to help the older adult successfully achieve these goals and objectives should be determined at this time. Depending on the type of program, a written action plan may be required.

An initial interview, interest profile, and observation determine a baseline of functioning. The horticultural therapist may also review the initial screenings or assessments given by the occupational, recreational, or physical therapists.

Assessments for nursing home residents are often lengthy and mandated by the state. This information, together with the established treatment issues, are used to write the goals. The goals are broad and based on the desired outcome. The objectives are the incremental steps necessary to achieve the goals. Several objectives may be required to meet one goal. It is important that objectives are written with time frames and methods of measurement so that progress can be evaluated.

Goals and objectives for older adults vary greatly, depending upon if the primary focus of treatment is health maintenance, rehabilitation from a traumatic incident, or coping with a chronic disease process or dementia-related issues. Horticultural therapy creatively addresses the issues facing older adults in a way that is motivating and enjoyable. In response to care, plants grow and change with the seasons to provide excellent tools with which to meet treatment goals. Many of the same basic horticulture and related activities are used with each of the previously mentioned treatment issues to meet a diverse array of specific individual needs.

Health Maintenance Through Community-Based Programs

Individuals striving to maintain health must have the opportunity for positive interaction with their peers in the community. Community-based treatment programs provide garden-club groups to address issues common to older adults who are in independent or semi-independent living situations. Some examples of goals and objectives that come out of such clubs are the following:

- *Goal:* To decrease feelings of isolation and loneliness and reestablish a network of friends with similar interests
  Objective: Individual will participate in the horticulture club at the senior center one day per week.
- *Goal:* To participate in mild exercise
  Objective: Individual will participate in weeding, cultivating, and pruning the raised-bed garden for twenty minutes three times per week.

- *Goal:* To increase awareness of nutritional needs of older adults
  Objective: Individual will participate in a "Cooking from the Garden" nutrition class one time per week for six weeks.

Community-based horticultural therapy programs provide the opportunity to participate in gardening activities, a supportive social environment in which to meet new people with mutual interests, strategies for increased participation in the community, adaptive tools and techniques to enable participation in gardening activities, and information on the nutritional benefits of using fresh vegetables and herbs in meal preparation. Case A, regarding Joan and the Senior Gardening Group, illustrates the operation, objectives, and activities of community-based programs.

Individual health needs determine the appropriate level of physical activity. For example, an individual with osteoarthritis should be encouraged to exercise the noninvolved joints and avoid stressing the involved joints. Conversely, it is important that an individual with rheumatoid arthritis exer-

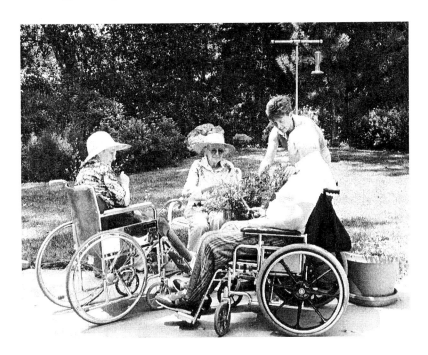

"Green Thumb" participants, with direction from a horticultural therapist, groom a container of plants at the Center Adult Day Care in Prescott, Arizona.

cise both the involved *and* noninvolved joints to prevent stiffness. Exercise promotes the retention of flexibility and range of motion.

Case A

The Senior Gardening Group meets monthly at an arboretum in Joan's community. Those attending range in age from sixty-five to ninety and live in their own homes, apartments, or independent-care facilities. Joan is basically healthy, but her hypertension and arthritis have slowed her down considerably. Most of Joan's friends have moved from the old neighborhood, and after her husband died she rarely left the house. Joan's daughter recommended the group as a way to get Joan to exercise more and get her back into the community. Although Joan didn't know anyone when her daughter first suggested she attend, she now looks forward to seeing her friends and exploring a variety of interesting horticulture topics and projects.

The Senior Gardening Group, led by a registered horticultural therapist, participates in programs on a variety of plant-related topics, including nutrition. The therapist also provides information and assistance on gardening techniques and tools adapted to accommodate a variety of changing physical needs. This is a necessary link in modifying both nutritional habits and level of activity. For example, raised beds and containers make the garden more accessible to individuals who use wheelchairs or who, like Joan, have difficulty getting down to the ground to work. The horticultural therapist showed Joan how to build up the handles of her trowel and cultivator since, due to arthritis, she was unable to close her hand tightly around the handles. The accessibility of the garden and the materials growing in it motivate her to at the very least ambulate to the garden. The growing of fresh vegetables supplements her diet, especially since it is hard to get to the store regularly for fresh produce. In addition, the use of herbs allows her to reduce her salt intake as requested by her physician.

While this group meets formally monthly, individuals within the group regularly get together independently to visit gardens in the community, share meals, or simply chat on the phone. In addition to the increase in physical exercise and better nutrition, Joan's social isolation has significantly decreased. She feels better about herself and knows there are people nearby that she can call when she needs help or companionship.

Chronic Disease and Traumatic Incident Rehabilitation

Goals are written to sustain or improve daily functioning levels and overall health and life satisfaction of individuals with chronic conditions.

Residents work in groups in both outdoor and indoor gardens, which allows them to impact as well as interact with their environments. The following goals and objectives may be included in programs for long-term-care patients. (See Figure 9.4.)

- *Goal:* To become less socially isolated and participate in purposeful activities
  Objective: Resident will participate in the horticulture group harvesting flowers for the dining room for twenty minutes two times per week.
- *Goal:* To maintain fine motor skills
  Objective: Resident will arrange cut flowers in bud vases two times per week.
- *Goal:* To improve feelings of satisfaction with living environment
  Objective: Resident will be responsible for watering the butterfly garden three times per week.
  Objective: Resident will participate in weeding the butterfly garden two times per week.

Rehabilitation following a traumatic incident focuses on the resulting orthopedic and neurologic issues. The overall treatment issue in the case of Sarah B, who broke her hip for example in Case B, is to return her to as independent a living situation as possible. The goal is to restore her physical strength and mobility, including sitting and standing balance. The following goal is a necessary part of the process in addressing the treatment issue. The objectives that follow are just two in a series of incremental steps to accomplish the goal.

- *Goal:* To restore physical strength and mobility
  Objective: Sarah will water indoor plants with a half-full watering can while standing for ten minutes.
  Objective: Sarah will water plants with a full watering can while standing independently for ten minutes.

Other examples of goals appropriate in a rehabilitation situation to build standing tolerance include and objectives progressing from unilateral tasks to bimanual tasks through activitiess such as soil mixing and flower arranging, respectively. (See Figure 9.5.)

The rehabilitation process necessitates challenging patients to address their deficits and/or employ compensatory strategies rather than simply assist them in completing the activity. The therapist should set the task up to allow the process to address the pertinent issues. A patient who is

FIGURE 9.4. Goals and Objectives for Long-Term-Care Patients

Treatment Issue: Decreased social interaction with peers

Horticultural Therapy Goal: To decrease the resident's social isolation

Objective 1:   Resident will initiate interaction with one person three times during the one-hour session with minimal cues for five consecutive sessions.

Objective 2:   Resident will interact with three people independently during the one-hour session for five consecutive sessions.

Procedure: *Week 1*

1. Review the procedures for harvesting and arranging cut flowers with the resident.

2. Review the goal of interacting with one other group member with the resident.

3. If the resident remains isolated after 15 minutes, provide one cue.

4. Repeat cues as needed until the resident initiates conversation three times with the same person.

5. Proceed to Objective 2 when the individual achieves the objective consistently.

Week 2

1. Follow steps 1 through 4 from week one.

2. If individual remains isolated, provide one cue and begin counting again.

3. Continue until the resident interacts with three different people during the one-hour session.

experiencing perceptual neglect should have the activity materials, sample projects, or task demonstration positioned toward the neglected side. This forces them to scan to get to the materials they want or observe the task taking place.

Case B

After the death of her husband ten years before, Sarah B sold her large house and garden and moved to an apartment in the Robin Hill Retirement Community. Sarah was seventy-one years old and had developed arthritis

FIGURE 9.5. Goals and Objectives for Rehabilitation

Treatment Issue: Improve living and self-care independence

Horticultural Therapy Goal: To decrease the amount of upper extremity support needed to stand independently.

Objective 1: The patient will perform unilateral tasks while standing with self-stabilizing support for twenty minutes.

Objective 2: The patient will perform bimanual tasks while standing without additional self-stabilizing support for twenty minutes.

Procedure: *Week 1*

1. Review the reason and procedure for scooping freshly mixed soil into storage bins with the patient.

2. Review the goal with the patient.

3. Demonstrate the proper technique for standing and supporting one's self with one hand, while scooping soil into the bin with the dominant hand.

4. Provide additional support if necessary.

5. If patient must sit and rest, begin to count again when patient resumes.

6. Repeat steps 1 through 5 until patient maintains a standing position with support for twenty minutes.

Week 2

1. Review the process of flower arranging with the patient.

2. Review the goals with the patient.

3. Demonstrate the technique for flower arranging, while standing and using both hands.

4. If the patient uses hand for self-stabilizing, begin count again.

5. Continue steps 1 through 4 until position has been maintained for twenty minutes without additional support.

in the joints of her right arm and hand. Fiercely independent, Sarah managed on her own for a long time. Then, as her arthritis progressed, it was recommended that she move into assisted living where she could get help with her daily routine.

Sarah had been an outgoing individual, a community leader with a busy schedule and social life. However, as she became more disabled, she became depressed and anxious about her situation. She tended to withdraw from other people and activities. One occupation she did enjoy, however, was gardening, at which she had excelled in earlier years. Her special favorites in her own garden had been herbs and vegetables.

A residents' garden with raised beds and adaptive equipment had been recently completed at Robin Hill. Still a reluctant joiner, she avoided the more structured garden club sessions; Sarah worked in the garden at her convenience. Despite the arthritis, she was able to work in the garden for short periods, using tools with special grips. The mild exercise and contact with soil and plants seemed to increase her mobility and raise her spirits. The garden was a place where she could express her creativity and reassert her independence.

The situation changed dramatically when Sarah slipped on wet pavement and broke her right hip. She was hospitalized for ten days, then returned to Robin Hill for an indefinite stay in the nursing wing of the retirement community. A treatment plan was developed for her recovery, the ideal outcome of which would be her return to her quarters in assisted living.

The Robin Hill staff had a serious commitment to therapeutic gardening and felt that Sarah would be an ideal beneficiary. It was decided that horticultural therapy would be an integral part of her rehabilitation program. During the first week, the therapist regularly brought activities to her room, which they worked on together until Sarah was able to use a wheelchair. This one-on-one contact allowed a relationship of mutual trust to develop. This preparation eased the transition into the structured horticultural therapy group of ten nursing patients. Goals for Sarah's rehabilitation, which were addressed by her participation in individual and group horticultural therapy, were (1) to restore physical strength and mobility, (2) to improve daily living skills, and (3) to return to a semi-independent living situation.

Physical benefits to Sarah from the regular activity provided by the gardening program included the following:

1. Sustaining and improving fine motor control by exercising arthritic hands and arm joints—Applied were tasks such as planting seeds and transplanting seedlings, harvesting and drying herbs for cooking and crafts, and weeding and deadheading flowers.
2. Maintaining upper-body strength and mobility during wheelchair confinement—Raised beds and containers, baskets on pulleys, and vertical gardens facilitated working from a sitting position, as did long-handled tools, foam-grip handles, which reinforced her weak grip.

Over the next month, with her hip fracture healing on schedule, Sarah became a staunch member of the gardening group. She soon graduated from the wheelchair to a walker.

Horticultural therapy's psychological benefits to Sarah were apparent as well. Depression about her condition and anxiety about the future were eased when she was sharing tasks and accomplishments with the group. She enjoyed her role as teacher when she passed along her expert knowledge of herb and vegetable gardening, the harvest of which was an important element in the year-round, indoor/outdoor program, incorporating culinary and nutrition education projects.

Sarah's enthusiastic participation in the horticultural therapy program resulted in renewed confidence in her ability to recover fully from surgery, have a degree of control over her environment, and regain some of her independence. She gained new interests and skills and an easy return to assisted living. During her recovery, Sarah gained valuable knowledge about managing her own health. She continued to garden with the group and independently, now understanding that horticulture provided valuable therapy for her arthritis and guaranteed a sense of accomplishment and well-being.

The Dementia Struggle

Horticultural therapy for patients with dementia provides sensory stimulation and mild exercise through creative use of safe gardening spaces, appropriate plant materials, and simple tasks. Emotional and cognitive goals are typically the primary treatment focus for patients with dementia. Sensory stimulation can improve orientation and trigger short- and long-term memory. The scenario of Tom in Case C illustrates the effectiveness of direct interaction with plants as a technique to increase social initiative and verbalization as well. In addition to the specifically targeted goals, the mild exercise included in a horticultural therapy program provides the opportunity to maintain range of motion and utilize both fine and gross motor skills. This is important for all older adults. Goals for an individual with dementia might include the following:

- *Goal:* To provide sensory and mental stimulation
  Objective: Patient will follow simple step-by-step instructions when planting a dish garden, for twenty minutes with minimal cues.
- *Goal:* To decrease aggressive behavior, anxiety, and agitation
  Objective: Patient will participate in mild physical exercise in the garden area a minimum of three times per week for thirty minutes.

- *Goal:* To improve orientation to reality
  Objective: Patient will participate in discussion about the fall season during a harvesting activity for twenty minutes.
- *Goal:* To stimulate long-term memory
  Objective: Patient will cultivate the old-fashioned cutting garden and participate in group discussion for twenty minutes with minimal cues, three times per week.

Gardens designed specifically to accommodate people with dementia provide a "clinic" for horticultural therapy to take place as well as accommodate the tendency to wander. Such gardens include encircling walking paths to enable individuals to return to the building without getting "stuck" or "lost." Plants are selected to promote reminiscing and pleasant memories and emotions. Columbine, peonies, hollyhocks, rosemary, lavender, geraniums, forsythia, lilacs, and chrysanthemums are a few examples. Plants known to be poisonous or injurious should be avoided. Physical problems associated with normal aging may also be present. Objectives addressing these needs may be incorporated as outlined above.

Case C

Soon after Tom M's eightieth birthday, his daughter consulted a gerontologist about his deteriorating mental state. Tom was becoming increasingly forgetful, withdrawn, and anxious. He often seemed lost in familiar surroundings. His family, in turn, felt he was becoming a stranger in his own home. Physically, Tom was frail, but still fairly active. Family members took turns looking after him, but of necessity, he was often left unattended.

The consulting physician diagnosed the onset of dementia. Among other strategies for Tom's care, the doctor recommended the Day Enrichment Program at Manor Park Nursing Center, which offered a full day of activities and lunch, twice a week. Not only would Tom's participation in the program provide him with the companionship and stimulation, his family would benefit from respite from his care.

The Day Enrichment Program offered many activities: music, arts and crafts, exercising, and horticulture. The main goals for group horticultural therapy activities were: (1) to provide sensory and mental stimulation, (2) to provide opportunities for physical exercise in a safe environment, and (3) to facilitate reality orientation. The center had a small, enclosed garden where the twelve clients could safely walk and tend the plants.

Programming was developed around a seasonal theme. Beginning in the late winter with seed selection and ordering from catalogues, and then planting, transplanting, and nurturing the young plants for setting out in the spring, therapeutic potential of the garden was realized through a varied display of annuals, vegetables, herbs, and everlastings. Plants were carefully selected for sensory appeal—tactile interest, taste and fragrance, color and contrast, and for practical use for crafts in the fall and winter months.

Gardens were familiar territory to Tom. Although it had been many years since he had been involved, the environment brought back pleasant childhood memories of his mother's garden. Memory stimulation played an important role in each day's achievements. Regular patterns of individual and group behavior were established and maintained, which helped participants focus on the tasks of daily living and orient them to the present.

Tom and the group were passive and unresponsive when first given a task to do, but soon were caught up in the enthusiasm of their teachers. They were able to follow directions when these were repeated often and when the activity was broken down into segments and followed step by step. Some responded well to verbal cues–others needed the visual demonstration to catch on. Repetition was needed consistently.

Handling living plants, enjoying their texture and odor, and focusing on their substantial reality dispelled for a while the lost and frustrated feelings that result from severe memory impairment. A sense of ownership and responsibility returned to Tom when he nurtured his plants and garden plot. Socialization skills were revived in all participants. Isolated and abstracted at the outset of the horticultural therapy session, they were soon talking among themselves and sharing their satisfaction with the end product, whether it was a plant repotted, a terrarium artfully arranged, or an arrangement made from flowers harvested in their garden. The therapeutic process that created this positive change in behavior is not well-understood, but the product is real and tangible, created by the client's own hands, and worthy of the pride and sense of accomplishment it engenders.

Tom also blossomed under the needed individual attention provided by the therapist and her volunteers. The positive effects of the mental, social, and sensory stimulation and the physical exercise provided by the program resulted in improvement in his home adjustment as well. The family was encouraged to provide him with indoor and outdoor gardening opportunities where he could "find himself again," reinforcing the skills he learned in the Day Enrichment Program.

Adaptation of Horticultural Activities to Meet Treatment Objectives

An activity is merely the process used to meet treatment objectives. The plant material is the source of stimulation and motivation. Both procedural and physical adaptations increase the potential for the patient to meet his or her goals. In other words, the same activity can be used to meet very different needs, depending upon the goals of each patient group.

An example to illustrate the many facets of a basic activity that the therapist can use to achieve the desired outcomes is the creation of tussie mussies. A tussie mussie is a small bouquet of herbs and flowers designed to express a particular sentiment. The therapist might have the group harvest the materials in a previous session, or might do it for the group if harvesting is not feasible or doesn't promote the desired individual objectives. If written skills are a goal, a written card detailing the message of the tussie mussie can be part of the process. If this is not a goal, a preprinted card with "Tussie Mussie, a Sentiment from the Heart" can be attached with the patient signature. The amount of time spent reminiscing about the history, fragrances, and type of individual plants can vary depending upon the need for memory stimulation and social interaction.

How an activity is presented has an impact on both physical and cognitive functioning. The patient can create the bouquet by holding the plant materials in his or her nondominant hand, or a bouquet holder resting on a weighted cup enables the patient to make it using one hand. Likewise, it might be too overwhelming to have all of the materials for the tussie mussie on the table at once. An alternative is to introduce items one to three at a time and in order of use. Decision-making opportunities can be complex (for example, choosing which message to send and which plants to use convey sentiments such as "happy anniversary," "welcome," or "I am thinking of you") or simple (for example, the whole group expresses similar messages such as "good cheer" or "best wishes" and the participants choose the ribbon color for the final touch).

Raised beds and containers bring gardens within reach of the patients while providing a community setting in which to interact. Long-handled, lightweight tools accommodate endurance and range of motion issues and enable participation in care and harvest of the group's gardens. Specialty tools are often not needed when ordinary household items will work. Lightweight ice cream scoops are a simple adaptation of an indoor trowel. The wide handle makes them easier for arthritic hands to hold. The small scoop is easier to direct into a small pot than a long trowel. To avoid confusion and ingesting inappropriate materials, utensils related to eating should never be used by patients with dementia.

In preparation and in conclusion of group activities, the passing of supplies works on range of motion and transporting watering cans (graded by weight, empty to full) addresses strengthening and mobility. Perceptual-motor tasks, such as pouring water or soil, and range of motion activities, such as reaching for supplies and wiping the table upon conclusion, are directly transferable activities of daily living. These processes can be incorporated into most group sessions depending upon the patient goals.

OBSERVATION, DOCUMENTATION,
AND CLIENT FEEDBACK

Requirements and methods of documentation will vary within each facility and type of program. It is essential to establish a system for recording client participation and measuring progress in order to determine if the objectives are being met. It is also important to provide feedback to the clients on their progress and communicate with the rest of the treatment team.

The information gathered at each session may be documented on a checklist, daily summary form, or progress note. Facilities often develop their own forms, coordinated between the therapeutic departments, to facilitate the documentation process. Programs that are rehabilitative in nature generally have more stringent reporting systems than those that seek to enhance the quality of life and maintain functional status.

Participants in a day program at Menorah Park Center for Aging in Cleveland, Ohio, garden and socialize in beds at a height appropriate for standing. Straw hats, clip-on sunglasses, and sunscreen provide protection from the sun.

Regardless of the method used, the therapist should review the patient's progress regularly. Immediately following group sessions is a good time to process how the group went. Review of the overall progress the individual is making and review of goal and objective appropriateness should occur during private discussion as well as at team meetings.

KEY ORGANIZATIONS

Administration on Aging, 330 Independence Avenue, SW, Washington, DC 20201.

Alzheimer's Disease, Education and Referral Center, National Institute on Aging, P.O. Box 8250, Silver Spring, MD 20907.

American Association of Retired Persons (AARP), 601 E Street, NW, Washington, DC 20049.

American Geriatrics Society, 770 Lexington Avenue, Suite 300, New York, NY 10021.

American Society on Aging, 833 Market Street, Room 516, San Francisco, CA 94103.

Gerontological Society of America, 1275 K Street, NW, Suite 350, Washington, DC 20005-4006.

National Institute on Aging, Public Information Office, Federal Bldg., Room 6C12, 9000 Rockville Pike, Bethesda, MD 20892.

National Institute of Mental Health, Information Resources and Inquiries Branch, Room 15C-05, 5600 Fishers Lane, Rockville, MD 29857.

GLOSSARY

Activities of Daily Living (ADLs) include two types. **Physical ADLs** are basic personal-care tasks including bathing, dressing, eating, transferring from a bed or chair, walking, getting outside, and using the toilet. **Instrumental ADLs** refer to more complex activities needed for independent living, such as handling personal finances, preparing meals, shopping, traveling, and doing housework.

Arthritis includes two basic types. **Osteoarthritis** is a noninflammatory, degenerative disease that destroys the protective barrier of cartilage in joints. As the cartilage wears away, bone ends become exposed and rub together, restricting movement. **Rheumatoid arthritis** is an autoimmune disease involving chronic inflammation that begins in the synovial membrane of joints and spreads to other joint tissue.

Chronic conditions are "long term, leave residual disability, require special training for rehabilitation, or may be expected to require a long period of supervision, observations, or care" (Atchley, 1994, p. 102).

Dementia describes a group of symptoms that are usually caused by changes in the normal activity of very sensitive brain cells. Two common types of dementia in older persons are Alzheimer's disease and multi-infarct dementia. **Alzheimer's Disease** is a chronic, degenerative, irreversible disorder that develops when nerve cells in the brain die. The disease affects the parts of the brain that control thought, memory, and language. The exact causes are not known, and a cure does not exist. **Multi-farct Dementia** is caused by a series of strokes that damage or destroy brain tissue. Untreated high blood pressure is the most important risk factor for multi-infarct dementia. Other main causes are high blood cholesterol, diabetes, and heart disease.

Diabetes is a chronic condition that occurs when a lack of insulin secretion causes an inappropriate elevation of blood glucose and associated alterations in lipid metabolism. Consequently, the body is unable to use insulin properly.

Heart disease develops when the normal flow of blood through the heart and to the body is impeded, the heart's pacemaker functions improperly, or the blood that supplies nutrients to the heart is blocked. **Stroke** is a sudden loss of consciousness due to a sudden interruption in blood flow to part of the brain.

Hypertension occurs when the heart must pump harder to force blood through the arteries to the lungs or to other parts of the body. **Blood pressure** is the force exerted by blood flowing against the walls of the blood vessels. A person is considered to have elevated blood pressure when the systollic pressure reading is above 140 mm or the diastolic pressure reading is above 90 mm.

Older persons refer to men and women age sixty-five years and above.

RESOURCES

Activities, Adaptation, and Aging. Binghamton, NY: The Haworth Press.

Ebel, Susan (1991). Designing State-Specific Horticultural Therapy Interventions for Patients with Alzheimer's Disease. *Journal of Therapeutic Horticulture.* VI:3.

Generations. San Francisco, CA: American Society on Aging.
Geriatrics. New York, NY: American Geriatrics Society.
The Gerontologist and *The Journal of Gerontology*. Washington, DC: Gerontological Society of America.
Journal of Gerontological Social Work. Binghamton, NY: The Haworth Press.
Kaplan, Maxine Jewel (1994). Use of Sensory Stimulation with Alzheimer Patients in a Garden Setting. In *People-Plant Relationships: Setting Research Priorities*. Joel Flagler and Raymond Poincelot (Eds.). Binghamton, NY: The Haworth Press.

REFERENCES

American Association of Retired Persons (AARP) (1995). *A Profile of Older Americans*. Washington, DC: AARP.
Atchley, R.C. (1994). *Social Forces and Aging*. Belmont, CA: Wadsworth Publishing Co.
National Institute on Aging (1993). *Age pages*. Washington, DC: U.S. Department of Health and Human Services.
National Institute on Aging Alzheimer's Disease, Education and Referral Center (No date). *Age Pages: Alzheimer's Disease and Multi-Infarct Dementia*. Washington, DC: U.S. Department of Health and Human Services.
Pfeiffer, G.J. (1989). *Taking Care of Today and Tomorrow*. Reston, VA: The Center for Corporate Health Promotion.
Reisberg, B. (1982). The Global Deterioration Scale for Assessment of Primary Degenerative Dementia. *American Journal of Psychology* 139:1138. Sept.

Chapter 10

Substance Abuse, Offender Rehabilitation, and Horticultural Therapy Practice

Jay Stone Rice
Linda L. Remy
Lisa Ann Whittlesey

INTRODUCTION

Public opinion regarding crime and criminals often is based upon inaccurate information and emotionally divisive political debates. The success of the horticultural therapist is predicated upon an empathic understanding of the particular needs of the population being served. The horticultural therapist utilizes adaptive gardening tools when working with disabled clients and raised beds when treating elderly clients. Horticultural therapists must understand and adapt to the unique needs of people who are imprisoned. Because the seeds of criminal behavior often are found in impoverished communities and violent or neglectful families, special attention is placed on the ecological context, i.e., the relationship between environmental deficits and impaired psychosocial development. Understanding the ecological context is essential for horticultural therapists to design and implement effective treatment programs for this population.

In this chapter we discuss the physical and emotional treatment needs of those incarcerated in our jails and prisons and identify inmates best suited for horticultural therapy. Our learning objectives are to

1. provide an ecological understanding of the physical, emotional, and social problems offenders endure;
2. consider specific horticultural therapy applications for this population;
3. explore the range of skills prisoners can develop from horticultural therapy; and

4. present case illustrations and relevant research findings on the efficacy of horticultural therapy as a treatment.

Pictures of prisoners tending plants convey hope and light amidst this gloomy, shadowed terrain of modern America. The significance of horticultural therapy for those incarcerated lies in its ability to counter alienation and isolation by reconnecting inmates to the community and the environment. An effective horticultural therapy program cultivates

- care for the earth,
- productive and positive work experience,
- suitable environments for reflection and self-understanding,
- care for the self,
- hope, and
- self-esteem.

IDENTIFICATION AND TREATMENT
OF CLIENT POPULATION

Demographic Characteristics

Prison Growth

The importance of developing and implementing comprehensive treatment programs for those incarcerated is apparent when the increase in this population is examined. In California, our nation's most populous state, prisons are the largest growth industry. A greater proportion of the California state budget is spent on prisons than on schools. Increased arrests for substance abusers and longer sentencing policies are contributing to the burgeoning population in our jails and prisons. By the end of 1994, there were 483,717 inmates in our nation's county jails, a 165 percent increase from 1980. There were 999,808 prisoners incarcerated in federal and state prisons, which represents a 213 percent increase since 1980. In addition, there were 2,962,166 people on probation and another 5,135,850 people on parole (Beck and Gilliard, 1995). The median sentence length (the fiftieth percentile) of federal and state prisoners in 1994 was forty-eight months (Beck and Gilliard, 1995).

Recidivism rates reflect the paucity of effective prison treatment programs and the deteriorating social conditions of the inner city communities, which spawn the majority of our incarcerated population. In 1991, over 60 percent of prisoners had been incarcerated in the past, and almost all of these prisoners had been in jail or prison for another charge in the five years prior to their current offense (Beck et al., 1992).

Who Is Incarcerated?

The terms "prisoner" and "inmate" are often used interchangeably. Generally, people serving time in a state or federal prison are called "prisoners" and those serving time in a city or county jail are called "inmates." This will be the distinction utilized in this chapter. The average age of people incarcerated in jails and prisons has been increasing due to longer sentencing and increased drug arrests. In 1991, 22 percent of federal and state prisoners were under the age of 25. Thirty percent of prisoners were between thirty-five and fifty-four years old (Snell, 1994). Men comprise the vast majority of those incarcerated. However, the rate of women in our federal and state prisons is increasing more rapidly than the male incarceration rate. By the end of 1994, women accounted for 6.1 percent of all state and federal prisoners (Beck and Gilliard, 1995) and 10 percent of the local jail inmate population (Perkins, Stephan, and Beck, 1995).

Racial and ethnic minorities comprise the majority of people incarcerated in America, and these percentages are rising. Between 1980 and 1992, the percentage of African American sentenced prisoners increased from 46.5 percent to 50.8 percent. At the end of 1993, African Americans were seven times more likely than whites to be incarcerated in state or federal prisons. Hispanic prisoners increased from 7.7 percent to 14.3 percent of the state and federal prison population in the years 1980 to 1993. On December 31,1993, nearly two-thirds of all sentenced prisoners were African American, Asian, Native American, or Hispanic (Beck and Gilliard, 1995).

Education and Employment

Many people come into the criminal justice system with an incomplete education and low-paying full- or part-time employment. About two-thirds of those incarcerated in 1991 had not completed high school, including 19 percent with an eighth grade education or less. Two-thirds of all people incarcerated in 1991 were employed during the month prior to their arrest, with half employed full-time. An estimated 38 percent of women and 13 percent of men had been receiving social security benefits, welfare, or charity before entering prison (Beck et al., 1992).

Limited education and employment are predictors of family and community instability. High unemployment in our inner cities is associated with limited access to community resources and a critical shortage of many vital health care, mental health, and social services.

Physical Problems

Many offenders enter the criminal justice system with chronic health problems. These are the result of living in impoverished neighborhoods with deleterious environmental conditions, and inadequate early nutrition and health care. Offender health is further compromised by substance abuse. Overall healthcare needs of those incarcerated have increased with the aging of the offender population, the increase in length of sentences, the rise in the percentage of pregnant female offenders, and the increase in infectious diseases such as HIV/AIDS and tuberculosis (Crawford, 1994). By early 1993, 11,565 inmate HIV/AIDS cases had been reported, including 3,500 fatalities (Hammett, 1994a). Tuberculosis (TB) has also risen in this population of poor inner-city minorities and injection-drug users. Jails and prisons further the spread of TB through overcrowding and poor ventilation (Hammett, 1994b).

An analysis of the San Francisco County Jail population determined that 81 percent of the inmates had one or more chronic health problems (Rice, 1993). Of further concern was that 16 percent of their children were already coping with chronic health problems. Nationally, a 1 to 10 ratio of children are so afflicted (National Center For Health Statistics, 1991).

Horticultural therapy often provides those incarcerated with their first experience of a healing natural environment. Learning how to cultivate life can be linked easily to information on nutrition and other important facets of self-care.

Mental Health Problems

Many people incarcerated in our jails and prisons have undiagnosed and untreated mental health problems. Those problems have roots in family instability, child abuse and neglect, and abject poverty. The high substance use found in this population may be considered an attempt to self-medicate in the absence of the social and environmental supports necessary to heal the emotional wounds from the past.

Family Instability

In 1991, about one-fifth of those incarcerated were married. Over one-half were never married. More than one-half did not live with both parents throughout their childhood. About one-fourth had parents who abused drugs, and a little less than one-third had a brother with a prison or jail record. Thirty-seven percent had at least one family member who had been incarcerated. Fifty-three percent of African American prisoners grew up in single-parent homes, compared to 40 percent of Hispanic prisoners and 33 percent of white prisoners. About 14 percent had lived in homes with no parent, while 17 percent had lived in a foster home, agency, or other institution at some time during childhood. In 1991, prisoners were the parents of 826,000 children under the age of nineteen. Forty-two percent of women and 32 percent of men had two or more children under eighteen. Six percent of women prisoners were pregnant when they entered prison (Beck et al., 1992).

Child Abuse and Neglect

The childhoods of those incarcerated often are characterized by deprivation, instability, and violence. Many offenders have been abused as children and have seen parents hit each other. These conditions suggest significant emotional deficits stemming from attachment failures, separation, and loss. As children, they did not have the continuity of care they needed to grow and thrive (Bowlby, 1969, 1973, 1980). Many did not receive the empathic parenting they required to form a cohesive and stable self experience (Baker and Baker, 1987; Kohut, 1984). As a result, many incarcerated people are angry, anxious, depressed, and distrust personal relationships. They often do not have a positive sense of their worth or their capabilities.

Substance Abuse

According to 1989 jail inmate self-reports, 44 percent reported using drugs in the month before their arrest, 30 percent reported using drugs daily in the month before their offense, and 27 percent reported using drugs at the time of their offense. In 1991, 49 percent of all state prisoners reported they were under the influence of drugs or alcohol or both at the time they committed their offense. Half the prisoners said they had taken illegal drugs in the month prior to arrest. Data from the Bureau of Justice Statistics show that 78 percent of jail inmates in 1989, 79 percent of state prisoners in 1991, 60 percent of federal prisoners in 1991, and 83 percent of youth in long-term juvenile detention facilities in 1987 had used drugs at some point in their lives (ONDCP, 1995). Despite the increasing population of offenders with substance abuse disorders, few receive treatment. A

1989 survey of 1,700 jails by the American Jail Association found only 28 percent had drug treatment programs. Of these, only 7 percent had a comprehensive range of services including group counseling, drug education, and transitional and aftercare planning (Peters, Kearns, and May, 1991).

Community Characteristics That Breed Social Problems

Sometimes community characteristics can so overwhelm the individual that incarceration is preferable to what must be faced on the streets.

> I remember the older guys would tell you how great reform school was. It was also a place where inner-city kids could have fun in a rural area. To me, reform school was a gas. Sometimes we would plan our juvenile activities so we could go to court in March or April and get sentenced to six months and spend spring and part of the summer in reform school. It was fun to get away from the city; you had your buddies there. (C. S. Dutton, quoted in Rothstein, 1990, p. B3)

Inner-City Environment

The majority of people in criminal institutions come from decaying inner-city communities. In order to appreciate the impact horticultural therapy may have on those incarcerated, it is important to consider the quality of their physical and social environments. These communities are noisier, denser, and have more exposure to environmental pollutants than affluent communities. Lee (1987) found a strong correlation between socioeconomic status (SES) and location of toxic waste dumps. Exposure to these conditions adversely affects the development of inner-city children. (Bronzaft, 1981; Bronzaft and McCarthy, 1975; Harvey et al., 1988; Moser, 1988; Needleman, 1985; Needleman et al., 1979; Rodin, 1976). Living in neighborhoods deemed ugly by the larger community adversely impacts how residents feel about themselves (Bronfenbrenner, 1979; Garbarino, 1992; Lewis, 1996; Proshansky, Fabian, and Kaminoff, 1983; Stainbrook, 1973).

Unemployment

These communities also are characterized by the absence of gainful employment. Many industries that employed the residents departed for suburban areas, other regions of the country, or other parts of the world.

Unemployment has a critically deleterious impact on the stability of family life. It fuels higher incidents of family violence and dissolution (Garbarino, 1992; Wilson, 1987). The absence of employment reduces the tax base, which limits the availability of community resources and supports and produces inadequate schools. The need for income propels an underground economy fueled by drugs. The absence of positive working role models skews the choices youths make and contributes to their involvement with the criminal justice system.

Access to Restorative Environments

Inner-city communities do not provide the physical or psychological space necessary for reflection. The physical environment is often denatured, and the few parks that exist are generally unsafe. Reflection is essential for understanding the influence of one's past and making appropriate future choices. Reflection requires a suitable environment and freedom from having to attend to critical survival needs (Kaplan and Kaplan, 1989). An institutional horticultural therapy program may afford participants their first opportunity for reflection in a safe environment.

HORTICULTURAL THERAPY

Offender Vocational and Treatment Programs

Offender vocational and treatment programs vary considerably nationwide. The types of programs offered are influenced by type and location of the institution and its political climate. The range of emphases in vocational programs include hard-labor punishment, producing state supplies and prison food, and specialized training programs in areas such as data processing, cooking, auto mechanics, and horticulture.

Horticultural programs in jails and prisons run a gamut from formal therapy programs to prison farming industries. In a 1990 phone survey of state prison systems, fifty-five prison authorities in forty-one states were contacted (Rice, 1993). Twenty-six prisons had horticultural vocational programs, fourteen prisons had working farms or ranches operating as correctional industries, four prisons had formal horticultural therapy programs, six prisons had informal horticultural therapy programs, one prison had a landscaping program, and four prisons had no program.

Program Goals

Horticultural therapy programs can be differentiated from horticultural vocational programs by their emphasis on the process of learning and the applicability of the material to other areas of the students' lives. It is not enough to provide inmates with vocational and academic skills. Wiley (1986) broadens the focus of correctional education:

> Vocational education for inmates . . . is simply insufficient when the emphasis rests solely on the acquisition of knowledge and skills. There must also be some sort of growth in values, beliefs, and attitudes that will enable the offender to return to and function in free society. (p. 2)

Horticultural therapy provides opportunities for experientially based learning. When an inmate gazes upon a garden growing by virtue of his or her effort, he or she will often feel pride in the accomplishment. This teaches that productive work enhances self-worth and self-esteem. Taking care of plants cultivates the experience of being responsible for other lives. Horticultural therapy can promote positive group interactions. For example, group activities transmit essential survival tools to inmates at the Williamson County's (Texas) Adult Probation Positive Experiential Learning (APPEL) program (Harris et al., 1993). The authors note,

> Group exercises, which typically make fewer physical demands, require participants to share responsibility and to solve problems as a team. Successful solutions to these problems depend upon the extent to which group members cooperate, trust and communicate with each other. Both individual and group experiences impart "lessons," which participants will later apply to problems in their personal lives. (p. 12)

Program Eligibility

In most instances, the jail or prison administration will determine which inmates are eligible for horticultural therapy. Since these programs usually take place outside, inmates will need outside clearance. Often times, this precludes people who have not been sentenced, have escaped, or have attempted to escape in the past. At San Francisco County Jail, there are no other offense-based restrictions. However, inmates who have been charged with domestic violence are required to take a violence prevention class prior to entry into the horticulture program. Some institutions will not give outside clearances to people who have committed violent crimes.

Therapeutic Relationships

Cultivating a therapeutic relationship within a correctional institution is an ongoing challenge to the efficacy of treatment. Horticultural therapists will gain better results by referring to program participants as students rather than inmates or prisoners. This places the emphasis on the present and fosters an identity associated with learning and growth. Prisons and jails often encourage staff to maintain formal and distant relationships with offenders, yet the relationship between teacher or therapist and student is characterized by respect, openness, and caring. These qualities are essential for a therapeutic relationship to develop. The horticultural therapist facilitates student development by communicating clear expectations of appropriate behaviors and attitudes. Students must then be held accountable for their actions.

A therapist must be sensitive to differences in life experience shaped by ethnic, class, social, and racial inequities. Teaching style and the quality of the therapeutic relationship often determines the impact of a horticultural therapy program. If the therapist is intimidating, judgmental, or unapproachable, the students may only begrudgingly do what they are asked to do and may not grow through the program. Horticultural therapists must be honest, receptive, and consistent in their treatment of students.

Program Design

Horticultural therapists should plan activities to promote specific outcomes. An effectively designed program

- outlines each activity's goals;
- sequences activities so that development is additive and cumulative;
- delineates goals that are attainable in the time allotted;
- maximizes individual accomplishments;
- teaches group participation, leadership, and responsibility; and
- describes relevance of activities to post-release life.

It is important to note that horticultural therapists working in this setting usually do not develop individual treatment goals and objectives. Generally, the offender population shares significant physical, emotional, and social needs in common. This coupled with the likelihood of inadequate program funding, understaffing, and high participant turnover, necessitates developing and implementing group treatment goals.

Therapeutic Metaphors

Therapeutic metaphors help students apply the lessons learned from gardening to their lives. A student's tomato plant may die because it was not watered sufficiently. The horticultural therapist might note that relationships suffer if they are not nurtured. The importance of weeding for plant growth and survival can be linked to the removal of harmful influences. If the garden is organic, the value of limiting chemicals can be discussed with a student who is a substance abuser. Contending with adverse weather conditions can be applied easily to the areas in the students' lives where they are affected by conditions beyond their control. If things grown in the garden are given to people in need, the students can be asked to reflect on how it feels to take care of and provide for others. The yearly natural cycle can be used to teach that there is a time for inner work on oneself and a time for outer flourishing.

Horticultural therapy provides a safe setting for challenging an offender's life choices. Is this type of experiential education really effective? According to Wood and Gillis (1979), "The subject's experience of successful problem solving stimulates his or her interest in solving other problems" (p. 7). Educational studies indicate that cooperative learning structures increase students' beliefs that their own efforts can make a difference in their success (Slavin, 1990).

HORTICULTURAL THERAPY ACTIVITIES

In this section, we consider specific horticultural therapy activities for people who are incarcerated. The skills and values each activity promotes, as well as their relevance for this population are reviewed. Case illustrations of jail and prison horticultural therapy programs are presented.

Garden Rows

Individual responsibility can be facilitated by providing each participant with a garden row to plant and tend. The horticultural therapist should determine the overall size of the garden row based upon the space available, number of participants, and time allocated for horticultural activities. The therapist should also establish guidelines for what can be planted, how certain vegetables should be planted, and specific instructions involving care and maintenance. Individual rows can be divided into separate growing segments. A fifty-foot row might be divided into five ten-foot sections. Each segment can be used for a specific type of crop. For example, the first segment could be utilized for root crops. The therapist can provide students with several choices such as radishes, carrots, or potatoes. The selection will depend upon location and time of year. In addition, including a "grower's choice" section encourages students to express individual differences and promotes their taking "ownership" of their row.

Several life issues can be addressed through this activity. Gardening requires delayed gratification. The soil is tilled and cultivated, the row is created, and the seeds are planted. Despite this considerable investment of labor, plants do not emerge immediately from the soil. Seeds and plants will survive only if they are tended. The students learn that some plants may die from overwatering, underfertilization, or other uncontrollable causes. Students who harvest the fruits of their labor from these row gardens also reap newfound feelings of pride and enhanced self-esteem.

Garden rows demonstrate:

- the value of perseverance and delayed gratification,
- the necessity of setting short-term goals, and
- the importance of adaptability in establishing new goals.

The significance of imparting these values and skills is best understood in light of the inadequate parenting many students received. Family instability prevents the students from learning that growth takes time, perseverance, and the ability to adjust ones goals. When a child's essential emotional and physical needs are inadequately met, they are more likely to act impulsively. Delayed gratification is possible when critical self needs have been nourished.

Horticultural therapists can use cluster gardens as a variation of this activity. Participants are divided into groups and given a small garden to tend. The therapist encourages and facilitates group interactions. The groups collectively decide the size of the garden, what crops to grow, how the garden will be organized, and who will be responsible for various activities.

Cluster gardens promote:

- cooperative work experience,
- communication skills,
- leadership, and
- positive group identity.

Students often come from neighborhoods that provide few social and communal supports. When working members leave to pursue jobs elsewhere, the social network deteriorates. Church and recreational youth group activities are reduced and schools provide fewer extracurricular programs. The gangs that develop in this vacuum often teach maladaptive social skills. Belonging to groups that transmit positive values, and the experience of being valued enhances an individual's self-development.

Propagation Activities

In this activity, students start new plants from vegetative and reproductive plant parts. The therapist sets guidelines for the number of cutting or seed starts, based upon the availability of greenhouse space and the quantity of stock plants. Students are encouraged to choose the types of plants they want to propagate. They are instructed in a variety of asexual propagation techniques such as tip cutting, leaf cutting, stem cutting, and air-layer propagation. Students label their flats of cuttings and are encouraged to closely observe when their plants have rooted. Once the plants have developed a root system, students pot them in larger containers. They then label their plants and place them in a designated area in the greenhouse or classroom.

Each student is responsible for watering, fertilizing, and caring for their own plants. The horticultural therapist can provide assistance and answer questions. However, it is important to afford students the opportunity to learn from their own choices and actions. For example, a student might ask if doubling the recommended amount of fertilizer will maximize growth. After explaining that overfertilization could potentially burn and kill the plant, the therapist should allow the student to choose how much fertilizer to use. This approach provides opportunities to experience the consequences of their actions.

The students also can be instructed in sexual plant propagation by seeds. Students are given seeds and are asked to plant, water, and care for the seedlings. As seedlings develop true leaves, the students are taught how to transplant them into larger containers. In commercial greenhouses, only the strong seedlings are transplanted while the weak ones are discarded. Participants may object to this practice, identifying themselves as weak seedlings that have been tossed aside. The horticultural therapist can encourage them to nurture all of the seedlings.

Propagation activities demonstrate the following:

- Overwatering or overfertilizing (compulsive behavior) can be harmful,
- New roots (new beginnings) grow from small cuttings, and
- Repotting plants (suitable environment) supports growth.

In essence, this activity teaches students how to grow. They first must confront the compulsive behaviors they have developed to cover their inner experience of pain and emptiness. The students learn through propagation that change occurs by taking small steps along a suitable path.

Landscape Design

In this activity, the students learn how to design a landscape from start to finish. The horticultural therapist provides basic information on landscape design. This includes elements of design, the design process, selection of plant materials, installation, and maintenance. The therapist selects the area to be landscaped and assists students in developing a site analysis. If a residential or commercial site is selected, the site analysis may be quite detailed. It can include marking power lines, phone lines, drainage patterns, existing vegetation, and other items. A simplified site analysis may be done for a specific bed area. After providing examples of design questions, the therapist can encourage students to work together to develop a list of questions to ask the client. Once the site and client analysis are completed, encourage each student to develop his or her own individual design for the client. Students can give oral presentations to the client detailing the landscape design.

There are several ways to expand this activity. The group can work with the client to develop a composite plan based on their individual designs. The students can use plants that they have propagated by seeds or cuttings in the landscape design. The students could design and implement a peaceful garden area or entrance area at their jail or prison.

Landscape design cultivates

- creativity and planning,
- group communication and interaction, and
- pride and ownership.

Growing up in impoverished social and physical environments breeds feelings of powerlessness. Feeling powerless to change one's environment in a positive manner can lead to destructive behavior. Landscape design provides students with a constructive way to express themselves in their environment. This physical activity imparts meaning and vitality to their sense of self.

Composting

Composting fosters environmental stewardship and personal responsibility. A group compost pile demonstrates how waste material such as weeds, manure, water, soil, and grass clippings can be turned into something useful and productive with hard work and time. The turning of the compost pile teaches the students that they need to turn over their old ways. Returning the finished compost back to soil is a powerful metaphor for giving back and making amends. This activity aptly illustrates principles discussed in twelve-step treatment programs.

Jail and prison horticultural therapy students are painfully aware that their lives have been wasted and that they are seen as social waste by the larger community. These feelings decimate their self experience. Carl Jung (1960) discerned that just as the decaying process in nature is integral to the emergence of new life, disintegration within the human psyche could lead to reintegration and growth. Composting provides students with a direct experience of this possibility for redemption.

Communication Development

All horticultural experiential activities can be enhanced by encouraging students to express themselves. For example, students can select and study a particular vegetable, grow it in their garden row, and give a presentation to the larger group on their vegetable. Becoming the "expert" on particular vegetables raises self-esteem. Posters and presentations can be placed in the food service areas to promote the benefits of vegetables and encourage healthy eating habits. Students often enjoy showing visitors "their" garden and sharing their gardening experience with correction officials and media reporters.

Communication skills can be cultivated by

- labeling plants in garden areas,
- designing an interiorscape plan for an office,
- measuring and calculating plant materials for a flower bed,
- making collages for group presentations, and
- leading garden tours.

In disorganized families and disadvantaged communities, the need to be seen, heard, and valued is often frustrated. This leads to maladaptive, negative, attention-seeking behavior. Horticultural therapy provides students with opportunities to feel valued by their peers, teachers, and the larger community. This has an immeasurable impact on their self-esteem.

Fostering Mentor Relationships

Utilizing seasoned students as mentors and positive role models adds another dimension to horticultural therapy programs. A mentor can be assigned to answer questions and assist students in planting their garden rows. This helps new students become comfortable with often unfamiliar horticultural activity. Mentoring also enables the experienced student to take on new challenges and responsibilities. The horticultural therapist must provide guidance on how to work with newer students in nonthreatening and nonjudgmental ways. Often this is achieved by example and role modeling. The therapist should closely observe and support the mentors until they feel secure and comfortable in this role.

Many students enter the horticultural therapy program with unmet self needs that fuel self-involved and antisocial behavior. They may never have experienced the fulfillment that comes from helping another person grow. Mentoring teaches them that their own painful experiences can be transformed into empathy and concern for others.

CASE STUDIES

Federal Prison Camp—Bryan, Texas

The Federal Prison Camp (FPC) in Bryan, Texas is a minimum security prison for women offenders. Its vocational education department offers a Master Gardener program sponsored by the Texas Agricultural Extension Service. The Master Gardener program is a volunteer organization with chapters in forty-five states. Their programs are open to all people regardless of socioeconomic level, race, color, sex, religion, or national origin (Texas Agricultural Extension Service, 1995).

At FPC, the class runs for fifteen weeks and is structured as horticultural therapy. Three classes are offered yearly. The class meets Monday through Friday from 7:30 a.m. to 3:30 p.m. The four-hour morning session provides experiential learning activities in the greenhouse and garden areas. The three-hour afternoon session includes classroom lectures, guest speakers, and laboratory activities. This program provides training in all aspects of the horticulture industry. Students work individually and within groups. Work groups are responsible for watering office plants and landscaping the prison grounds. These groups are rotated every four to six weeks to provide students with the experience of working in different settings with a variety of people.

Upon completion of the program, students receive a Master Gardener certificate and forty continuing education units from Texas A&M University. Selected students from this program are placed into a job training program run by the horticulture department at Texas A&M University.

Students receive additional training in all areas of horticulture and assist in greenhouse, lab, and data-collection activities.

The Master Gardener's nationwide network has proven valuable for ex-offenders. It provides job placement, positive leisure-time activities, and community-based support upon release. Many of the students remain active in this program when they return to their communities. Some students have left FPC and secured employment in horticulture. Five participants found jobs in the retail greenhouse industry and another now works for a wholesale florist. One woman completed her GED in prison and finished the Master Gardener program upon release. Currently, she is enrolled in a university horticulture program. Many former students have opened their own businesses in the horticulture industry.

The following comments were taken from student evaluations, cards, and letters received from 1990 to 1995. "You have given me the grounding I truly needed to see how instant gratification isn't what I want. To grow something has taught me to appreciate life more than I thought possible." "I want to thank you for all that you have done for me while I was incarcerated. Most of all thank you for treating me as a person." "You taught me a lot more than horticulture—you taught me self-respect, self-worth, and dignity. More than any staff member here, I credit you with my successes." One student wrote a poem, which is excerpted below.

> In search of ways to do quality time,
> We tilled the land, studied the vine.
> We watched God's process of making things grow,
> And after a while, He was cultivating us (little did we know).
>
> We can't speed it up, can't hurry the process
> But with each day we serve, we're making progress.
> Toward becoming like stately evergreen trees,
> Standing tall, unwavering, proud to be seen.

Currently, FPC's horticultural therapists are evaluating the program's impact on student locus of control and self-esteem (Migura and Zajicek, 1995). Further research will assess the long-term effectiveness of the Master Gardener program in cultivating employability, personal accountability, effective communication, group collaboration, and a positive communal experience. Research results will further the development and utilization of horticultural therapy programs. Currently, other prison facilities and drug treatment centers in Texas are developing Master Gardener programs at their facilities.

Graterford State Correctional Institution—Graterford, Pennsylvania

Graterford is a maximum security prison for men. In the mid-1980s Matt Epps, a prisoner, asked if he could start a garden. He later wrote a proposal for a prison horticulture program. The horticulture project began in 1991 and is coordinated by the Montgomery County Agricultural Extension Agent, Penn State University Master Gardeners, and other community volunteers. The program has grown from an initial twenty students to 360 students currently tilling forty plots (Pruyne, 1994–1995). Almost the entire prison grounds are covered with plants and vegetables. Plants, seeds, topsoil, and mulch are donated to the program. Students save money from other prison jobs to purchase supplies such as stones, bricks, and fountains. The students receive vocational certificates, which facilitate their getting nursery and greenhouse jobs upon their release. A special grant has been secured to build a greenhouse that will extend the program to a yearround one. Nancy Bosold, a Montgomery County Extension agent, comments on student involvement: "They get excited over things like flower arranging and growing produce that wins prizes at the county fair. They help each other with weeding, mulching, and they share their vegetables with one another" (Pruyne, 1994–1995, p. 36).

New Jersey Department of Corrections

In 1992, the New Jersey Department of Corrections, in partnership with Rutgers University, established a horticultural training program for youth offenders, ages fourteen through seventeen. Groups were selected from several residential and day programs, including Voorhees, Florence Crittenden, Union, and Middlesex. Students were brought to the Rutgers University, Cooks College Campus one day weekly three times a month. Classes begin with a forty-five-minute lecture, which is followed by a ninety-minute experiential activity in the greenhouse. The students have lunch at the campus student center and then participate in a structured outdoor activity or field trip. The outdoor activities include pruning trees and shrubs, transplanting nursery stock, planting garden plots, and operating landscape equipment. Each of the participating institutions has a garden and/or greenhouse for the students to work in upon their return. This twenty-five-week horticulture therapy program provides horticultural skills for future employment and fosters positive behavior. Preliminary findings indicated that the youth experience personal satisfaction and pride in their horticultural projects (Flagler, 1993).

San Francisco County Jail—San Francisco, California

In 1984, the San Francisco Sheriff's Department began an organic horticultural therapy program for county jail inmates called the Garden Project. The purpose of this program is to transmit meaningful work skills to students, while cultivating a heightened awareness of self in relation to community and nature (Rice and Remy, 1994). In 1989, the department used money set aside for public art at their newly built program facility to construct a greenhouse. Initially, the horticulture program was offered as an optional vocational education program.

In 1991, the Sheriff's Department designated the horticultural program a mandatory core treatment component for almost every eligible inmate. There are three two-hour class daily, two are for men and one is for women. Classes generally begin with a group discussion on an aspect of gardening that can be applied to the lives of the students. This is followed by specific training in organic gardening and work in the fields. Food and flowers grown by the program are donated to homeless shelters and home-bound AIDS patients. Trees started by the Garden Project are planted in San Francisco neighborhoods.

Recognizing the need for support post-release, the Garden Project branched out into the community with the creation of an after-care garden. When grant or other donated money is available, the students are paid $68 weekly for sixteen hours of work. Within the program facility and upon release, the students receive life-skill and employment training, such as mock job interviews, resume development, family-life education, and anger management. The post-release garden grows organic produce, which is sold to fine restaurants, health food stores, and a bakery. The Sheriff's Department has secured a contract with the Department of Public Works to employ graduates from the program in an urban tree corp. The tree corp plants and cares for trees in predominantly low-income neighborhoods.

One participant in the Garden Project stated, "The garden gives me a chance to think about my problems and what I did, and what I want to do better in my life. Coming here gave me the opportunity to start thinking about these things" (Gordon, 1990, p. 1). Another participant describes the greenhouse in the following way:

> All of these trays are seeds that are germinating. Mainly what seeds need in order to germinate is water and light, and someone to pay attention to them and make sure that they don't have any problems . . . This program is like that—it starts a germination in your mind, to try and instill the tools you need to grow and to make changes in your life. (Richards, 1991, p. 1)

An exploratory experimental study of the garden project determined that county jail inmates had experienced a high percentage of loss and violence during childhood. These factors contributed to early involvement with alcohol and drugs, which in turn led to involvement with the criminal justice system. Inmates were assessed for psychocial functioning prior to random assignment to the garden project or other jail programs. They were reassessed prior to release and at least three months post-release. Treatment effects included reduced post-release substance abuse by Garden Project participants, lower depression in Garden Project participants with emotionally detached mothers, and sustained hope and desire for help throughout treatment and follow-up (Rice and Remy, 1994).

FUTURE PRACTICES

Challenges

Many of those incarcerated enter the criminal justice system with unresolved early traumatic life experiences. In order to successfully leave this system and reintegrate with the larger community, they must understand and work through these painful early experiences. It is difficult to stop avoiding or numbing one's pain. Having positive life experiences that provide hope for the future makes facing one's pain possible. Horticultural therapy can be an effective first step for treating this population. The effectiveness of offender horticultural therapy programs will be limited if they are not integrated with other treatment modalities, such as drug treatment, psychological counseling, and family-life education.

Many people leave jails and prisons penniless, homeless, and estranged from friends and family. Horticultural therapy has been shown to be an effective vocational education program, providing meaningful and useful employment skills for people who are incarcerated in our nation's jails and prisons. Horticultural therapy students will need post-release planning to enable them to find employment and other needed services upon their release.

Remedies

Too often, troubled youths and adults in our culture do not receive services until they enter the criminal justice system. Juvenile delinquents are provided with counseling, art therapy, and recreational activities that are not available in their own communities. Drug-dependent mothers receive drug treatment in jail that is minimally available on the streets (Remy et al., 1983). If horticultural therapy programs are an effective correctional treatment modality, how much more of an impact might these programs have for people who are struggling within their home communities? Community gardens can be used to teach children, build intergenerational bridges, and help detoxify the social environment (Garbarino, 1995).

In San Francisco, the county jail horticultural therapy program started an aftercare program for those leaving the jail. This outside garden quickly began to draw participants from the community who had not been previously incarcerated. A day care program was established, so that parents could work and bring their children along with them. Parent and child equally benefit from their interaction with a beautiful natural setting. Horticultural therapists working with this population must forge links with community residents and agencies interested in initiating preventive urban gardening programs.

While horticultural therapy has been practiced in this country for over 100 years, it is not a widely known treatment modality. Horticultural therapists must publicize their programs in local newspapers, magazines, and on television. Generally, the quality of media coverage of offender horticultural therapy programs has been impressive. The image of prisoners growing things portrays a hopeful message that people can change and grow.

Horticultural therapists must present their work to local social welfare, mental health, and criminal justice professionals. Bringing these professionals out to the garden or greenhouse to observe horticultural therapy firsthand, conveys the vitality and viability of this treatment modality. Horticultural therapists must also arrange for comprehensive evaluation of their programs. Wherever possible, research should meet the standard of quality established by other social and physical sciences. Horticultural therapists can evaluate their own programs with the help of qualified researchers. They might also form a collaborative relationship with researchers interested in studying horticultural therapy. It is critically important for research findings to be disseminated for broader acceptance, funding, and program development. (See Chapter 16 for additional information on conducting horticultural therapy research.)

An ecological treatment intervention takes into account the multiple intersecting factors that contribute to health or disease. For horticultural therapists to treat those who are incarcerated, they must form links with other community agencies and professionals working with this population. A case management model stresses the importance of designing programs that link students with the transitional support and assistance they will need post-release. Research has shown case management maximizes appropriate utilization of services, promotes reintegration of clients into the community, and improves the quality of clients lives (Sherwood and Morris, 1983; Goering et al., 1988; Franklin et al., 1987). Without this bridge to the community, treatment gains are likely to be lost post-release, when former inmates have to struggle once again to survive in extremely adverse life conditions.

The importance of addressing the context as well as the individual is inherent in horticultural therapy. This ecological perspective generates a moral imperative for horticultural therapists to be advocates for developing healthy psychological, social, and physical environments that enhance human survivability and growth.

Graterford Prison, Pennsylvania. (Photo by Stacie Bird)

Graterford Prison, Pennsylvania. (Photo by Stacie Bird)

An inmate tends the garden at Graterford Prison, Pennsylvania. (Photo by Stacie Bird)

KEY ORGANIZATIONS

American Horticultural Therapy Association, 362A Christopher Avenue, Gaithersburg, MD 20879, 301-948-3010.

American Orthopsychiatric Association, 330 Seventh Avenue, 18th Floor, New York, NY 10001, 212-564-5930.

Center for Mental Heath Studies (CMHS), 5600 Fishers Lane, Department of Health and Human Services, Rockville, MD 20857, 301-443-2792.

Children's Defense Fund, 25 E Street, NW, Washington, DC 20001, 800-233-1200.

Fortune Society, 39 W 19th Street, New York, NY 10114-0726, 212-206-7070.

National Institute for Corrections (NIC), 1960 Industrial Circle, Suite A, U.S. Department of Justice, Longmont, CO 80501, 800-995-6429.

GLOSSARY

attachment: John Bowlby defined "attachment" as an enduring affectional relationship between child and caregiver, the purpose of which is to protect and nurture the child. Attachment behaviors are inborn abilities forming a basic motivational behavior system that characterizes interactions between mother (or other primary caregiver) and child. The goals of the behavior system are to maintain proximity to the caregiver and thereby ensure the child's survival. Note that need for affectional relationships persists into adulthood and forms the core of committed adult relationships.

case management: Case management provides links to often uncoordinated social services via an interpersonal, supportive relationship. Support systems help people adapt to life transitions, crises, and stresses. Using guidance and feedback, case managers provide opportunities to develop problem-solving capabilities and emotional mastery.

ecological context: An ecological context asserts the interdependence of organism and environment. Urie Bronfenbrenner believes an ecological understanding of human development enables researchers and therapists to recognize and address environmental factors that obstruct or enable humans to express their full potential. An ecological orientation predicts an interaction between developmental and environmental influences on self experience.

empathy: Empathy is the capacity to understand another person's experience from their perspective.

inmate: A person serving a sentence in a city or county jail is referred to as an inmate. Usually the sentence is one year or less.

loss: Attachment theory suggests the developmental consequences of loss of a primary caregiver are anger and anxiety.

parole: Generally, parole refers to supervision post-release from a state or federal prison (although some local jails use this term as well). There are usually conditions of parole such as nonfraternization with other offenders, no use of drugs, and no weapons. Parole may be initiated after a sentence has been reduced.

prisoner: A person serving a sentence in a state or federal prison, usually for a year or more, is referred to as a prisoner. In some cases, a prisoner can be returned to prison for a shorter period for violating his or her parole.

probation: Probation usually refers to supervision post-release from a city or county jail. There are usually conditions of probation such as nonfraternization with other offenders, no use of drugs, and no weapons. Probation may be initiated after a sentence has been reduced.

psychosocial development: This is an understanding of development as influenced by early life experience and the interaction of individual psychological processes with social influences.

recidivism: This refers to reinvolvement with the criminal justice system upon release from jail or prison.

self: According to Heinz Kohut, the self represents the core of the individual's psychological universe. Self capacities regulate self-esteem and promote an inner sense of identity. They continue to develop throughout the life cycle.

self needs: A child's stable and cohesive self emerges through empathic attunement, positive mirroring of its essential nature, availability of others for idealization, and twinning, which represents the child's need to feel alikeness with others.

separation: Attachment theory suggests the developmental consequences of separation from a primary caregiver are sadness and depression.

traumatic life events: Kohut asserts repeated and traumatic empathic failures in childhood contribute to the experience of self-fragmentation, thereby impeding the experience of wholeness. Depression and hostility, as well as substance abuse and antisocial behavior, may be manifestations of self-fragmentation.

KEY BOOKS, JOURNALS, AUDIOVISUAL MATERIAL

Agriculture and Human Values. Agriculture, Food, and Human Values Society, P.O. Box 14938, Gainesville, FL 32604, 904-378-0386.

Cobb, E. (1977). *The ecology of imagination in childhood.* New York: Columbia University Press.

Ecopsychology Newsletter. The Ecopsychology Institute, P.O. Box 7487, Berkeley, CA 94704-0487.

Fortune News. The Fortune Society, 39 West 19th Street, New York, NY 10011.

Garbarino, J. (1992). *Toward a sustainable society: An economic, social and environmental agenda for our children's future.* Chicago: Nobel Press.

Gore, A. (1992). *Earth in the balance: Ecology and the human spirit.* Boston: Houghton Mifflin.

Journal of Offender Rehabilitation. The Haworth Press, 10 Alice Street, Binghamton, NY 13904-1580, 800-342-9678.

Kaplan, R. and Kaplan, S. (1989). *The experience of nature: A psychological perspective.* Cambridge, England: Cambridge University Press.

Lewis, C. A. (1996). *Green nature/human nature: The meaning of plants in our lives.* Champaign/Urbana, IL: University of Illinois Press

Seeds of Change (Video). The Garden Project, 35 South Park, San Francisco, CA 94107, 415-243-8034.

REFERENCES

Baker, H. S. and Baker, M. N. (1987). Heinz Kohut's self psychology: An overview. *American Journal of Psychiatry, 144,* 1-9.

Beck, A. J. and Gilliard, D. K. (1995, August). *Prisoners in 1994.* Washington, DC: Bureau of Justice Statistics, U.S. Department of Justice.

Beck, A. J., Gilliard, D., Greenfeld, L., Harlow, C., Hester, T., Jankowski, L., Snell, T., Stephan, J., and Moton, D. (1993, March). *Survey of state prison inmates, 1991.* Bureau of Justice Statistics, U.S. Department of Justice.

Bowlby, J. (1969). *Attachment and loss: Vol. 1. Attachment.* New York: Basic Books.

Bowlby, J. (1973). *Attachment and loss: Volume 2. Separation: Anxiety and anger.* New York: Basic Books.

Bowlby, J. (1980). *Attachment and loss: Volume 3. Loss: Sadness and depression.* New York: Basic Books.

Bronfenbrenner, U. (1979). *The ecology of human development: Experiments by nature and design.* Cambridge, MA: Harvard University Press.

Bronzaft, A. (1981). The effect of noise abatement program on reading ability. *Journal of Environmental Psychology*, 215-222.

Bronzaft, A. and McCarthy, D. (1975). The effect of elevated train noise on reading ability. *Environment and Behavior, 7*, 517-518.

Crawford, C. A. (1994). Health care needs in corrections: NIJ responds. *National Institute of Justice Journal, 228*, 31-38.

Flagler, J. (1993). Correctional youth and the green industry. *Journal of Therapeutic Horticulture, Z*, 49-55.

Franklin, J. L., Solovitz, B., Mason, M., Clemons, J. R., and Miller, G. E. (1987). An evaluation of case management. *American Journal of Public Health, 77(6)*, 674-678.

Garbarino, J. (1992). *Toward a sustainable society: An economic social and environmental agenda for our children's future.* Chicago: Nobel Press.

Garbarino, J. (1995). *Raising children in a socially toxic environment.* San Francisco: Jossey-Bass.

Goering, P. N., Wasylenki, A., Farkas, M., Lancee, W. J., and Ballantyne, R. (1988). What difference does case management make? *Hospital and Community Psychiatry, 49(3)*, 272-276.

Gordon, R. (1990, July 31). A garden of healing: Prisoners reap new outlook. *San Francisco Independent, 35*, pp. 1-2.

Hammett, T. (1994a, January). *1992 update: HIV/AIDS in correctional facilities.* Washington, DC and Atlanta: U.S. Department of Justice, National Institute of Justice, and U.S. Department of Health and Human Services, Centers for Disease Control and Prevention.

Hammett, T. (1994b, January). *Tuberculosis in correctional facilities.* Washington, DC and Atlanta: U.S. Department of Justice, National Institute of Justice, and U.S. Department of Health and Human Services, Centers for Disease Control and Prevention.

Harris, P. M., Mealy, L., Matthews, H., Lucas, R., and Moczygemba, M. (1993). A wilderness challenge program as correctional treatment. *Journal of Offender Rehabilitation, 19*, 149-164.

Harvey, P. G., Hamlin, M. W., Kumar, R., Morgan, G., Spurgeon, A., and Delves, H. T. (1988). Relationship between blood lead, behaviour, psychometric and neuropsychological test performance in young children. *British Journal of Developmental Psychology, 6*, 145-156.

Jung, C. G. (1960). Stages of life. In R. F. C. Hull (Trans.), *The collected works of C. G. Jung: Structure and dynamics of the psyche* (Vol. 8). Princeton, NJ: Princeton University Press, Bolligen Series XX. (Original work published in 1947).

Kaplan, R., and Kaplan, S. (1989). *The experience of nature: A psychological perspective.* Cambridge, England: Cambridge University Press.

Kohut, H. (1984). A. Goldberg (Ed.), In *How does analysis cure?* with the collaboration of P. E. Stepansky). Chicago: University of Chicago Press.

Lee, C. (1987). *Toxic waste and race in the United States.* New York: Commission for Racial Justice, United Church of Christ.

Lewis, C. A. (1996). *Green nature/human nature: The meaning of plants in our lives.* Champaign/Urbana, IL: University of Illinois Press.

Migura, M., and Zajicek, J. (1995). *Effects of vocational horticulture education on the locus of control, self-esteem, and life satisfaction of women inmates at a federal prison camp.* Unpublished master's thesis, Texas A&M University, Texas.

Moser, G. (1988). Urban stress and helping behavior: Effects of environmental overload and noise in behavior. *Journal of Environmental Psychology, 8,* 287-298.

National Center for Health Statistics. (NCHS). (1991, December). Current estimates from the national health interview study, 1990 *Vital and Health Statistics.* Washington, DC: U. S. Department of Health and Human Services, Public Health Service Centers, Centers For Disease Control.

Needleman, H. L. (1985). The neurobehavioral effects of low-level exposure to lead in childhood. *International Journal of Mental Health, 14,* 66-77.

Needleman, H. L., Gunnoe, C., Leviton, A., Reed, R., Peresie, H., Maher, C., and Barrett, P. (1979). Deficits in psychological and classroom performance in children with elevated dentine lead levels. *New England Journal of Medicine, 300,* 689-695.

ONDCP. (1995). *Drug and crime facts. 1994.* Washington, DC: Bureau of Justice Statistics, U.S. Department of Justice.

Perkins, C. A., Stephan, J. J., and Beck, A. J. (1995, April). *Jail and jail inmates 1993-1994: Census of jails and surveys of jails.* Washington, DC: Bureau of Justice Statistics, U.S. Department of Justice.

Peters, R. H., Kearns, W. D., and May, R. L. (1991). *Drug treatment services in jail: Results of a national survey.* Washington, DC: Bureau of Justice Assistance, U.S. Department of Justice.

Proshansky, H. M., Fabian, A. K., and Kaminoff, R. (1983). Place-identity: Physical world socialization of the self. *Journal of Environmental Psychology, 3,* 57-83.

Pruyne, R. (1994-1995). Gardening for the health of it: Hort therapy benefits those with special needs. *Penn State Agriculture, Fall/Winter,* 34-40.

Remy, L. L., Haskell, S., Glass, R., and Wiltse, K. (1983). *300 Families in San Francisco: An evaluation of services to San Francisco's most vulnerable parents and children.* San Francisco: Coleman Children and Youth Services.

Rice, J. S. (1993). *Self-development and horticultural therapy in a jail setting.* Unpublished doctoral dissertation, San Francisco School of Psychology, San Francisco.

Rice, J. S. and Remy, L. L. (1994). Cultivating self-development in urban jail inmates. In M. Francis, P. Lindsey, and J. S. Rice (Eds.) *The healing dimensions of people-plant relations: A research symposium.* Davis, CA: University of California, Davis, Center for Design Research, pp. 229-256.

Richards, T. (1991, Spring). Bright wisdom and new life at the county jail. *Toward a Human Future, 4(2),* 1-5.

Rodin, J. (1976). Crowding, perceived choice and response to controllable and uncontrollable outcomes. *Journal of Experimental Social Psychology,* 564-578.

Rothstein, M. (1990, April 19). Dutton: A star of "Piano Lesson" began life anew on stage. *The New York Times,* p. B3.

Sherwood, S. and Morris, J. N. (1983). The Pennsylvania domiciliary care experiment: Impact on quality of life. *American Journal of Public Health, 73(6),* 646-653.

Slavin, R. E. (1990). *Cooperative learning: Theory, research and practice.* Englewood Cliffs, NJ: Prentice-Hall.

Snell, T. L. (1994, June). *Correctional populations in the United States, 1992.* Washington, DC: Bureau of Justice Statistics, U.S. Department of Justice.

Stainbrook, E. (1973). Man's psychic need for nature. *National Parks and Conservation Magazine, 47(9),* 22-23.

Texas Agricultural Extension Service. (1995). *1994 annual report: Texas Master Gardener program.* College Station, TX: Texas A&M University.

Wiley, L. J. (1986). *The effect of teaching style on the development of moral judgment in prison inmates.* Unpublished doctoral dissertation, Texas A&M University, Texas.

Wilson, W. J. (1987). The truly disadvantaged: The inner city, the underclass, and public policy. Chicago: University of Chicago Press.

Wood, D. E. and Gillis, J. C. (1979). *Adventure education.* Washington, DC: National Education Association.

PART THREE:
SETTINGS FOR HORTICULTURAL THERAPY PRACTICE

Chapter 11

Outdoor Space and Adaptive Gardening: Design, Techniques, and Tools

Jean Kavanagh

Outdoor facilities for therapeutic gardening reinforce and evoke a wide array of positive responses through growing, harvesting, processing, observing, and experiencing plants in their garden environment. These landscapes are called therapeutic gardens. They are special landscapes intended to provide an outdoor location for horticultural therapy. A garden setting can increase the horticultural therapy benefits realized by patients in clinical settings as a natural outcome of expanded opportunities for programmed and prescribed activities. The number and quality of horticultural therapy gardens have increased as their unique benefits are identified. In hospitals, hospices, clinics, residential care institutions, and other health care settings, the therapeutic garden is rapidly becoming a place of healing for patients, visitors, and staff.

THE HORTICULTURAL THERAPY GARDEN

A purposeful horticultural therapy garden is an outdoor setting for horticultural therapy, which provides distinct opportunities for the therapist to shape the modality and for the patient to experience a variety of stimuli drawing from the richness of garden environments. As a specialized landscape, the horticultural therapy garden may include many different facilities and activities within its boundaries. However, the common goal of improving the physical, intellectual, emotional, and spiritual well-being of garden participants through horticultural activities and experiences encourages some common features.

The purpose of any therapeutic garden is to maximize the number, quality, and intensity of interactions with plant materials in the garden

287

landscape. The benefits of such interactions with plants can be realized through active participation in the horticultural tasks of the garden or through passive appreciation of garden spaces and features. These gardens supporting horticultural therapy goals and objectives often are constructed and maintained to use plants to achieve one or more of the four following goals:

1. **Sustain clinical therapy** by providing large and small spaces, group and individual places, and in-ground and container opportunities within the garden used in an ongoing horticultural therapy program
2. **Empower individuals by generating** new confidence, new skills, and new understanding through encounters in the garden
3. **Emphasize the restorative qualities** that permit the garden to be a place of respite, recovery, and simple enjoyment
4. **Provide a landscape amenity** that can be used by staff and visitors as a place of beauty

Design factors in gardens and outdoor spaces can accentuate a person's occupation or preoccupation with plant and plant-based experiences. The factors most likely to influence this concentration are those which are common to many such gardens. Table 11.1 presents descriptive characteristics of horticultural therapy landscapes that were adopted by the Board of Directors of the American Horticultural Therapy Association. The list identifies the shared traits of landscapes serving different populations in a variety of programs, climates, and materials.

A systematic schedule of garden events increases use and recognition of a therapeutic garden as a site intended for casual as well as formal participation. Programmed activities in a horticultural therapy garden promote and encourage a variety of experiences, including

1. individual, horticulturally oriented tasks employing the plant materials and the living environment of the garden clinical therapy;
2. group activities and opportunities for socialization during horticultural activities, events, and classes; and
3. the demonstration of horticultural practices and new sources of horticultural knowledge or expertise to patients, staff, and visitors.

UNIVERSAL DESIGN

At one time, outdoor horticultural therapy settings were designed to meet the specific abilities and disabilities of a relatively limited resident or

TABLE 11.1. Therapeutic Garden Characteristics

---

1. **Features are modified specifically to improve accessibility to plants and gardening techniques.** A variety of features and elements within the garden enable each person's positive interactions and experiences with plants, with plant characteristics, and with the ability to engage in garden activities regardless of ability or disability.

2. **Activities are scheduled and programmed.** A series of events, clinical sessions, and/or horticultural activities is scheduled to invite, welcome, and occupy people in the garden.

3. **Perimeters are well-defined.** The boundaries of the garden are clearly marked and the spaces within the garden clearly defined for patients, students, and visitors.

4. **There is a profusion of plant and people interactions.** Garden users consciously experience healthy, vigorous plants as the predominate setting and as the focus of the garden, its activities, and its spaces.

5. **Conditions are comfortable and supportive.** Plants, maintenance practice, chemicals, materials, lighting, and other conditions in the therapeutic garden are managed and maintained to eliminate hazards to the well-being of patients, visitors, and staff.

6. **Universal design principles are used.** The garden is usable and enjoyable for people with all levels of abilities and disabilities.

7. **The garden is a recognizable place.** The garden or portions of it can be unique and recognizable environments easily distinguished from surrounding context.

---

visitor population. These gardens presented many barriers to users with other disabilities. Today, a therapeutic garden is most often intended to enable and to serve people having a wide variety of abilities and disabilities. Such gardens are designed using the principles of universal design, which enable all garden visitors and users to share a common garden experience and to enjoy the garden at their own pace, on their own time, in their own way. Table 11.2 enumerates the principles expressed by design educators and professionals participating in the Universal Design Education Project of the Adaptive Environments Center in Boston, Massachusetts, and developed by the Center for Universal Design in the School of Design at North Carolina State University.

TABLE 11.2. Universal Design Principles

---

1. **Equitable use.** The design does not disadvantage or stigmatize any group of users. *Do all potential users have access to all features of the design?*

2. **Flexibility in use.** The design accommodates a wide range of individual preferences and abilities. *Could the design be used safely with a single fist or elbow or with the aid of mobility or other assistive devices?*

3. **Simple, intuitive use.** Use of the design is easy to understand, regardless of the user's experience, knowledge, language skills, or current concentration level. *Could the design be used safely and effectively by a novice or newcomer without instructions?*

4. **Perceptible information.** The design communicates necessary information effectively to the user, regardless of ambient conditions or the user's sensory abilities. *Could the design be used safely and effectively with eyes closed? Ears plugged?*

5. **Tolerance for error.** The design minimizes hazards and adverse consequences of accidental or unintended actions. *Does the user have the ability to reverse any action?*

6. **Appropriate and anticipated physical effort.** The design can be used comfortably and consistently by individuals with a range of abilities with an expected level of fatigue appropriate to the activity. *Is the level of effort needed to safely and effectively use the design predictable, appropriate, and consistent with the context, environment, and user expectations?*

7. **Size and space for approach and use.** Appropriate size and space is provided for approach, reach, manipulation, and use, regardless of the user's body size, posture, or mobility. *Can the design be used easily and effectively by a small wheelchair user as well as by a tall ambulatory person?*

FACTORS IN DESIGN AND USE

A horticultural therapy garden accentuates plants as the focus of activities and experience. Nevertheless, the garden must remain a functional, practical, and safe landscape setting in which people of all abilities and disabilities can realize therapeutic benefits. Many factors influence the eventual usefulness and the continuing potential of the designed garden to provoke this horticultural occupation and preoccupation in patients, staff, and visitors. Many of these factors are interrelated or profoundly interact with clinical goals and objectives. How can this information be developed and introduced in the design process?

SITE ANALYSIS

Garden design efforts will be most effective if the complexity of these relationships is examined and identified at the beginning of the project. This stage, often called the **site analysis phase**, thoroughly studies the proposed program of activities and the existing and possible characteristics of the site. A team including, at a minimum, the design professional, the administrator, and the clinical staff will be most efficient in this study. Including patients in the deliberations where possible provides important insights and understandings.

The site analysis can result in a great deal of information to track and evaluate. Putting that information in a chart form can help to keep this task manageable. Plotting horticultural therapy objectives against site characteristics as a matrix helps to list and to visualize programmatic objectives as they are and will be influenced by the site. A sample of a matrix used to coordinate interrelated factors is illustrated in Table 11.3.

The issues examined in the site analysis process include (1) defining goals and objectives; (2) site selection; (3) site inventory; (4) site assessment; and (5) proposed design values and attributes.

GOALS AND OBJECTIVES

It is best to bring administrative and clinical objectives into the design-discussions steps in the planning and design process. This allows clinical objectives, institutional goals, and potentially serious site factors to be identified, discussed, and included in considerations prior to the design of the garden. The purpose of this effort is to develop a conscious understanding of the proposed uses for the garden and of the site's native environmental assets and liabilities. Despite common garden features and

TABLE 11.3. Goals Assessment Matrix Format

Proposed Facilities	Related Site Factors						
	Clay Soil	Steep Slopes	Bldg. Shade	Tree Shade	Staff Access	Staff View	Adjac. Streets
Independent Use		Yes*			Yes*	Yes	Yes
Clear Path System	Maybe	Yes*	Maybe		Yes*	Yes	Yes
Ground-Level Beds	Maybe	Yes*	Maybe	Yes	Yes	Yes	
Raised Beds	Yes	Yes	Maybe	Yes	Maybe	Maybe	
Large Containers		Yes	Maybe	Yes		Yes	
Small Pots			Maybe	Yes*			
Central Work Space		Yes	Maybe	Yes*	Maybe	Maybe	Yes
Cookout Area		Yes	Maybe	Yes	Yes	Yes*	Yes

therapeutic goals, most horticultural therapy programs will identify some different characteristics, uses and activities as goals and objectives for their gardens.

Prior to designing or constructing a garden, it is important that the clinical staff and administration consciously determine the clinical goals and objectives the garden will support. Objectives might include some or all of the following:

- Individual client responsibility for garden plots
- Maximizing seasonal difference in program tasks and activities
- Cultivating craft materials for the horticultural therapy program
- Production and sales of decorative or edible plant materials by the institution or by the clients
- Developing a "demonstration" garden to identify garden techniques, materials, and relationships
- Serving as models of excellence in landscape design, and providing job training or a sheltered workshop setting

• Providing a place of refuge, beauty, and solitude for patients, staff, and visitors

Objectives can further vary according to emphasis on whether the horticultural therapy program is to generate income for the institution, on whether plant materials will be permanent or perennial, and if the garden will be used year-round.

Site Selection

Site selection is a critical step in the development of any garden. Even when a horticultural therapy garden has been located primarily because of its proximity to an entrance, a clinical center, or to another indoor facility such as a greenhouse or commons room, the site will influence every factor in the garden. An ideal garden site is selected only after it has been determined that the location has excellent soil, optimum solar orientation, a mild microclimate conducive to the prospering of both plants and people, relatively level topography, and readily available utilities such as water and electricity. Such optimal conditions are rarely available to the clinical gardener. Instead, the site for the garden is often bounded by tall buildings, exposed to extremes of heat and cold, and has soils that are derived from excavations elsewhere on the property. Most garden sites must be designed and developed with special care to overcome their negative factors in use and design.

Site Inventory

Each site feature capable of influencing eventual design can be identified and located in a process called **site inventory**. It is important that the inventory determine precisely what factors of the chosen site are likely to affect the design or use of the garden. Typical climate considerations include regional weather patterns, local weather patterns and extremes, seasonal precipitation, and on-site sun and shade patterns. Soil type, soil fertility and/or structure, geologic features and stability, hydrology and stream influences, and topography and slope factors are crucial factors in accessibility, plant health, site selection, and structural design. Particularly interesting vegetation, including existing large trees and shrubs and on-site patterns of cultivated and volunteer vegetation, provide clues to existing conditions affecting the potential success of the garden. Existing buildings, structures, paths, parking, and other site improvements can directly affect the development of the site.

A complete inventory of site factors cannot guarantee a good design, but it does encourage knowledgeable decision making by thoroughly examining the assets and problems associated with the site and the planned horticultural, social, developmental, and economic uses.

Site Assessment

A **site assessment** is a procedure for evaluating the relative importance of each potential site inventory factor in the eventual garden design. For instance, if universal access and ease of movement throughout the garden are objectives generated for many clients with mobility difficulties, the importance of suitable topography and slope becomes increasingly critical to an economic design and construction. Degree of slope will further increase in importance if the soil is not stable, since erosion potential increases when unstable soil is cultivated on steep slopes. An assessment is used to define the most important or critical interactions of program and site.

Although separate tasks with different information, both the site inventory and the site assessment can be complex and highly detailed or quite simple. While it is possible to keep all of this information in one person's memory, communication is greatly improved if written formats such as Table 11.3 are used to share findings. This simple tool for handling the multiple variables encountered in the steps of garden planning is called a matrix.

A **matrix** is a type of chart used for recording interrelated information. In a matrix, rows of one type of variable are graphed against a vertical column of different but potentially related factors. A matrix format makes the connections between the site inventory and the goals and objectives for the garden evident at a glance. It records the answers to the question "Will this factor affect a specific use or therapeutic garden facility on this site?"

A further level of evaluation in a matrix can be added by using symbols to indicate the degree of significance for problematic or highly supportive characteristics. In such instances, symbols such as an asterisk (*) or a filled circle (●) are often used to indicate a high degree of significance or influence between proposed uses and existing site factors.

PROPOSED DESIGN VALUES AND ATTRIBUTES

The garden will be expected to sustain the activities and movement patterns needed for horticultural therapy through conscious design of materials specified for paths, walls, and other structures; of appropriate spaces; of regionally adapted structural techniques; and of thoughtful low-maintenance plant selections. Each of these attributes can be recognized as an important component of spatial quality, representing horticultural therapy values in horticultural therapy gardens around the country. Values and attributes that ordinarily support clinical programming and clinical activity in garden spatial arrangements are general statements of effects and qualities sought in the garden. These can include any of those listed in Tables 11.1 and 11.2 and usually incorporate the following attributes:

1. Masses of vegetation, which increase the horticultural dominance of plants within the garden.
2. A simple and legible organization of spaces and paths, which permit the visitor to understand the garden and to navigate within it easily.
3. Design and construction features that improve or increase the comfort, safety, and security of the garden and its convenient use to the care providers, patients, and visitors in the garden.
4. Paved, nonslip paths and gathering spaces surfaced with colorful, nonglare, textured, but not rough or irregular, surface materials.

A horticultural therapy garden is the result of careful planning and study. The time invested in developing a good, accurate, and sensible statement of goals and objectives, in conducting an inventory and assessment of on-site factors, and in developing a clinical consensus about the values and characteristics of the proposed garden will be well-invested. These efforts help to bring about a horticultural therapy landscape that directly relates to the proposed horticultural therapy activities.

DESIGN AND USE OF SPACE

The early studies and subsequent discussions between the administrators, staff, and potential users of the garden will address the four central factors in therapeutic garden design: (1) clinical objectives for individuals to realize as benefits from the garden activity; (2) experiential objectives describing the sensory and activity options garden participants will encounter in the garden; (3) maintenance objectives addressing the budgetary and staff support anticipated for the garden and its ongoing activities; and (4) spatial and material characteristics that constitute the aesthetic, the accessibility, the legibility and the organizational components of the garden experience. As the design process continues, the designer and the horticultural therapy representative must collaborate to develop increasingly detailed proposals for specific elements in the garden that meet these four objectives.

The designer will engage in a site analysis at a professional level, using information provided by the therapy staff and administration. It is important for the horticultural therapist to

- communicate the relationship between clinical objectives of horticultural therapy and the methods, materials, and activities for achieving these benefits in the garden environment.
- provide a succinct but complete list of proposed garden events and activities to the designer early in the process.

When professional design services are engaged to design the horticultural therapy garden, the designers will rely on the horticultural therapist for information about the patients and the garden facility, features, and forms anticipated to serve the horticultural therapy program. Few designers have expertise in horticultural therapy and even those with experience in therapeutic landscapes welcome an understanding of how a particular clinical program will use the garden designed for them. As a working team member, the clinician becomes a professional "filter," ensuring that horticultural therapy values are represented in and strengthened by the proposed landscape and its experiential, maintenance, spatial, and material characteristics.

Clinical Objectives

A horticultural therapy garden will be designed with specific clinical outcomes in mind for the patients who use it. The horticultural therapy staff working in conjunction with related therapy staff and administrators must outline the types of activities and the levels of participation expected from the users of the garden. If the garden were only intended to be a place of refuge, shelter, and solitude, little active participation in horticultural activities would be anticipated. On the other hand, if the purpose of the garden is to engage the patients in the actual plant selection, planting, and maintenance of the garden, the level of participation will be quite high. If the garden is to adequately reflect and serve the horticultural therapy program in general, the designers and the therapy staff must share a common idea about the activities and purposes to which the garden will be put. Because each element in the garden derives from the stated goals and objectives for the horticultural therapy program and its garden, and from the site inventory and assessment, the clinical, experiential, and maintenance objectives for the proposed garden are essential to include in early considerations.

Experiential Objectives

A horticultural therapy garden may be a place for hard work and satisfaction or may be a place in which patients, visitors, and staff are encouraged to experience their senses, to exploit the sensations that one can find in a garden. Horticultural therapy staff should clearly state a level of sensations that are proposed for their garden. If the designer is to maximize the senses of smell, touch, hearing, and perhaps even taste, then the staff must be adamant in seeking these qualities in each and every aspect of the design proposal. If the staff is firm in their belief that edible plants in the garden should be discouraged, then they must clearly advise the designer to avoid plant materials and plant relationships that present fruits and other potential edible plant parts in enticing ways. The staff could ensure, for instance, that no berry-like fruits would be included in the garden.

Maintenance Objectives

Every horticultural therapy garden should be constructed to meet the budgetary and maintenance objectives of the institution, its administrators, and the horticulture therapy program and staff. The single most common error made in the investigative phases of program development and site analysis for horticultural therapy gardens is the underestimation of professional maintenance needs. Most programs anticipate that their patient users will contribute to the maintenance pool, but find that these patients are much less capable of providing the complete range of maintenance needed to maintain the garden. This usually means that after the garden is constructed and budget solidified, the staff must assume more responsibility for maintenance or must find available budgetary excesses to hire maintenance staff to fulfill that role. If the garden is designed for its maintenance objectives from the outset, it is more likely to be maintained in an economical manner. Building for low maintenance may also increase the initial construction budget. However, this increase may be offset by years of reduced maintenance costs.

Spatial and Material Characteristics

Horticultural therapy gardens vary in their appearance. Some are extremely sophisticated, controlled, elaborately designed installations, while others are ad hoc collections of found containers, donated plants, and the fruits of many a seed packet with the most rudimentarily designed garden elements. Both extremes of the horticultural therapy garden are valid examples of gardens, of effective spaces, and of spatial and material decision making. The spatial and the material qualities of the garden will be based on the design team's understanding that factors and relationships can increase or limit the participation of individuals with disabilities in the garden experience. The team achieves this understanding by seeking answers to questions about the proposed garden and its visitors.

1. Who will use the garden?
2. Why will they use it?
3. How will the horticultural therapy staff interact with the garden?
4. How will patients, visitors, or staff discover and experience the garden?
5. Will some modes of experience or sensing in the garden be less important than others?
6. How will the garden be maintained?
7. Who will actually maintain the garden?
8. How many spaces, places, or types of activities will be needed in the garden?
9. What materials will be used in the construction?
10. What types of plants will be grown in the garden?

From the outset of design efforts, it is important for the horticultural therapy staff and administrators to determine and express their concept of the spatial qualities and the materials that a patient or visitor will encounter within the garden. Some decisions, such as spatial organization or size, will determine the ease with which first-time visitors can understand and move throughout the garden. Material selections and choices will have a strong impact on the final cost of construction but will also be a strong influence on the level of maintenance required, ease and comfort of use, practicality of independent use, patient cost and change, accessibility, and other qualitative components of the garden experience.

While there are few rules for guiding the effective spatial development of a garden, staff and administration will recognize that the orientation, way-finding, and usability characteristics of the garden will be profoundly affected by the number and nature of the spaces within the garden and by the path of travel or circulation route that connects them. An objective to keep in mind is that of optimizing variation and diversity within the garden by providing linear path systems that loop back to the origination point, since circling through the garden is more likely to maximize the experiential, spatial, and clinical experiences of garden visitors.

ADAPTING GARDEN SPACE AND FACILITIES

The horticultural therapy garden acts as a catalyst in the healing process. People having a broad range of abilities and disabilities benefit from a horticultural therapy garden. Because of this potential for so many different outcomes, an effective therapeutic garden is designed to be flexible and to be diverse in its horticultural opportunities. This flexibility is reflected in plant-growing methods that enable people of all abilities to engage in horticultural activities. Two general types of planting situations are employed in therapeutic gardens: (1) traditional planting and (2) container planting. Rarely is a garden restricted to only one type of planting situation.

Traditional planting, in which the plants are grown in a large soil mass, such as that of a prepared, traditional, in-ground planting area or a raised planting bed, relies on garden facilities. Table 11.4 describes the typical characteristics of common planting situations. Traditional garden practices take advantage of native soils or of introduced soils specially formulated for the region. As therapeutic gardens, these represent a range from the most basic to the most elaborate examples of therapeutic landscape planting in a real garden setting. Container horticulture is a viable alternative to ground-level or raised-bed gardening. Container gardening plant care is similar to the type of care normally given house plants. By using containers, any negative aspects of the native soil can be avoided because containers allow a complete and permanent change from the physical and chemical properties of the native, on-site soil.

Container planting provides horticultural opportunities in which plants are grown in a variety of purchased or recycled pots, containers, window boxes, and tabletop planters.

TRADITIONAL PLANTING

The horticultural therapy garden usually employs a mix of traditional planting situations including at-grade or **ground-level** beds and borders; **slightly raised** beds or borders, which may be encased with some stable material as a kind of curb or edging; and **raised beds** or borders, which can be completely supported by retaining walls made of a variety of materials. Table 11.5 compares characteristics of planting bed heights.

TABLE 11.4. Typical Characteristics of Planting Areas

	GROUND-LEVEL PLANTING BEDS	12+" RAISED PLANTERS	VERY LARGE CONTAINERS	LARGE POTS AND CONTAINERS	SMALL POTS AND CONTAINERS
TIME FRAME	Fairly immediate. Dig, prepare soil, and plant (soil + plants = garden).	Several months min. Raised wall structures must be designed, built, and soil filled before planting.	Immediate (container + soil mix + plants = container garden).	Immediate (container + soil mix + plants = container garden).	Immediate (pot + soil mix + plant = garden).
COSTS	Inexpensive. Primarily horticultural supply costs. Requires no additional structures.	Expensive. Materials, labor, and maintenance costs for walls and related structures in addition to horticultural supplies.	Containers are costly in large sizes but can last many years. Use of various construction products and recycled nonhorticultural containers can reduce costs.	Moderate cost for new containers but can reuse old containers.	Relatively inexpensive. Can use many different types of containers to hold soil medium.
SOIL	Existing soil. May need modifications. Natural hydrology.	Introduced mix. Can improve on natural soil conditions and can control soil to meet special plant needs. Soil mix characteristics are important to water retention or loss.			
WATER NEEDS	Normal for region. Mulch helps soil retain moisture. Irrigation system desirable in some regions.	Seasonal treatment common to soil type and region. Mulch helps keep soil uniformly moist. Irrigation system desirable in many regions.	Hand water by staff or patients. Drip system is desirable. Care must be taken to avoid over or under watering. Sensitive to summer heat and drying and to winter cold.		
CLIMATE	Tolerant of summer heat/winter cold.	Soil can dry easily. Summer heat/winter cold sensitive.	Tolerates some winter cold.	Sensitive to summer heat and to winter cold.	
	Surface mulch protects from sudden temperature changes. Sun/shade affects plant selections.	Ease in moving containers around improves flexibility in meeting varying sun/shade requirements. Porous materials may be damaged by freezing temperature.			

	1	2	3	4	5
DEFINED SPACE	Plants and edging mark perimeter of spaces. Nearby shrubs and trees can enclose.	Retaining wall and masses enclose space. Structures provide clear definition of garden spaces.	A few containers can mark garden space. Can look like permanent arrangement/ planting.	Many used for larger plants can mark garden space. Impromptu arrangement/temporary exhibit encloses space.	Too small to mark garden space without supporting structure. Impromptu arrangement/temporary exhibit
ACCESSIBILITY	Difficult from seated position. Requires mobility and strength. Preferred by some. Adapted tools can improve reach.	Greatly improved. Preferred for mobility or stamina limitations. *Soil and plants can be within reach and near eye-level of seated gardener.*	Container height variations available.	Best with support/ structure. Can be set on walls, sills, tables, or shelves of various heights for flexibility.	Good but requires support/structure. Easily transported to tables, trays and shelves at any level. Good for individual involvement.
SPECIAL OPPORTUNITIES	Can mix permanent and seasonal planting. Flexible—can expand and reduce planting areas as needed.	Height of raised beds can vary (18″-36″) to suit special needs/ preferences. Seating, shade, and other structures can be built as part of retaining walls.	Containers can be visually attractive design elements in the garden landscape.	Groups can be re-arranged easily. Containers can be visually attractive design elements in the garden landscape.	Individual potted plants can leave with a patient.
VARIATIONS	Bed shape, length, and width can vary. Pattern of beds and planting rows can add interest. Soil can easily be raised 6″-12″ height.	Raised structure shape, length and width can vary. Wide range of materials possible (wood, stone, brick, concrete, etc.)	Container variations, such as purchased, hand-made, "found" or recycled containers, half-barrels, drain tiles, flue tiles, cannisters or paper cups. Hanging planters, column and vertical planters, or various sizes possible.		
LIMITATIONS	Existing soil may be poor, rocky, shallow or infertile. Need paths into wide or deep beds.	Width of raised bed limited to normal reach (arm length minus wall width).	Weight of filled container may exceed structural capacity of roofs/decks. Frequent watering needs.	More frequent watering required. Some type of structure/support for containers is best.	More frequent watering required. Small pots can tip easily. More frequent transplanting needed.

TABLE 11.5. Comparison of Attributes Associated with Planting-Bed Heights

ATTRI-BUTE	CULTIVATION	CLIMATE	ACCESS
GROUND-LEVEL BED	**Local soil dominant.** Fertility depends on soil composition. Heavy clays, rocky, shallow soils, or wet conditions can make cultivation difficult for people who use assistive device or who lack stamina or upper body strength. Can modify existing soils over time by adding needed organic matter, nutrients, etc. as standard practice.	**Regional climate patterns prevail.** Row covers and other structures warm soil in spring. Can rely on regional rainfall or choose to install irrigation system.	**Accessibility limited.** Soil and plants are below eye-level. Strength and agility needed to garden. Changes in topography further decrease access. Improve accessibility with path design and surface material choice. Preferred garden level by some despite access difficulties. Most commonly underestimated need in horticultural therapy gardens.
BEDS RAISED 6" TO 12" HIGH	**Local soil can be amended or partially replaced.** Slightly raised bed permits the addition of additives to substantially modify or replace troublesome local soils. Soil modification efforts can be realized immediately.	**Regional climate patterns can be modified slightly.** Raised soil level warm earlier in season. Row covers and other structures warm soil in spring. Can rely on regional rainfall or choose to install irrigation system.	**Accessibility improved for some.** Increase in height and soil improvements reduce strength and agility needed to garden. Curbing or edging of raised beds improves way-finding. Paths improve accessibility with appropriate surface material choice. Narrow path width reduces access for people who use assistive devices.
BEDS RAISED 12" OR MORE	**Local soil can be completely replaced.** Raised bed permit the addition of new ideal soil mix to replace troublesome local soils. Improved soil qualities can be realized immediately. Construction costs high and directly affected by material choices and installation methods. High degree of construction skill/experience required. Professional construction recommended.	**Regional climate patterns modified significantly.** Raised soils warm earlier in season but can be prone to overheat in summer and freeze in winter. Row covers and other structures warm soil in spring. Could rely on regional rainfall but installation of an irrigation system is recommended.	**Accessibility improved for most people.** Soil and plants at eye-level. Height and optimum soil improvements reduce strength and agility needed to garden. Varying wall/soil levels increase choices in the garden. Changes in topography can be incorporated in terrace system. Retaining walls of raised beds improve way-finding. Reach across walls reduces working garden area. Accessibility improved with wide path and surface choice between retaining walls.

COMFORT	SPACE	SUMMARY
Comfort elements must be specially constructed. Permanent structures such as benches, supports, windscreens, and shade structures must be added and may be the only structures in garden. Temporary structures must be stored and moved around garden site.	**Spatially relates to surrounding landscape.** Permanent shrub, tree, and plant masses possible to exploit. Can include large area and expand as needed if land is available. Difficult to completely define and enclose associated group meeting or gathering space without above eye-level plantings or structures.	**Assets include convenience and economy.** Soil preparation is relatively simple. Inexpensive option. Little or no construction is required so few material or construction costs are incurred.
Comfort elements must be specially constructed. Permanent structures such as benches, supports, windscreens, and shade structures must be added but may be integrated with raised-bed curb or edging. Temporary structures must be stored and moved around garden site.	**Some spatial and visual separation from surrounding landscape.** Permanent shrub, tree, and plant masses less likely to be included. Can expand planting areas with minimal construction. Easier to define and enclose associated meeting or gathering spaces with slightly raised plantings and edging structures.	**Assets include simplified soil modification, improved accessibility, and economy.** More successful horticulture in areas of poor or shallow soils. Improves drainage and encourages earlier soil warming in spring. Cost of construction related to material choices and installation methods. Modest construction skills required. Raised soil level improves accessibility.
Comfort elements can be incorporated in structures. Permanent structures such as benches, windscreens, supports, and shade structures can be readily integrated with raised-bed wall construction. Wider retaining walls can be constructed as seating walls. Temporary structures must be stored and moved around garden site.	**Distinct spatial and visual separation from context of surrounding landscape.** Permanent adjacent or existing shrub, tree, and plant masses less likely to be included. Can only expand planting areas with new raised bed construction. Walls clearly define and enclose associated meeting or gathering spaces. Patterns of raised beds mixed with ground-level beds provide easily recognizable spatial organization.	**Assets include beauty, ideal soil conditions, and improved accessibility.** Retaining walls can be attractive elements in the landscape. Can direct attention to plant materials, landforms, and structural relationships. Successful horticulture in poor soil areas because ideal soil and drainage can be created. Soil mass tends to warm earlier in spring and to extend growing season in fall. Raised soil level improves accessibility.

Ground-Level Bed Plantings

Ground-level beds are planting areas at the existing soil grade. They often are thought to be the least convenient and the least enabling of growing areas. The need for in-ground beds can be significantly underestimated and the opportunities for ground-level gardening underrepresented. Because ground-level beds are inexpensive, versatile, and enduring methods of gardening, it is the facility most likely to be added in further expansions of existing garden programs.

The typical vegetable garden we first knew as children was composed of ground-level beds. This garden is traditionally created by removing turf and weeds from an area, digging the soil, and possibly incorporating a soil additive such as compost to improve the soil prior to planting or seeding. Paving the path areas around the ground level beds improves accessibility for people who use wheel chairs and other assistive devices. Since the soils throughout a region will be similar, as will the climate, rainfall, pests, native vegetation, and environmental factors, the growing conditions in a ground-level bed are determined by the native environment.

Assets of ground-level therapeutic gardens are the following:

1. The low cost of garden construction without structures, since no professional construction experience other than that needed for the paved or surfaced walks within the garden is needed.
2. The speed with which the garden can be constructed and developed, since no lengthy delays for design or structures are encountered.
3. The ability to stress a strong spatial relationship to context in which nearby woods, shrubs, and landscape plantings appear to be part of the garden.
4. The ease of garden expansion where space is available.

Problems associated with ground-level beds include the following:

1. All horticultural activities involving cultivation, seeding, planting, and maintenance must take place at ground level, causing some difficulties for people using wheel chairs, people with joint impairments, or people with other mobility impairments who cannot reach down to the soil eighteen to twenty inches below a chair's seat.
2. Plants, flowers, fruits, and foliage can be well below or well above eye level.
3. Significant strength and agility is required to garden and to maintain the bed.
4. Too few ground-level beds are available because therapeutic landscape planners are prone to underestimate the need for ground-level

beds in favor of an increase in raised beds during the early programming and analysis phases.

5. Negative characteristics of the native soils can dominate horticultural conditions and practices and can limit horticultural activities and selections.

Ground-Level Beds Raised Six to Twelve Inches

Soil can be raised above the original ground level an additional six to twelve inches. The added fill material results in a slightly raised bed with minimal structural edging to retain it. This type of planting situation, described best in *Square Foot Gardening* (Rodale Press, 1981), is valued because no engineered structure or wall is required to retain the soil. A brick, stone, concrete curb, or timber edge is used to retain the garden soil at a chosen height above the surrounding soil. This is essentially a modified ground-level bed recommended for improved accessibility and for use in areas where heavy clays and other problem soil conditions can be reduced or eliminated by the increase in soil depth. Improvements from soil modifications of this kind can be realized immediately.

The assets of slightly raised beds include the following:

1. Low cost of construction, since no professional construction experience other than that needed for the edging is needed.
2. Spatial and visual connections to the surrounding landscape despite the increased separation of the raised soil surface and the visual quality of the edging.
3. Desired expansions are a simple project of defining and enclosing the garden.
4. Improved accessibility and easy cultivation for some people with disabilities.
5. Low retaining structures, which are an orientation device for people with a vision impairment.
6. The slight elevation of soil, which can cause earlier spring warming and extend the growing season.

Problems associated with slightly raised soils include the following:

1. The cost of soil additives or the new soil used in raising soil levels purchased and transported from another location can increase construction costs.
2. The vertical edge holding the soil in place forms a six- to twelve-inch obstruction well below eye level, which can be a tripping hazard unless clearly marked or highly visible.

3. The raised-bed retaining structures form a barrier on either side of the path and can potentially limit maneuvering space in narrow areas.

Beds Raised Twelve Inches or More

Raised beds are soil masses contained within retaining walls twelve inches or more in height. The common heights for such walls are sixteen to eighteen inches, twenty to twenty-four inches, or twenty-eight to thirty-two inches, with some raised-bed planting areas being as high as thirty-six or forty-two inches. Generally, it is most appropriate to provide a variety of raised bed heights in any single garden. Since the height of a raised bed affects its accessibility, a person who uses a wheelchair may prefer an eighteen- or twenty-four-inch wall height for its improved reach. An individual who prefers to stand while gardening may find the thirty-six- or forty-two-inch raised-bed areas more comfortable because he or she can lean against the garden walls for support. A diversity of needs can be met by constructing raised beds at different heights.

The retaining walls of raised beds are also versatile structures in the horticultural therapy garden. Because the width of a retaining wall can form a wide seat, the wall itself can serve as bench seating within the garden. Wood or metal benches can be attached to the walls. The walls can be indented to form seating or gardening alcoves useful for people who use wheelchairs. As a direct result of the color, texture, surface finish, shape, durability, and uniformity of materials, the physical qualities of a retaining wall determine its comfort as a bench and its convenience, comfort, and accessibility as a raised bed. Common construction materials include brick, stone, concrete, concrete block, and various forms of wood or timber.

The assets of raised beds include the following:

1. Retaining walls can be used to incorporate changes in topography into, for example, a terrace-like design in which the uphill side of the garden meets the grade of the land above while the downhill side is held in place by a retaining wall forming a raised bed near an accessible walk or gathering area.
2. An optimum soil mix can be introduced to the raised bed to further reduce the strength and agility needed to garden.
3. The walls of the raised beds can provide seating in the garden and can clearly define associated meeting and gathering spaces.

4. Opportunities arise to incorporate changes in topography into terrace-like spaces in which a retaining wall forms a raised bed on the downhill side.
5. The walls of raised beds provide visual separation from the surrounding landscape which can reduce the impacts of nonsupportive adjacent activities.
6. The wall structures can be elements of beauty, bringing designed forms and materials into the garden.

The problems associated with raised beds include the following:

1. Designing, constructing, and maintaining raised beds and retaining walls comes at a high cost.
2. There is a need for professional design and construction to ensure structural integrity and stability because raised beds and their structural retaining walls are exposed to rain, frosts, and other weather effects.
3. The depth of the retaining wall reduces the length of the gardener's reach into the bed.
4. Wood products, though relatively economical as construction materials, require increased levels of regular and seasonal maintenance to eliminate excessive weathering and splintering.

CONTAINER PLANTING

Container gardens are composed of collections of pots, decorative containers, window boxes, tabletop planters, and other purchased, found, or contrived methods of holding small amounts of soil in place for plant growth. Containers can vary from highly elaborate, decorative, and expensive ceramic pots to simple plastic bags with soil, scaled and cut to permit the insertion of small plants. There is no fundamental definition of what a container is, but rather what it does. A container permits the practice of horticulture in areas without open-soil opportunities.

In many horticultural therapy gardens, plants cannot be grown in open soil. The garden may be on a rooftop, in an almost completely paved courtyard area, or in a temporary location from which the plants and program will move to a new location within a foreseeable future. In such locations, a container garden may be the only option. Container gardens are also selected as a horticultural therapy garden type because of the intimate quality and the personal ownership they evoke in the garden.

Container gardens often provide a high degree of accessibility because containers can easily be raised, lowered, or sized to meet specific access needs.

Pots and Other Containers

Containers for gardening vary widely in expense, size, attractiveness, and durability–in fact, in every way imaginable. It is possible to build containers; to buy containers retail, wholesale, or by mail order; or to find suitable containers that are well-adapted for a use in a therapy program in food processing, laundry operations, and other commercial processes. Containers can be found to meet the budgets and the expectations of every program.

The critical factors in the selection of a container are cost, convenience, durability, and its ability to meet plant requirements. A special potting soil mix will be introduced to every pot. Lack of water is the greatest threat to success in outdoor container gardening. In regions with very hot, dry air throughout the growing season, it will be advisable to use an organic or polymer soil additive to help maintain even soil moisture between waterings. Pots and containers can benefit from an irrigation system in which small emitters or drip lines are placed within each pot. Table 11.6 describes the relative characteristics of containers.

Very Large Pots and Containers

Very large pots and containers are most commonly purchased. While costs vary, they are usually expensive but quite durable if the material selected is adapted to the regional climate of the garden. Containers over fourteen-inch diameter are for heavy soil and are therefore inconvenient to move and once filled with soil, are unlikely to be moved to new locations at all. It is therefore important to evaluate the planter's materials and permanent location to ensure that light and rainfall characteristics will support plant life. The original siting for this size of container must be conducive to the well-being of the growing plants and of working patients, staff, and visitors.

There is no optimum size or most appropriate style of planter. Containers are commonly selected for their appearance, their particular size, and for the horticultural capabilities needed to support the continued growth of a particular plant or collection of plants. Since accessibility will be an issue, very large raised planters must be high enough to reach easily from a seated position (sixteen to twenty-four inches in height) or must be securely elevated on some type of structure or "foot," which raises the pot to within the reach of a seated person. Smaller containers can be lifted and carried to work spaces; very large containers cannot.

The effects of excessive rainfall during wet springs and summers can be minimized by ensuring that drainage through the container is positive and that no water collects in the container over long periods of time. The lack of water, however, can only be remedied by the application of water.

Large pots and containers are those which are large enough to support a "mature" plant but which still are possible for a single individual to move. Ten-inch through fourteen-inch-wide containers are within this range although they can be very heavy. These containers are likely to be the eventual location of young plants as they are potted up to increasingly larger sizes. Large containers can provide seasonal accents with brightly colored, intensely fragrant, or tactually interesting plant material as a changing display within the masses of plants in the garden.

Since a collection of large containers in a container garden usually requires some type of display or support structure to exhibit the growing plants and to provide them with adequate light and air, facilities associated with such container gardens will include purchased or constructed shelving, walls, ledges, benches, and work tables of all types. A container of this size must be lifted to a raised surface if it is to be cultivated by a person who uses a wheelchair or who is unable to stand or to kneel during the horticultural activity. If the plant is to be moved during horticultural activities, its total weight must be relatively light and a table or bench-height support must be conveniently located in the horticultural therapy garden. If a plant is to remain on display during horticultural therapy activities, it best to grow and display it in its permanent location within the reach of the patient.

Pots of this size are relatively inexpensive and are very convenient to use. Horticultural preferences vary; there is no hard and fast ideal material for containers. They may not be as durable as larger sizes, because they are often made of less durable materials. Trees and large shrubs cannot live in containers indefinitely of this size and must be potted into even larger sizes as they grow. Since binding the root mass encourages flowering in some species and almost completely eliminates it in others, the pot size requirements of each plant must be met. Containers of all sizes dry out more quickly than in-ground or raised beds, particularly if the container is made of a porous material with an unglazed or unsealed surface. Maintenance may be higher on smaller pots as the more restricted soil mass will become dry more frequently than pots holding a larger volume of soil.

Small pots and containers have a valued place in the horticultural therapy garden. A small pot or container can be easily picked up, held, carried, and manipulated by a person with little upper body strength. Containers can range from eight-ounce paper cups to ten-inch-diameter

TABLE 11.6. Materials for Constructing Planting Areas, Containers, and Walks

	QUARRY, FIELD, AND OTHER STONE	SMOOTH/EXPOSED AGG. CONCRETE	BRICK AND OTHER PRE-CAST UNITS
GROUND-LEVEL PLANTING BEDS	Low maintenance if placed properly. Color, size, and surface textures vary with type of stone and regional geology. Heavy weight, especially useful, durable, attractive, perceptible planting bed edge.	Very low maintenance if installed properly. Can be formed into slightly raised curb edge of planting beds. Glare problems with light tones. Colorants added to reduce glare and create a more "natural" look.	Low maintenance. Retains good looks and condition. Porous, unfired brick, tile, and clay units damaged by frosts. Units moderate to heavy weight. Curved forms possible. Colors and tones available. High installation cost.
	Use as permanent, at-grade edging or slightly raised curb marking between paths and planting areas/beds.		
12+" RAISED PLANTERS	Handsome, durable, low maintenance, and natural looking. Can be constructed with or without mortar. Color, size, and surface textures vary with type of stone and regional geology. Heavy weight difficult to manage easily. Professional construction recommended. Wall widths suitable for seating but reduce arm reach to garden areas. Available from many sources.	Durable, low maintenance if constructed properly. Curves, rectangles, and other shapes use wooden forms as molds for curving. Thinner walls improve reach to garden areas. Wider wall areas can be used as seating. Planting bed walls need metal bar reinforcement. Glare problems with light tones. Colorants reduce glare and create a more "natural" look.	Low maintenance walls retain original good looks and condition. 8+" wall width improves reach into garden areas. Wider walls can be built for seating where needed. Fired brick, tile, and clay units avoid winter frost damage. Curved forms constructed easily. Units can be laid in many patterns. Colors and tones available. High cost materials and construction. Ordinarily nontoxic
LARGE PERMANENT CONTAINERS	Not ordinarily used in constructing freestanding containers.	Durable, low maintenance. Curves, rectangles, surface decorations, and other shapes available. Thin walls and reduced container depth improve reach. Glare with light tones. Colorants reduce glare and create a "natural" look. Textural variations possible. Very heavy but "lightweight" models available. Plants overwinter in larger soil masses. Group to form mass.	Not ordinarily used in constructing containers.
WALKS AND OTHER PAVED AREAS	Uneven surface. Usually not accessible, even, slip resistant walk surface. Often difficult for people with mobility impairments. High installation cost. **Compacted crushed gravel** and other loose surfaces not recommended.	Stable, durable, accessible, even, slip-resistant surface for walks and paths for all levels of ability. Moderate installation cost. Low maintenance. Colorants added to reduce glare. **Pressed concrete** patterns attractive but uneven surface.	Attractive material with some surface irregularities. Many color selections. Contrast elements and patterns possible. High installation cost.

TREATED AND ROT-RESISTANT WOOD	ARTIFICIAL MATERIALS	CLAY FLUE LINERS AND DRAIN TILES
Medium-high maintenance reduces splintering. Lightweight, easily installed edge for beds. Weathers to a rough but attractive surface. Can rot easily and must be replaced frequently. Regional cost variations. Chemical preservatives not recommended.	**Recycled plastic "wood."** Durable, lightweight, attractive, material edging. Little maintenance. Little weatherizing. Color, texture and size choices. **Faux stone: Durable, attractive, uniform edging. Little maintenance. Color, texture, shape and size choices.**	Not ordinarily used in ground-level plantings. Can be used to retain clumps and rampant growing plants in restricted areas.
High maintenance retains good looks and reduces splinters. Thin walls improve reach to garden areas. Weathers to a natural, rough but attractive surface. Can build bench seats on wall units. Preconstructed units can be moved into place. Wood against soil decays and must be replaced regularly. Material costs vary widely. Chemically treated woods not recommended for growing plants used as food.	**Recycled plastic "wood."** Very durable, attractive, material for retaining walls/raised beds. Very low maintenance. No decay with soil contact. Choice of color, texture, and dimension. Seat possible to construct. **Faux stone:** Very durable, attractive, uniform material for retaining walls/raised beds. Little maintenance. Choice of color, texture, shape, size. Rough for seating.	Not ordinarily used in constructing raised planting beds.
High maintenance retains good looks and reduces splinters. Thin walls improve reach to garden areas. Weathers to a rough but attractive surface. Large preconstructed units can be moved into place. Wood can decay at soil contact. Costs vary regionally. Chemically treated wood can be allergenic or carcinogenic.	**Recycled plastic "wood."** Very durable, attractive, material for planters. Little maintenance. Weathering minimal. Choice of color and texture. **Faux stone:** Can be cast as low maintenance container as form of concrete planter.	Durable, attractive, low maintenance, small scale units. Pre-fabricated cylinders and rectangles (flue liners, drain tiles, and various tube forms). Thin walls and limited width, optimize reach. Heavy but moveable empty. Can be damaged by frosts. Can absorb heat in summers. Sizes vary. Culture as container plants. Group to form mass.
Common decking material. Uncommon paver. Splinters possible. Tends to decay at points of contact with soil. Some common wood preservatives thought to be allergenic or carcinogenic.	**Recycled plastic "wood."** Often slippery when wet. Uncommon as paver. **Faux stone** or "brick" surfaces more regular and even than natural stone if laid with care. **Impact-absorbing surfaces** available. Usually high cost.	Not suitable paved surface. Unglazed materials not recommended in regions subject to prolonged frosts.

exquisite works of the ceramic arts. Small container gardens can be contrived from found objects as simple as grapefruit juice cans or rolled newspaper cones. Plants may be sown and grown in small containers only until they require or are large enough to be moved to a larger container.

A wide variety of plants ranging from the most tolerant of tropical or "desert" plants to African violets, gesneriads, ivies, lilies, small shrubs and other small outdoor plants can be grown in very small containers. However, even the bean seed planted in a paper cup familiar to almost every kindergarten child must eventually grow out of doors or be planted in a larger container if it is to sustain growth.

The cost of small containers varies but is usually significantly lower than the cost associated with larger containers. While small containers are convenient, the conditions and the locations needed to support them are much more limiting. The containers themselves must be placed at a height that protects them from accidental bumping by passers by. A structure of some sort is needed to avoid mechanical damage to pot or plant. Table or bench structures, such as those found inside greenhouses, and various other structures that expose the plants to view and to needed sunlight and rain, can be constructed or purchased. Retaining walls and other structures within the garden can provide a temporary setting for small plants, however, small plants must be placed in protected locations protected from extremes of temperature, light, and air. The soil in small containers dries out extremely quickly in hot, dry climates or during summer drought periods and may have to be watered more than once a day. In more moist climates, the small containers can become waterlogged during wet seasons.

The great benefit of the small container is that it can be "owned" by the patient who cares for it. It can be moved to an interior room and returned to the garden. It can be brought into a classroom, meeting rooms, or outside meeting spaces of the garden for cultivation, display, and appreciation.

Window boxes, hanging plants, and other constructed containers are favored where the garden contains or is adjacent to fences, overhead structures, or buildings. Window boxes and other suspended containers can minimize or reduce the structural and architectural impacts of the building on the garden's image and environment. Cost will vary according to the materials selected, the quality, and the nature of the container purchased. The durability of hanging containers varies with the material. Window boxes and other wooden containers will undergo a fair degree of weathering and rot, but can be expected to survive and function for some years. Once placed and planted, it is extremely difficult to move hanging or suspended containers of moderate or larger size. However, they still offer opportunities for horticultural activities that may otherwise be lacking in regions or locations with restricted growing possibilities. Typically, container plants are grown in striking compositions of color, texture, and fragrance, taking advantage of the broadest range of horticultural variation and diversity.

The dimensions of selected containers are affected by (1) the available space, (2) the structural support that can be provided the containers, and (3) the stated objectives of the horticultural therapy program. As with other horticultural endeavors, basic horticultural requirements of the plant material must be met. The most significant of these requirements in container-grown plant materials is likely to be water, as plants in hanging or suspended containers often are more exposed to drying winds and baking sun.

Many **window boxes** and constructed containers fixed to vertical surfaces are made of wood or unglazed terra cotta. If the wood is left in its natural state, it will weather to a soft gray color, which brings more attention to the plant material. Painted or sealed wood, which must be treated regularly to retain its characteristics, can add color and personality to the garden scene. As with any wood product in the garden, care must be taken to minimize or avoid splinters and other problems associated with weathering wood. In some climates, terra cotta is susceptible to severe frost damage.

Window boxes, hanging plants, and other constructed containers are often shallow and so may have a larger surface-area ration than most other planting situations. The resulting increased moisture loss can cause them to dry even more quickly than average container plants. A drip watering system will enable the gardener to focus on the plants and to reduce watering as a constant activity during hot dry summers. On the other hand, if watering is just the activity that engages patients, staff, or visitors, a certain portion of these containers could remain unirrigated. Perennial plant material is rarely used in the window boxes or containers of this sort as the soil and plants freeze readily in the winter season.

Window boxes can be secured to the walls of buildings without suitable window sills. Window boxes are sometimes used in roof gardens where the structural elements in the roof are incapable of supporting a full range of plant growth. Distributing some of the weight of the garden soil to the walls of the building permits additional plants to be grown in the horticultural therapy roof garden.

Hanging planters and hanging pots can be moved outside in season and returned to an interior room or greenhouse as cold weather approaches. They can be planted with annuals for the growing season or support tropical plants grown year-round. Locate these containers in as shaded an area as they can tolerate with healthy growth. Plants needing sunny, exposed sites in ground-level beds are not as tolerant of extremely high light situations when grown in hanging containers. Hanging plant cultivation is utilized most where floor area is limited. Individual plants can be suspended from hooks, wood framing, or wires. Watering is made more convenient if plants can be raised and lowered from a seated position using a rope and pulley system.

Vertical planters are commonly constructed as a type of freestanding planter. A constructed vertical planter has one or both of the large front and back vertical surfaces open to the air and planted with young plants inserted through a plastic or mesh layer holding the soil in place. Irrigation distributed through the vertical soil mass helps to keep the entire soil mass moistened. Vertical planters are often composed of stacked wire-fronted "drawers" or boxes held in a bookcase-like unit that can be removed for maintenance. Although vertical containers are extremely heavy, they still must be anchored in place to avoid the toppling in high wind and other dangerous situations created by their high center of gravity. They can be bolted or secured at the base or to a wall capable of withstanding the forces generated by a six-foot column of soil.

Flowers, vegetables, fruits, and any other type of plant material capable of supporting its own weight can be grown in vertical garden planters. The wall-like character of vertical planters standing alone can enclose spaces of special character or special use within the horticultural therapy garden (Table 11.6.).

LEGAL ISSUES

Horticultural therapy garden administrators are responsible for compliance with a wide variety of laws, codes, and regulations. Although therapeutic landscapes will be provided the legal protection afforded units of the larger organization or institution, other issues remain of particular interest in the outdoor setting. Legal issues directly related to garden design in particular concern issues of accessibility, employment, and liability.

Accessibility and Employment: The ADA

The Americans with Disabilities Act of 1990 (the ADA) establishes the legal framework for the current accessibility and enabling guidelines as a component of the Civil Rights Act of 1964. This law mandates access and programmatic inclusion in public landscapes. Horticultural therapy landscapes "open" to public access for exhibition, activities, and amenities are directly affected by several provisions of the act that call for access, modification of structures and programs, and if necessary, alternative delivery of information and experiences. Even gardens operated as "private" facilities usually must meet ADA guidelines in employment and personnel actions.

Merely providing access into and through the horticultural therapy garden is probably not sufficient to avoid potential ADA violations. Since the process and primary experience of horticultural products is a significant component of the landscape, garden access alone is inadequate. Enabling design and/or universal design must be employed. However, current ADA guidelines do not establish optimum design standards, especially in the outdoors. New, greatly improved outdoor facilities and landscape design guidelines are being developed for publication. The new guidelines appear to be based on universal design goals and objectives. They prescribe access for all people of all abilities to landscape experiences, features, and recreation amenities.

Horticultural therapists are attuned to shaping and extending the garden and horticultural activities for people with disabilities. There is often a natural fit between both ADA and universal design goals and objectives and the operation of a horticultural therapy garden. It is not reasonable to construct a garden or garden features that restrict or deny participation to a patient, visitor, or staff member with a disability not common in the program population. However, many horticultural therapy gardens of the past have actually been remarkably inaccessible. The movement toward enabling gardens is meeting that challenge directly by introducing a change from developing new gardens for defined client groups to a more widespread effort to serve patients with all levels of abilities.

Liability

When financially possible, responsibility for the design of the horticultural therapy garden should be delegated to a design professional. A landscape architect is the professional whose training and experience provides the best preparation for the garden design effort. Other design, horticulture, or landscape installation experts can also be consulted in the development of the design with varying degrees of success. However, liability to the institution usually increases in cases where the designer is not professionally insured, accredited, or licensed. Responsibility for daily operations and maintenance rests with designated personnel who must continuously monitor materials and conditions in the garden just as they would in any interior portion of the institution. Program administrators are best advised to consult their legal counsel for any concerns about the horticultural therapy garden. Following legal counsel and exercising normal care and watchfulness should suffice to limit potential liability.

BIBLIOGRAPHY

*Bartholomew, Mel. 1981. *Square Foot Gardening.* Emmaus, PA: Rodale Press.

Boisset, Caroline and Greene, Fayal. 1993. *The Garden Source Book.* New York, NY: Crown Publishers.

Brookes, John. 1984. *The Garden Book: Designing, Creating, and Maintaining Your Garden.* New York, NY: Crown Trade Paperbacks.

Brookes, John. 1994. *Garden Design Workbook: A Practical Step by Step Course.* New York, NY: Dorling Kindersley.

*Rothert, Gene. 1994. *The Enabling Garden.* Dallas, TX: Taylor Publishing Company.

*Stoneham, Jane and Thoday, Peter. 1994. *Landscape Design for the Elderly and Disabled.* Wappingers Falls, NY: Garden Art Press.

*Tufts, Craig and Loewer, Peter. 1995. *The National Wildlife Federation's Gardening for Wildlife.* Emmaus, PA: Rodale Press.

*Willis, Barbara (Ed.). 1990. *Rodale's Illustrated Encyclopedia of Garden and Landscaping Techniques.* Emmaus, PA: Rodale Press.

---

*Indicates texts of particular value to the design and maintenance of horticultural therapy gardens.

Chapter 12

Inside Space and Adaptive Gardening: Design, Techniques, and Tools

Douglas L. Airhart
Kathleen M. Airhart

INTRODUCTION

Many successful horticulture therapy (HT) programs utilize and emphasize indoor spaces and activities such as craft projects, floral design, and plant propagation to provide rewarding and effective therapeutic experiences (Hewson, 1994). A greenhouse is not mandatory for successful programming, although some type of plant-growing space can improve the choices of plant materials available for activities. Many programs need to start small or stay small due to funding issues. A plant room or horticultural therapy room is the place to start.

The purpose of this chapter is to present ideas about how indoor spaces can be used for HT programs. The learning objectives of this chapter are to present considerations for equipment and materials for programming, to provide options for different levels of funding and program activity, and to describe basic accessibility and safety concerns for interior spaces. This chapter discusses (1) adaptive indoor space, including activity space, storage space, growing space, and greenhouse space; (2) adaptive tools and equipment; (3) safety precautions; and (4) accessibility criteria.

ADAPTIVE INDOOR SPACE

The three primary requirements are activity space, storage space, and office space for privacy and storage of confidential records and other materials. Many programs have supplemental space for displaying and selling items to raise funds, extra growing space for plants, or possible greenhouse and headhouse spaces, but these are not absolute requirements for success. If a greenhouse facility is desired, the administration must be prepared to hire staff with specific training in greenhouse management, because a person who can grow houseplants on a window sill is not automatically able to manage a greenhouse facility. A poorly managed or maintained plant room or greenhouse will not be a benefit, but will become a serious problem and possibly a liability.

Activity Space

Size

The size for activity and work areas will depend on the number of patients being served, the type of projects planned, and accessibility requirements. For an initial estimate, complete a sample activity project to determine the optimum space needed. Consider factors such as convenient placement of raw materials, ease of maneuverability in the work space, and space to put the finished product. Use that amount of area for each additional person expected in order to estimate the total space needed. Accessibility standards may require that you multiply by an even larger factor, e.g., wheelchairs and prone carts require greater space at work stations to turn and maneuver.

Work Areas

Work areas typically will need even more space per project to accommodate the larger tools, carts, pots, and plant materials you will be using. More space will be required if bulky or messy materials (e.g., potting mix, stock plants for propagation cuttings, filling flats and containers, or foil wraps) are being used.

Tables and Seating

The best choice for tables are those that are adjustable to different heights for use while in a sitting or in a standing position, or to accommodate wheelchair users. Without adjustable tables, various tables of different work heights and knee-space heights will be required. The comfort range for any individual is that the tabletop be level with or slightly below the elbow when the forearm is extended parallel to the floor. All seating should be adjustable for sitting activities. The individual comfort range is a height that positions the thigh parallel to the floor and allows the feet to be placed flat on the floor.

One innovative program mounted a tabletop to a hydraulic barber-chair base to create a workspace adaptable to sitting or standing positions (Figure 12.1). Inexpensive, lightweight but durable molded plastic chairs, with high or low backs, with or without armrests, and that stack one upon another to reduce storage space, are available at discount stores. Hospital supply warehouses offer tables with durable finishes, optional raised edges to contain spills, central pedestals with adjustable heights, and optional wheel mounts to move them more easily. These are more adaptable than the standard four-legged tables.

Arrangements

To supervise group activities or presentations, an effective position is boardroom fashion with the leader at one end (the head of the table) to be able to view all participants. Adequate space is required around the work-

FIGURE 12.1. Hydraulic barber chair base adapted to tabletop to provide adjustable heights for different client needs. (Enid A. Haupt Glass Garden, Rusk Institute, New York Institute for Rehabilitation; photo by Douglas Airhart, HTM)

table area to allow the leader to maneuver between patients and behind chairs safely and comfortably. One program devised an open-ring configuration with a permanent aisleway that placed the supervisor in the center, using a swivel stool to allow pivoting to view each client. Each perimeter workstation had two recessed basins fitted with removable covers to store materials and tools or to hold debris.

Designs

Tables and work areas of different designs and configurations will be needed, depending on whether they are designed to accommodate an individual, groups of different sizes, cotreatment activities with more than one therapist involved, or for vocational job training. A good method to determine your actual needs is to visit other sites with similar user situations to observe activity sessions and to ask those therapists what they would do to improve their areas.

Vocational programs may need only potting benches and other work stations for instruction and practice, a blackboard for group instruction, a bulletin board for work assignments and a scheduling calendar. A therapeutic situation may require less individual workspaces but additional tables to increase group interaction.

Storage Space

Each program will have different supplies, tools, and equipment inventories that will demand certain space, location, and security precautions for adequate storage. The amount of space and configuration of shelves and pegboard wall storage needed can be determined by observing the display areas at your suppliers. Ask to tour their warehouse to gain a better appraisal of the space needed.

Bulk Items

The majority of horticultural supplies are sold in case lots, with 50, 100, or up to 250 pots and flats per case. You may purchase potting mix components (e.g., peatmoss, perlite, vermiculite, trace elements, lime) in bulk and mix it on site or purchase more convenient premixed bales and bags. Potting mix bales are becoming smaller and more manageable. Allow adequate space between shelves to accommodate case heights or bale dimensions.

Inventory

Supplies should be inventoried and evaluated monthly to determine reordering schedules. Utility cabinets with drawers, cupboards, and open or covered and possibly lockable shelves may be required. Some programs have received salvage cabinets from public schools or residence halls that are being renovated. A small refrigerator will help preserve rooting hormones and seeds; a moderate size refrigerator can store cut flowers and greens for short periods.

Safety

Tools, especially sharp tools, should be itemized and inventoried regularly and kept in locked storage. An effective method is to use wall-mounted pegboard storage with the outline of each tool painted on the pegboard where the tool belongs. A quick glance will identify missing tools. Color coding tools with sharp edges or points and more hazardous items might improve the daily evaluation. Some facilities require the sharp tools to be inventoried and locked *after each group.*

Growing Spaces

Light Carts

A light cart or other fluorescent light system can provide good growing conditions for a wide selection of plants (Luse and Mandeville, 1977). It is also an excellent environment for rooting cuttings of plants. Most foliage plants and many flowering plants can be grown or produced successfully by people with little experience. It allows a successful growing experience for clients or patients and it can create a pleasant viewing environment almost anywhere in a facility, not just in treatment rooms or activity centers. Many light carts can be purchased from horticultural suppliers (Figure 12.2), but their expense may be prohibitive to beginning programs (Rothert and Daubert, 1981).

Construction. For utilitarian use, shop-light fixtures can be hung from chains over tables, supported from adjustable wall-shelf brackets, or even placed on bricks or blocks to provide growing space below. Another simple unit can be made by placing a light fixture over an unused aquarium, with spacers to allow for some air circulation. This configuration will help create a humid environment around the plants, decreasing the need for watering or maintenance, and protect the plants in situations where vandalism might occur.

FIGURE 12.2. Indoor light cart constructed of plastic pipe with a single, accessible shelf. The lamp fixture may be raised or lowered by changing the length of chain. (Bryn Mawr Rehab; photo by Karin Fleming, HTR)

Typically, purchased units have three tiers of lights and shelves, with equal spacing between shelves. Flowering plants are placed closer to the lights to encourage flowers by using stands or spacers such as inverted pots. With a little ingenuity, a variety of attractive and useful units may be built by facility maintenance staff or volunteers to meet special situations. The space may be enlarged to allow easy access for touching and caring for the plants. The units might be made to be more attractive for placing in public visiting areas, unused hallway spaces, or corners of residential meeting rooms. These units will look less institutional and their use will make the space more personal, such as clients, patients, or visitors might install in their own homes.

Lighting. The actual number of fixtures you need will depend on the number of plants you have to grow or the number of light cart units you need. For most situations, four foot or longer fluorescent fixtures are recommended. Shorter-length lamps may be used, but the light intensity decreases near the ends of the lamps, reducing the adequate growing space. Generally, a combination of cool white and warm white lamps will provide adequate intensities for most plants (Luse and Mandeville, 1977). The more expensive grow lamps or lamps that emit selected wavelengths are not needed to grow most plants. What is needed is long duration of light, such as fourteen to eighteen hours per day, which can be accomplished by using automated cycle timers. Inexpensive household timers may be satisfactory, but durable commercial units from horticultural suppliers are suggested.

The next important consideration is placing the plants in close proximity to the lamps to ensure adequate intensity. Plants requiring lower light intensities should be placed on the outer edges and ends of the lighted area. Higher intensities can be achieved for individual plants in single units by placing plants on spacers such as inverted pots. Fixtures suspended from chains might have one end suspended at a different height. This arrangement can accommodate plants of different heights or different intensity requirements.

Spotlight bulbs are not efficient for growing large numbers of plants in light carts, but they can be used to highlight some plants on a light cart in public areas, or the attractiveness of specimen interior plants. Some flowering plants will respond to red light energy emitted by incandescent bulbs. Adding one or two incandescent bulbs between two banks of fluorescent fixtures can improve the response of flowering plants grown in light carts.

Growing Regimes. The main consideration for these units is heat. Incandescent bulbs are known to emit heat. Fluorescent lamps are relatively cool, but the fixture ballasts emit a great amount of heat. In situations with multiple shelves of fixtures or multiple light units in one room, the emitted heat must be ventilated or at least circulated with a fan. When more than six fixtures are closely arranged, consider rewiring the system of units to place the ballasts in a remote area where the heat is not damaging or can be more easily ventilated.

The second consideration is watering. The light intensity, duration of lighting, and heat will increase the demand for water, causing plants to dry out more often. Be sure to check the moisture level of the growing mix before watering. If the mix is wet, even if the plant is wilted, do not apply water. Check the roots for damage first. If the mix is sufficiently dry to need watering, apply water until it begins to run through the drainage holes in the pot. If the pot has a drainage container, be sure to drain excess water from it. The best situation is to have pots placed on gravel in the basin to prevent having the pot bottom be submerged in water after watering.

Humidity is the third consideration. Typically, institutions and interior rooms have low humidity levels that are not optimum for plant growth. The light intensity and heat emitted by light carts will tend to decrease humidity levels, and the plant population and the gravel-filled basins will tend to raise humidity. Low humidity may cause some leaf curling or premature flower dropping, which can be corrected. High humidity levels can cause disease problems. If you notice water condensing on plant leaves at night, or if some plants are growing roots from their stems, the humidity is too high. If an aquarium is being used, the humidity will be high. A good indicator of excess humidity is condensation on the interior glass of the aquarium, in which case, increase air circulation until the moisture evaporates.

Acclimatization. Although light intensities in light carts seem high, they are relatively low compared to exterior light levels. If the plants being grown in the light cart are to be used in exterior situations, they must be acclimatized to the exterior environment. Acclimatization is a process of gradually preparing a plant for a change of environment. This process takes at least a week, and more time is better.

Begin by moving the plants outdoors during cloudy weather, or placing them under shade outdoors, for an hour or so the first day, then increase the time period daily. By the end of the first week, begin placing the plants in the expected light environment (full sun, morning sun, afternoon sun, shade) for a short duration, gradually increasing the duration on a daily basis. This process should protect plants from being stunted or damaged after being transplanted into the growing environment.

Windows

Windows may serve as adequate spaces for growing plants (Yeomans, 1992). Sometimes the existing sill can hold small pots and saucers, or you may decide to install shelves for holding plants adjacent to the window (Figure 12.3). Plants grown in windows tend to lean toward the light and will need to be turned frequently to keep their growth balanced. If you wish to grow flowering plants, supplemental lighting with incandescent bulbs is recommended. Be sure to check for and protect your plants from drafts from leaky windows. A source of water should be close by to reduce carrying distance and the chance of spillage. Cultural and sanitation practices are especially required to control pests and diseases in smaller growing spaces.

FIGURE 12.3. Shelves installed in bay window to hold art. (Photo by D. Airhart)

Greenhouses

Many signs and shapes of greenhouses are available that can be used in HT programs (Ball, 1991). When considering a greenhouse for an HT program, remember there are costs for planning, building permits, the structure, utilities and their connections, plus the costs for operation, management, and maintenance (Nelson, 1991). A basic list of equipment needed for greenhouses is presented in Table 12.1.

Site Considerations

The greenhouse site must be level and well-drained, with few objects to cast shade on it, and water, electricity, and fuel supply must be available. Proximity to high-intensity lights, spotlights, or vehicular traffic lights at night creates problems with growing specialty crops such as poinsettias unless shade cloth is used to protect them from outside sources of light. Likelihood of vandalism should be considered when selecting the glazing material.

TABLE 12.1. Suggested Equipment for Therapeutic Greenhouses

20X hand lens
absorbent for chemical spills (cat litter, vermiculite)
aspirated thermostatic controls
assorted plastic kitchen measuring tools
box fans for air circulation
disinfectant solution
eye goggles
fertilizer proportioner
first aid kit
flashlight
fogging nozzle
gas furnace pilot lighter (butane, extension wand)
good quality hoses
Lemon Joy®, Ivory® detergent, Safers Soap®
measuring equipment
minimum-maximum thermometers
MSDS file for all chemicals
nitrile gloves
pest identification pictures
pesticide sprayer
pH meter
rain gear or tyvek® suit
record-keeping file or calendar
respirator
rubber boots
scale
soil thermometer
solubridge (soluble salts meter)
temperature sensing alarm system
trash cans
two-wheel cart
watering wand and breaker nozzle
warning/reentry signs

Source: Crater, 1993.

Work Areas

Work areas for filling pots and transplanting (Figure 12.4) must be included in the design or be nearby. Many HT programs have a simple eight-by-ten foot lean-to structure attached to a building work room, and anticipate no future expansion. Some vocational programs have multiple greenhouse units of varying size and utility. Your program can be adjusted

FIGURE 12.4. Work station inside greenhouse with visibility to activity room for supervision. (Bryn Mawr Rehab; photo by Karin Fleming, HTR)

to accommodate an existing greenhouse, or the structure can be built to the size needed to meet your program goals.

Structure

Greenhouses can provide an ideal growing environment and protect plants from harsh natural weather conditions, allowing plants to grow quickly and year-round (Corr, 1993b). The main environmental factor, light, can be maximized by using few overhead structural supports and keeping the greenhouse covering clear and clean. There are many styles available, but the Quonset style and pipe-frame or truss-frame greenhouses are most common (Nelson, 1991).

Quonset, or hoop, greenhouses are relatively inexpensive, energy efficient, and transmit light well. They may be built with hoops reaching the ground or with raised vertical sides that improve usable interior space. Most are covered with two layers of polyethylene sheets and inflated with a small fan to provide an insulating dead air space. Greenhouse-grade polyethylene with UV inhibitors is suggested to ensure longevity. Even so, the plastic must be replaced every three years. Vandals, mowers, and projectiles can easily create holes in the sides.

Truss and pipe-frame style greenhouses usually have more headroom and are expensive to build, but they can last many years. They are typically less energy efficient if covered with glass, but light transmission is excellent. Newer glazing materials such as polycarbonates (Lexan®) and acrylic sheets (Exolite®) can improve energy efficiency with only minor loss of light transmission, and they are not as fragile.

Headhouse

A headhouse area, which can be constructed in many ways, is generally attached to the greenhouse. This provides space for working, storage, lockers and break room, restrooms, office, and possibly sales or shipping. If attached to the northern side or end of the greenhouse, it can improve winter heat efficiency by acting as a wind break. Most HT programs should anticipate and plan for a headhouse area equal to 15 percent of the greenhouse area.

Benches

Benches are not required to grow greenhouse crops, but they make it easier to reach the plants. Air circulation under the bench decreases the chance of disease and pests, and make chores like weeding and spraying less difficult. Simple and inexpensive benches may be constructed using cinder blocks for support. Pallets or skids placed on them make short benches or snow fencing on boards running between the blocks may be used for longer benches. More permanent wooden frames can be made using welded wire fencing. Prefabricated plastic sectional benches and benches of expanded metal can be purchased from many greenhouse suppliers.

Management

Trained staff is required to operate greenhouses effectively (Larson, 1992). Existing staff may attend training sessions to learn to operate a small greenhouse. Qualified pretrained staff is needed for larger greenhouse operations. Registered horticultural therapists should be able to manage most greenhouse crop production and pest control (Airhart, Airhart, and Tristan, 1995) situations and provide effective programming. A brief summary of basic management strategies is presented in Table 12.2.

ADAPTIVE TOOLS AND EQUIPMENT

Tools and equipment help us garden more effectively or more efficiently, allowing us more enjoyment of our efforts (Straus, 1991). If a

TABLE 12.2. Greenhouse "Always" and "Never" Rules

Always be careful with pesticides.
Never allow weeds to develop in or around the greenhouse.
Always keep good records.
Never assume your equipment is working properly.
Always keep the end of the hose off the floor.
Never allow any compound containing 2,4-D into the greenhouse.
Always automate whenever possible.
Never begin a crop without considering the market for sales.
Always control insects and diseases before they become a major problem.
Never use creosote or pentachlorophenol; use copper naphthenate.
Always check to be sure any pesticide is safe for greenhouse use before applying it.
Never reuse pots or flats without sterilizing them.
Always allow air intake for gas, oil, coal, or wood-burning heaters.
Never drop leaves, flowers, or buds onto the floor.
Always maintain adequate air circulation.
Never sprinkle; always irrigate.
Always test a new growing mix before planting in it.
Never discard plants or dump plant debris near the greenhouse.
Always wash after using tobacco products when working in the greenhouse.
Never stop learning.

Source: Corr, 1993a.

specific tool is required to accomplish a task, not having the tool is a serious hardship to training and a disadvantage in therapy. Horticultural therapists can assist clients to learn skills or work with less effort by selecting an appropriate tool or selecting tools that can be adapted to the task for the client (Freeman, 1985; Thabault and Ocone, 1982).

Tools

There are three main factors to consider when choosing tools: the weight, the handle length, and the type of grip on the handle (Airhart, 1990). Some tool manufacturers attempt to incorporate all three considerations when designing tools modified to make them easier to use or for special applications (Relf, 1983). Others have designed lightweight extension handles for regular tools, such as grippers, large grabbers, scissors, and watering nozzles that can have application for inside use. A sample tool list for HT is presented in Table 12.3.

Modern alloy metals, thin steel, and polypropylene materials have become the material of choice for many tools and handles, reducing weight, and increasing maneuverability. The use of interchangeable tool heads that snap-lock into the same handle makes it convenient to perform

TABLE 12.3. Sample Tool List (Inventory will vary with program size)

---

adaptive cutters
aprons
brooms, brushes, and dust pans
canvas gloves
cultivators, short and long
dibbles
extended handle cut-and-hold pruners
growing mix scoops
hand pruners by-pass and anvil blade
kneelers
labels
lapboards
nitrile gloves
pencils
scissors
sun hats and sunglasses
trowels
watering cans
watering wands and breaker nozzles

---

multiple tasks without transporting excess weight, and helps reduce storage clutter. The handles are available in long or short lengths for more adaptability to tasks.

Tools designed with shorter handles to encourage children to garden have proven adaptable for wheelchair users because the tools are not as cumbersome. Elderly clients favor these tools because the tools are light and durable, but the handle may require grip cushions. Individuals with visual impairments may need short handles on tools to be able to locate and place the point or prong into position using two hands. A few tool handles, such as rakes and the interchangeable polypropylene handles, are being extended to reduce the need for bending while using the tool. Other modified tools, such as rakes with adjustable tines and ratchet pruners, provide adaptations to their users. Another favorite tool is a hose nozzle with trigger valve or on-off position that prevents excess usage or spillage between pots, benches, or different plant locations.

Ergonomic grips, bent to complement the wrist angle to reduce strain, are available on hand trowels, small shovels and rakes, and hand pruners. Pistol-grip tools allow the index finger to wrap around a hook on the handle to provide stability. An adjustable "D-grip" attachment fastened partway down a handle improves the wrist angle while grasping and also increases leverage. Some therapists have simply applied a short length of rubber or Styrofoam pipe insulation to a tool handle to make it more comfortable to use, or bigger and easier to grab with weak hands. Universal cuffs, long used by physical therapists to adapt household utensils, can help adapt the use of hand tools. Individuals with use of a single arm can fashion an armband grip or use a strip of Velcro material to secure the handle to the forearm for stability.

Public reception of these modifications has been favorable, and more improvements should be forthcoming. A horticultural therapist must adapt ideas, tasks, and tools of horticulture to the goals of improving self esteem and providing specific activities.

Equipment

The amount and selection of equipment will be influenced by the size of your growing space, the type of program application (e.g., therapeutic or vocational), and number of staff.

As with tools, equipment made from lighter metal alloy and polypropylene material is available. The portability of equipment is particularly important because the worksite is not always stationary nor permanent. Two-wheel and four-wheel carts are favored by some programs over wheelbarrows due to their stability in use, and some can be attached as trailers to wheelchairs. One program developed a trolley cart with shelves and basins for a variety of indoor uses, equipped with large-diameter wheels that improve maneuverability over bumpy surfaces (Boutard and Airhart, 1982). The Morrison Horticultural Center in Denver has growing tables with lockable wheels on the legs that allow portability. Many programs use wheeled grain-storage bins or wheeled bases for garbage cans to contain potting mixes or other bulk materials.

PATIENT OR CLIENT SAFETY PRECAUTIONS

Every human service facility should have written policies and staff designated to maintain the safety of all clients in training or patients attending sessions. Consult first with this individual to develop the proper safety procedures for the use and maintenance of tools and equipment expected to be used in your program.

Basic Safety Procedures

Staff and volunteers should be trained regularly in safe use and storage of tools and equipment. A list of safety procedures should be posted prominently in the tool and equipment areas.

Tool Procedures

Tools are to be used only by patients under the direct supervision of staff during any session. Patients should be instructed in the proper use of tools, and especially should be assessed by staff regarding use of sharp objects prior to obtaining them. Patients must be seated or kneeling while using short-handled tools. Walking with tools in hands is not acceptable; tools should holstered or in an apron pocket. Tools should be distributed by staff and counted after each session to ensure patient safety and stored in a locked closet in the designated restricted storage area. An inventory of tools should be made by staff after each session.

Safety Checklist

A program safety checklist is presented in Table 12.4. Programs with greenhouses are required to comply with Worker Protection Standards (Faust, 1995), as summarized in Table 12.5.

Notification of Support Services (Fire, Police, Security)

The local fire marshall, police chief and ambulance services director, and any off-site security officer should be invited to tour your facilities. They will appreciate your advanced preparation and concern and will have good suggestions to improve safety considerations for you and their personnel. Be sure they have *all* phone numbers needed to contact you and your immediate assistants in an emergency. Prepare and give these individuals a map of your site and your inventory list of chemicals. Distinctly highlight on the map the locations of any storage areas with flammable, caustic, corrosive, or toxic materials and have these areas well-marked on the exterior entry doors as well. Copies of Material Safety Data Sheets (MSDS) are available from your suppliers and are required to be posted in an accessible notebook.

Pesticides and Hazardous Chemicals

All staff, volunteers, clients, patients, and guests must be protected from accidental exposure to pesticides. Without proper training (possibly requiring staff to obtain a state applicator's license), the use of chemicals should be avoided or at least minimized. The use of specific biological agents is preferable (Airhart, Airhart, and Tristan, 1995). Homemade formulas such as garlic, onion, or pepper sprays may be dangerous if applied incorrectly.

TABLE 12.4. Checklist for Site Safety Specification

1. Storage closet
    all equipment opposite its label
    floor kept as free from materials as possible
    utilize the shelves and pegboard
    neat
    no items closer than twelve inches to the ceiling
    notify director of any inventory needs

2. Activity area
    supplies *cleared* and *cleaned* nightly
    materials put away and organized nightly
    all spills/stains to be cleaned up nightly
    neat
    tables *cleared* and *cleaned* nightly
    shelves and cabinets neatened and dusted weekly
    clean cabinets annually
    all sharp instruments stored out of patient access daily
    wash and disinfect lapboards after each use
    gloves used once should be placed in laundry receptacles
    aprons placed in laundry receptacles as needed
    gloves and aprons washed weekly

3. Chemical cabinet
    return all chemical and flammable materials and organize nightly
    fire cabinet emptied and cleaned quarterly
    notify director of any inventory needs
    cabinet remains closed and locked always, and after each use
    material safety data sheets posted on cabinet and in fire safety manual

4. Greenhouse
    post "Wet Floor" sign when watering
    immediately squeegee any wet areas on floor
    maintain plants and equipment in a neat and orderly fashion
    sweep floor daily
    apply disinfectant solution to floor weekly
    report pest problems to director immediately

TABLE 12.5. Ten Steps to Pesticide Safety Compliance: Worker Protection Standards (Faust, 1995)

---

1. The Centralized Bulletin Board—WPS Quick Reference Poster, Emergency Medical Information; Pesticide Application List; Right to Know and Employee Safety-Health Protection; Material Safety Data Sheets.

2. Pesticide Safety Training—Prior to any handling; within first week of work.

3. Decontamination Sites—Water to washing and eyeflushing; coveralls, drinking water, eyeflushing water; less than one-quarter mile; worker vs. handler considerations.

4. Emergency Assistance—Provide transportation and written information.

5. EPA-Approved Warning Signs.

6. Monitor Handlers—Every two hours for Restricted Use pesticides.

7. Restricted Entry Interval.

8. Early-Entry Exceptions.

9. Ventilation Criteria for Greenhouses—Check the label.

10. Personal Protective Equipment—Check the label.

11. Chemical Sprays Procedure—seal doorways to patient areas with plastic sheeting; close vents and disengage heating and cooling units; applications only between 7:00 p.m. and 7:00 a.m. (nonpatient hours); notify maintenance and security personnel; post spray notification signs and reentry information on all doorways.

ACCESSIBILITY

Americans with Disabilities Act Regulations

The Americans with Disabilities Act (ADA) of 1990 has had far-reaching implications to employers, businesses, educational institutions, and especially to an estimated forty-three million persons with disabilities in all aspects of life (USEEOC, 1991). The purpose of the ADA is protection for persons with disabilities against discrimination in economic, educational, and vocational opportunities.

The law establishes a three-part definition of disability:

1. A person with a physical or mental impairment that substantially limits one or more major life activities
2. A person with record of such an impairment
3. A person who is regarded as having such an impairment

A major life activity includes caring for oneself, performing manual tasks, walking, seeing, hearing, breathing, speaking, learning, or working. A person with record of such an impairment could include those with a medical diagnosis in remission or those persons recovering from substance abuse or mental illness. Persons regarded as having a disability refers to those who are discriminated against, and is based solely on the perceptions of others; for example, a person who is HIV positive but shows no symptoms of the disability, or a person with a facial disfigurement that does not affect job performance in any way. Family members of persons with disabilities are also covered by this law.

This law affects all areas of public participation, including employment, public services, public accommodations, and telecommunications. Accommodations may be made accessible through relocation of services or building modification.

The ADA requires that places of public accommodation be physically accessible. The standards of accessibility may vary for new, existing, and altered facilities. All new construction since 1993 must be accessible and usable by persons with disabilities. Existing facilities are required to remove structural barriers if easily accomplished and with little expense. An altered building must be made readily accessible and usable by persons with disabilities to the maximum extent feasible. Alteration and construction must conform to the specifications of ADA Accessibility Guidelines (ADAAG).

Most medical care facilities, such as hospitals, nursing homes, and rehabilitation centers, have been required to be accessible since 1973 (Section 504 of the Rehabilitation Act). With the passage of the ADA, the definition of facilities was broadened to include all facilities that had been previously exempt (those not receiving federal funds). For a medical care facility to be required to meet the specifications of ADAAG, it must be a facility that meets at least one of the following criteria:

1. A facility where people receive physical or medical treatment or care
2. A facility where people may need assistance in responding to an emergency
3. A facility where a period of stay may exceed twenty-four hours

Accessible parts of the facility must include food service areas, activity areas, gift shops, libraries, and greenhouses.

Prior to 1990, greenhouses were not typically designed to accommodate persons with physical disabilities. With ADA, this is not only a desired outcome but also a necessity required by law. The primary consideration in designing a greenhouse or HT room should include space allowance, accessible routes, reach ranges, protruding objects, ground and floor surfaces, activity areas, ramps, doors, and accessible restroom facilities. Accessibility of materials should be addressed by following appropriate space allowances and reach-range guidelines. The design and specifications of each of these criteria should be based on the average functioning of a nonambulatory individual. If these specifications are followed, most disabilities will be compensated. For more information about accessible design in public facilities, the Americans with Disabilities Act Accessibility Guidelines (USEEOC, 1991) should be carefully studied.

Space Allowance

A standard wheelchair height is twenty inches to the seat, twenty-nine inches to the armrest, and thirty-six inches to the handlebar pushers. Wheelchair width averages twenty-five to thirty inches and lengths of forty to forty-two inches are the norm (Rothert, 1994). The minimum ground space required to accommodate a single wheelchair and occupant is thirty by forty-eight inches, however, the minimum clear width of a designated accessible route must be no less than thirty-six inches wide. For two wheelchairs to pass in a designated accessible route, the width must be no less than sixty inches. In aisles less than sixty inches wide, a passing space (sixty by sixty inches) must be designated no less than every 200 feet. The average turning space required for 360 degree turns is sixty by sixty inches (Figures 12.5 and 12.6).

Reach Ranges

Reach ranges vary between forward and side reach and should take into consideration personal stature and ability. If clear floor space allows only forward approach to an object, the maximum high forward reach is forty-eight inches and the minimum low forward reach is fifteen inches (Figure 12.7). A person who is reaching over a table with his or her legs extended beneath has an average reach of twenty inches if he or she has maximum clearance for armrests of the chair (Figure 12.8). If clear floor space allows only parallel approach, the maximum high-side reach is fifty-four inches and

FIGURE 12.5. Wheelchair turning space (USEEOC, 1991)

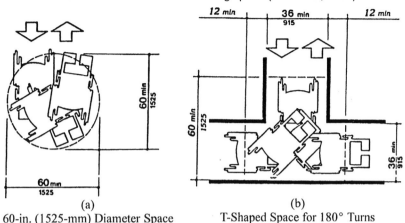

(a)
60-in. (1525-mm) Diameter Space

(b)
T-Shaped Space for 180° Turns

low-side reach is nine inches. If a table or shelf is approached from the side, a maximum twenty-four-inch side reach is allowed if the object is at least the height of the armrest, but not greater than thirty-four inches (Figure 12.9).

Accessible Routes

Ground and floor surfaces along accessible routes and in accessible rooms and spaces must be stable, firm, and slip resistant. If gratings are located in walking routes, the spaces must be no greater than one-half inch in any one direction. If there is a rise in any walking route with a slope greater than 1:20, a ramp must be installed. The least possible slope should be designed into any ramp, but not to exceed a slope of 1:12 with a maximum length of thirty inches per rise. Dimensions exceeding thirty inches must include a level resting area (Figure 12.10).

Protruding Objects

For objects on walls, the zone of hazard is from twenty-seven to eighty inches above the floor. They may not protrude more than four inches (Figures 12.11a and b). Drinking fountains, telephones, and trash receptacles should *not* be placed in main egress areas.

Seating Space

When planning seating spaces in activity areas for persons in wheelchairs, accessible paths of travel and clearance heights must be provided.

FIGURE 12.6. Minimum clear floor space for wheelchairs (USEEOC, 1991)

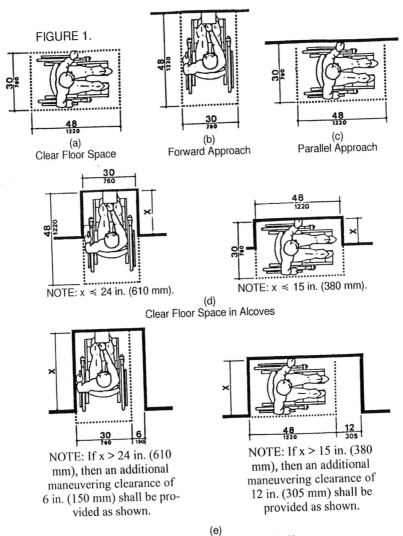

FIGURE 1.

(a)
Clear Floor Space

(b)
Forward Approach

(c)
Parallel Approach

NOTE: x ≤ 24 in. (610 mm).

NOTE: x ≤ 15 in. (380 mm).

(d)
Clear Floor Space in Alcoves

NOTE: If x > 24 in. (610 mm), then an additional maneuvering clearance of 6 in. (150 mm) shall be provided as shown.

NOTE: If x > 15 in. (380 mm), then an additional maneuvering clearance of 12 in. (305 mm) shall be provided as shown.

(e)
Additional Maneuvering Clearances for Alcoves

FIGURE 12.7. High forward reach limit (USEEOC, 1991)

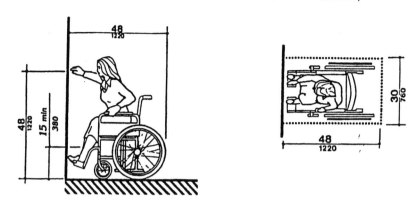

FIGURE 12.8. Maximum forward reach over an obstruction (USEEOC, 1991)

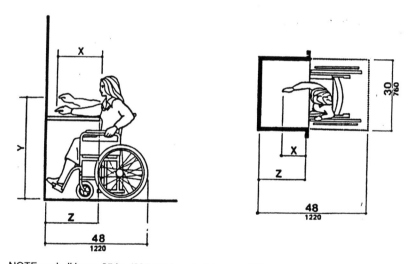

NOTE: x shall be ⩽ 25 in. (635 mm); x shall be ⩾ x. When x < 20 in. (510 mm), then y shall be 48 in. (1220 mm) maximum. When x is 20 to 25 in. (510 to 635 mm), then y shall be 44 in. (1120 mm) maximum.

To maneuver around an empty table, thirty-six inches of clearance is required. In order to pass a person already seated at a table, an additional thirty-six inches clearance space is required. Knee space beneath a table

FIGURE 12.9. Side reach (USEEOC, 1991)

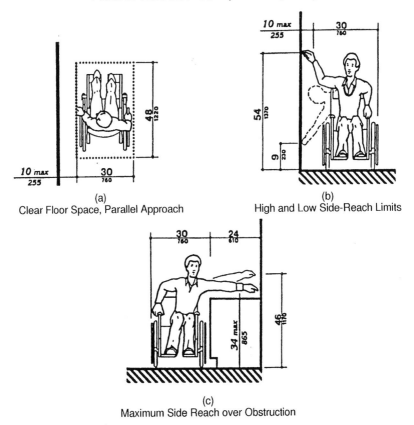

(a)
Clear Floor Space, Parallel Approach

(b)
High and Low Side-Reach Limits

(c)
Maximum Side Reach over Obstruction

requires no less than nineteen inches in depth and thirty inches in width (Figure 12.12). Clearance space for knees should be no less than twenty-seven inches and twenty-eight to thirty-four inches to clear arm rests. An ideal situation is a table on a pedestal that may be raised and lowered to meet individual needs.

Doors

Door widths must have a minimum clear opening of thirty-two inches when the door is opened to ninety degrees (Figure 12.13). The maximum force for pushing or pulling open a door should not exceed a five-pound force effort. If automatic doors are used, no more than a fifteen-pound

FIGURE 12.10. Sides of curb ramps (USEEOC, 1991)

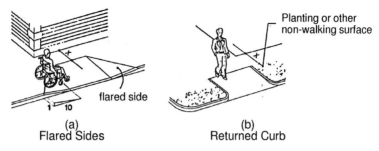

(a)
Flared Sides

(b)
Returned Curb

*If x is less than 48 in.,
then the slope of the flared side
shall not exceed 1:12.*

FIGURE 12.11a. Walking parallel to a wall (USEEOC, 1991)

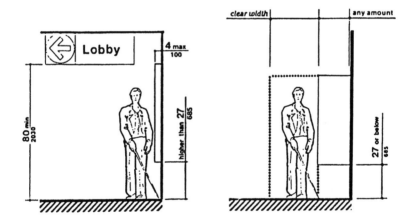

force should be required to stop the door movement. Thresholds at door-ways should not exceed three-fourths inch in height with a slope of less than 1:2. Door handles must have a shape which is easy to grasp with one hand and which does not require tight pinching or twisting of the wrist to operate. Handles must be mounted no higher than forty-eight inches above the floor.

Restrooms

Restrooms should include the standard sixty-inch minimum turning space between facilities. If this is not feasible, several clear floor space

FIGURE 12.11b. Walking perpendicular to a wall (USEEOC, 1991)

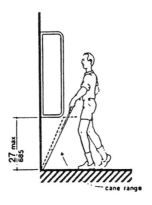

FIGURE 12.12. Minimum clearance for seating and tables (USEEOC, 1991)

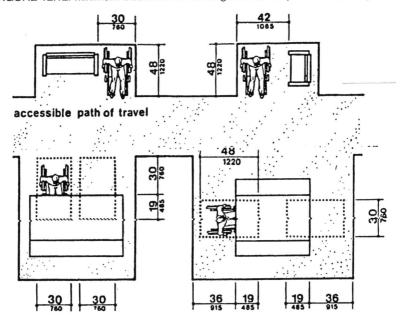

options may be considered for the toilet area, including a forward approach of forty-eight inches width and sixty-six inches depth (Figure 12.14). An outswinging door is preferable with the toilet seat hung twenty inches from the floor. Handrails on the side or rear of the toilet should be located thirty-three to thirty-six inches from the floor. Siderails should be run the full length of the seating area and extend to seventeen inches from the front of the toilet (Figure 12.15).

FIGURE 12.13. Clear doorway width and depth (USEEOC, 1991)

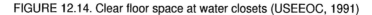

FIGURE 12.14. Clear floor space at water closets (USEEOC, 1991)

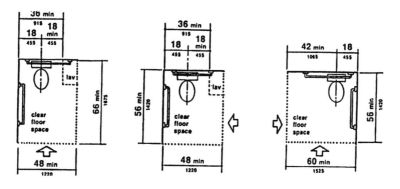

FIGURE 12.15. Toilet Stalls (USEEOC, 1991)

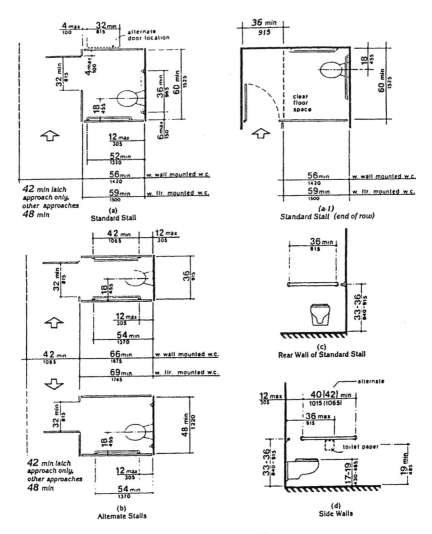

CASE STUDIES

Bryn Mawr Rehab

The HT program at Bryn Mawr Rehab hospital in Malvern, Pennsylvania demonstrates the adaptability of horticultural therapy to its environment. Initial plans for HT were incorporated into the business plan of the hospital four years prior to its implementation. Preliminary work included forming a multidepartmental HT committee, defining the program, projecting volumes, establishing space requirements, developing an operating budget, visiting sites, fundraising, and developing building plans. This preliminary background work provided hospital personnel with a better understanding and commitment for such a unique program at its facility.

Departments participating in the planning phases of program development included physical therapy (PT), occupational therapy (OT), speech therapy, therapeutic recreation (TR), cognitive retraining, vocational education, the outpatient program, and a long-term residential facility. Each departmental representative helped to establish goals that could be met through horticultural therapy. A basis for cotreatment with horticultural therapy was established, which would support all existing rehabilitation programs and services and provide a new and exciting therapeutic milieu for patients and staff.

Indoor Plant Room

The first HT groups met in a room shared by therapeutic recreation, Day Hospital, and HT (Figure 12.16). The room was twenty-five-by-fifteen feet with windows along the west wall. A fifteen-foot-long three-tiered shelf was built and placed along the window as the primary indoor plant area. Grow lights were affixed to the top tier to aid in seed germination. Large potted plants also sat along the floor area in front of the window.

Most activities took place around a large table with the therapist and staff providing plant materials to participants. Cuttings, plant maintenance, seeding, and repotting were HT activities led by OT, TR, and day treatment on a weekly basis. Vocational clients and PT patients provided the more strenuous physical labor of hauling water, mixing potting mix, and rotating plants on shelves. Outdoor raised flower boxes contributed to the activities that could take place in this confined space.

While therapists worked on goals for their patients, such as fine and gross motor control and eye-hand coordination, patients enjoyed a functional activity that provided real-life meaning after their discharge.

Greenhouse

During the time that the indoor plant activities were taking place, planning and fundraising for the greenhouse continued to completion, and the greenhouse construction began. The accessible greenhouse was 1,500 square

FIGURE 12.16. Indoor light cart constructed from metal shelving with spacing at various heights to accommodate different plants. Plastic trays collect drainage water and provide humidity. Large westward window with vertical blinds provides natural light to room and light cart. (Bryn Mawr Rehab photo)

feet with an adjacent 1,500 square feet for combined use of activities, sales, storage, and office space to involve as many as 250 patients per week (Figure 12.17).

All aisles were six feet wide to allow two wheelchairs to pass and turn. The cement floor sloped down from the middle to the walls for drainage, but the benches were level, twenty-eight inches from the floor at the center aisle and thirty-four inches at the wall. The four-feet-wide peninsular benches could be reached from either side. The perimeter benches were three feet wide due to access from only one side. Each peninsular bay had a water spigot and retractable hose with a trigger-valve water wand and breaker nozzle. A metal sheath provided storage off the floor for each water wand. A partial wall of glass block segregated the greenhouse from the propagation and pot-washing areas, the rear access door, and the loading area. The pot-washing area proved critical to operations, allowing dirty pots and flats to be sanitized for reuse by soaking overnight and rinsing the next morning.

FIGURE 12.17. Bryn Mawr Rehabilitation greenhouse plan and display area.

Schematic plan of therapeutic features of Bryn Mawr Rehab HT program: activity area for tables, lockable storage room, flower cooler, work/display counters and shelves, sales area, staff offices, restroom.

Schematic plan of therapeutic features of Bryn Mawr Rehab greenhouse: wide aisles, peninsular benches, raised bed planters, work area, propagation area, pot-washing and storage area, loading area.

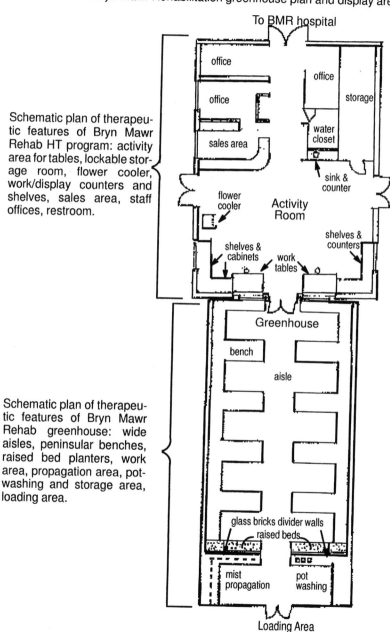

To BMR hospital

office

office

office

storage

water closet

sales area

sink & counter

flower cooler

Activity Room

shelves & counters

shelves & cabinets

work tables

Greenhouse

bench

aisle

glass bricks divider walls

raised beds

mist propagation

pot washing

Loading Area

Plants maintained in the greenhouse were designated as stock plants, patient activity plants, and sale plants. Group activities were increased with the new space, and individual therapists were invited to bring patients to the greenhouse to work one-on-one on a plant activity with the HT's assistance. The PTs and OTs had patients reaching, bending, walking, lifting, and developing fine motor skills through planting activities.

Activity/Sales Area

The activity area was designed with accessible storage space for craft supplies, four adjustable height tables that could accommodate up to twenty-four patients, and a display area for patient projects (Figure 12.18).

Volunteers, including many former patients, conducted the daily functions of the sales area, which provided income for funding the HT program. Handmade floral items, fresh flowers, and greenhouse plants were popular sales items year-round. Seasonal displays and sales of poinsettias,

FIGURE 12.18. Height-adjustable tables in use at Bryn Mawr Rehab. Accessible storage space and craft display area visible on rear wall. (Bryn Mawr Rehab; photo by Karin Fleming, HTR)

bedding plants, and chrysanthemums guaranteed income all year from the supportive hospital staff and patrons.

The HT program grew from an idea to a small adjunctive program, and then to a full department that conducted six patient groups *per day* throughout the year. The involvement of two HTs, a group of dedicated volunteers, and countless hospital staff enabled many patients to enjoy the therapeutic benefits of HT.

Menorah Park Nursing Home

Looking Ahead is a small, privately funded program for brain injury clients. It is a part of a larger program called the New Beginnings Adult Day Care facility located at the Menorah Park Jewish Orthodox Nursing Home in Cleveland, Ohio. New Beginnings is designed to serve as a day treatment program for adult dementia patients for purposes of respite care for family and the safety of the client. The nursing home also provides full-scale residential living, assisted living, and independent living centers for older adults. The population of Looking Ahead consists of approximately ten nonresidential persons from various parts of the county, ranging in age from young adult through age fifty, who have survived brain injury due to closed or open trauma, disease, and stroke. The director of the program, an occupational therapist, oversees this group and is assisted by a part-time horticultural therapist and a full-time OT assistant.

The Day Treatment Center and Looking Ahead are located within the nursing home and share horticultural therapy space with staff, therapists, and group meetings for various client activities. The room is T-shaped with two entrances and has low-level lighting through tinted glass located along one wall. Though the area seems sunny, it is not an ideal situation for growing plants. However, through the ingenuity of the horticultural therapist, the room has been transformed into a fully functioning horticultural therapy activity center with up to one hundred plants in need of care. Wall space is well-utilized, with a counter, a sink, a storage/potting table, plant shelves in front of the window, and two light carts for propagation purposes. Potting mix is stored in twenty-gallon trash bins, and the room has a tile floor for easy maintenance. In the center of the room is a treatment table that services the needs of both ambulatory and wheelchair clients.

The main purpose of the horticultural therapy program at Looking Ahead is to support the goals of each individual in the Brain Injury program. Their goals may include improving fine motor skills, increasing gross motor function, enhancing cognition and memory, and beginning prevocational goals. These goals are achieved through horticultural therapy activities such as handwriting labels, propagating new plants, filling pots, transplanting, locating materials, and washing pots and tools. Tasks are assigned to clients according to their relative goals in the program.

The horticultural therapist visits the program once every two weeks to assist the day-treatment staff in developing goals and maintaining horticultural activities. Plant needs (need for more propagation, transplanting, fertilization, or new plant materials) are assessed, and sales and distribution of plants to residents are planned. Clients are responsible for maintaining the plants during the week, watering and grooming them, and tending to other daily needs. Clients are each assigned a section to water on a regular basis under the supervision of the OT assistant.

Fundraising plant sales are held several times per year. The clients are responsible for pricing, labeling, advertising, making signs, setting up tables, and taking special orders. Clients also make live centerpieces complete with bows and wrapping for special occasions. All money earned by these projects goes back to funding the program. A welcoming gesture of a free potted plant is made by the clients in the HT program to each new resident of the nursing home.

Looking Ahead is very successful in that it provides an outlet for creativity and a feeling of productivity among its clients. It has been funded to expand to include a patio garden that will provide activities for its residential nursing home patients. The success of the program continues to hinge on special gifts, donations, and volunteer help. Goals to expand the program include increasing the hours of the HT for specific activities and providing vocational experiences, at a neighboring arboretum, for graduates of the brain injury program.

RESOURCES

Accessibility Issues

Access Board, 800-USA-ABLE.

Department of Justice, 800-514-0301.

Supplier Catalogues

E.C. Geiger, Box 285, Route 63, Harleysville, PA 19438-0332, 215-256-6511.

Florist Products, Inc., 2242 North Palmer Drive, Schaumburg, IL 60173, 800-828-2242.

Green Thumb, P.O. Box 5980, Virginia Beach, VA 23455-5980, Fax 804-496-9061.

Hummert International, 4500 Earth City Expressway, Earth City, MO 63045, 314-739-4500.

Langenback, Department L4600, P.O.Box 1140, El Segundo, CA 90245-6140, 800-362-1991.

Smith & Hawkin, 25 Corte Madera, Mill Valley, CA 94941, 415-383-4050.

Turner Greenhouses, Highway 13, Goldsboro, NC 27533, 800-672-4770.

Verilux, P.O. Box 7633, Vallejo, CA 94590, 800-786-6850.

Walt Nicke, 36 McLeod Lane, P.O. Box 433, Topsfield, MA 01983, 508-887-3388.

Trade Publications

Garden Talk, Walt Nicke Company, P.O. Box 433, Topsfield, MA 01983.

GM Pro/Greenhouse Management & Production, P.O. Box 1868, Fort Worth, TX 76101-1868, 800-434-6776.

Greenhouse Business, McCormick Communications Group Ltd., P.O. Box 698, Park Ridge, IL 60068-0698, 708-823-5650

Grower Talks, P.O. Box 532, 1 North River Lane, Suite 206, Geneva, IL 60134, 708-208-9350.

Greenhouse Grower, Meister Publishing Co., 37733 Euclid Avenue, Willoughby, OH 44094, 216-942-2000.

Professional Organizations

American Horticultural Therapy Association, 362A Christopher Avenue, Gaithersburg, MD 20879, 301-948-3010.

American Society for Horticultural Science, 600 Cameron Street, Alexandria, VA 22314-2562, 703-836-4606.

International Light Gardening Society of America, 128 West 58th Street, New York, NY 10019.

Ohio Florists' Association, 2130 Stella Court, Suite 200, Columbus, OH 43215, 614-487-1117.

Professional Plant Growers Association, P.O. Box 27517, Lansing, MI 48909, 517-694-7700.

Society of American Florists, 1601 Duke Street, Alexandria, VA 22314, 800-336-4743

REFERENCES

Airhart, Douglas. 1990. Adaptive tools and ideas to make gardening easier. 114th Annual Meeting of the American Association on Mental Retardation. Atlanta, GA. May 1990.

Airhart, Douglas, Kathleen Airhart, and John Tristan. 1995. Implementing pest-control strategies for vocational and therapeutic greenhouses. *HortTechnology,* Vol. 5, Number 2:182-184.

Ball, Vic. 1991. *Ball red book,* fifteenth edition. Reston, VA: Reston Publishing Co.

Boutard, Sherry and Douglas Airhart. 1982. *Trolley cart for adaptive use.* Stockbridge, MA: Berkshire Garden Center.

Corr, Brian. 1993a. *Greenhouse "always do-s" and "never do-s".* Knoxville, TN: Agricultural Extension Service, University of Tennessee.

Corr, Brian. 1993b. *Greenhouse philosophy and structures.* Knoxville, TN: Agricultural Extension Service, University of Tennessee.

Crater, Douglas. 1993. *Retail greenhouse management.* Knoxville, TN: Ornamental Horticulture and Landscape Design, University of Tennessee.

Faust, Jim. 1995. 10 steps to WPS compliance. *Tennessee Flower Growers Association Bulletin,* Vol. 4, No. 2.

Freeman, Jennifer. 1985. Helping Hands. *Horticulture* August:54-58.

Hewson, Mitchell L. 1994. *Horticulture as therapy.* Homewood Health Center, 150 Delhi Street, Guelph, Ontario N1E 6K9, Canada.

Larson, Roy. 1992. *Introduction to floriculture,* second edition. New York, NY: Academic Press, Inc.

Luse, Sue Beven and Mary Mandeville. 1977. Gardening under lights. Fifth Annual Conference, American Horticultural Therapy Association, Marianna, Florida.

Nelson, Paul V. 1991. *Greenhouse operation and management,* fourth edition. Reston, VA: Reston Publications.

Relf, Diane. 1983. *Not your usual tool.* Newsletter. American Horticultural Therapy Association, Vol. 10 (1): 1, 4.

Rothert, Eugene Jr. and James Daubert. 1981. *Horticultural therapy for nursing homes, senior centers, retirement living.* Glencoe, IL: Chicago Horticultural Society.

Rothert, Gene. 1994. *The enabling garden: A guide to lifelong gardening.* Dallas, TX: Taylor Publishing.

Straus, Martha.1991. *Adapting gardening to meet your changing needs.* Friends Hospital, Philadelphia, PA 19124.

Thabault, George and Lynn Ocone. 1982. *Tools and techniques for easier gardening*. Gardens For All, 180 Flynn Avenue, Burlington, VT 05401.

U.S. Equal Employment Opportunity Commission (USEEOC) and the U.S. Department of Justice. 1991. *The Americans with Disabilities Act Handbook*. Document No. 869-010-000-96-1, US Government Printing Office, Washington, DC 20402.

Yeomans, Kathleen. 1992. *The able gardener: Overcoming barriers of age and physical limitations*. Pownal, VT: Storey Communications.

Chapter 13

Botanical Gardening: Design, Techniques, and Tools

Matthew Frazel

INTRODUCTION

Botanic gardens and arboreta have arisen over the last twenty years to provide training and knowledge about the use of horticulture in therapeutic settings. As U.S. public gardens have developed more programs for their municipal constituencies, therapeutic horticulture programs under the auspices of public gardens have gained in ascendancy. This chapter will explore this phenomenon.

STATEMENT OF THE ISSUES

Where can someone in a large city in the United States receive training or even an exposure to the concepts and practice of horticultural therapy (HT)? Because horticultural therapy degree programs may not exist at schools and universities in certain areas, it makes sense that the botanic gardens and arboreta of most major U.S. cities should offer some training or orientation to the concepts and practice of horticultural therapy.

Most urban-centered public gardens are developing greater awareness of their roles and responsibilities to their communities. Sometimes this is due to the fact that public garden funding stems from a municipality. In other cases, the motivation has more to do with a desire to increase visitorship and the impact of programs on an urban environment. In this context, horticultural therapy programs under the auspices of public gardens take on added significance and meaning. This chapter will examine how three different public institutions have responded to an awareness of the community, to the needs of people with disabilities, and to the need to provide training in the concepts and principles of horticultural therapy.

The Role and Value of Public Gardens as a Horticultural Therapy Resource

History and Development

In the medieval world, monastic gardens and libraries served to guard and nourish knowledge through a period that became known as the Dark Ages. This knowledge bank encompassed such broad areas as medicine, botany/horticulture, music, literacy, and the study of languages. In the contemporary world, gardens usually take the form of municipal or university-based botanical institutions and some private gardens open to the public.

Today public gardens provide the urban city dweller with a place of quiet and repose. The beauty of a garden serves as a restorative *bank* for the contemporary urbanite. Increasingly however, such gardens have taken on newer missions and embraced programs that go beyond a strict plant-focused approach. The reality of public funding for gardens has challenged many institutions to more closely link their programs to the public that supports them. Programs of this sort include information services about gardening for the home, school programs teaching the elements of science literacy, neighborhood garden initiatives, and pertinent to this chapter, horticultural therapy programs.

In this context, the modern public garden can act as a rich source of training, information, and programming in the specialized use of horticulture in therapeutic settings. An urban-based public garden can serve as a forum for the dissemination of the ideas and methods associated with therapeutic horticulture. The value of a public garden as a place for restoration of the city-dweller can encompass the restoration of spirit, body, and psyche, a restoration that flows from involvement in therapeutic horticulture programs.

The role a public garden develops in response to this specialized call of the community varies from garden to garden. Examining the unique aspect and history of each public garden reveals how some gardens have developed such programs and others have not. The strengths and emphasis of one garden may be a specific plant collection or an orientation toward display gardens. Other gardens have developed a program-based or audience focus and produce a wide variety of programs aimed at various segments of the public.

Two main factors contributed to the rise of horticultural therapy programs at public gardens:

1. The environmental movement: This movement arose in the 1960s and 1970s, contributing to a greater awareness of the importance of green and open spaces in cities, and also spawned a tremendous interest in gardening. By the mid 1980s, according to a survey conducted by Gallop Organization Inc. for the National Gardening Association, gardening was among the top three most popular outdoor leisure activities. Over 44 percent of the nation's households garden in some form or another. The environmental movement brought a greater focus on eating healthier, more naturally grown fruits and vegetables. This led city-dwellers to turn to their local expert sources of information about plants: public gardens.

2. The disability rights and civil rights movement: As the civil rights movement evolved in the 1960s, issues pertaining to people with disabilities came to the fore. During the late 1970s and 1980s, disability rights issues became more broadly voiced throughout society. This culminated in 1990 with the passing of the Americans with Disabilities Act. As awareness grew regarding disability rights, public gardens were challenged to offer more programs that were accessible to all people regardless of their level of ability.

The coming together of these two influences (the environmental movement and the disability rights movement) led to a response on the part of publicly funded public gardens to more directly address their population base in the form of programs, classes, services, and tours. As public gardens began to turn more to their community settings in an effort to *humanize* their gardens, horticultural therapy programs were a natural component of this development.

The formal development of horticultural therapy in the United States in the early 1970s is linked to a number of hallmark public garden-based horticultural therapy programs. Under the auspices of botanical garden institutions, horticultural therapy programs began to develop at gardens in New York City, Chicago, Denver, Cleveland, and in North Carolina. Such programs were often spearheaded by a single determined staff member with a vision. Grant funding initially supported a number of these programs. Broad, institutional support developed after audiences were identified and programs began having an impact. These programs led gardens to reach out to groups not traditionally served and have elevated the profile of public gardens in their communities and cities. (See Table 13.1.)

TABLE 13.1. Horticultural Therapy Services Offered by Public Gardens in the United States in 1995

Indoor Horticultural Therapy Projects

- growing orchids
- repotting older plants
- terrariums
- growing windowsill herbs
- kitchen gardening
- desert dish gardens
- cooking with herbs
- learning to take plant cuttings

- making fresh holiday flower arrangement
- growing African violets
- planting bulbs for indoor bloom
- ordering seeds for the summer garden
- planting seeds for the outdoor garden

Outdoor Horticultural Therapy Projects

- preparing a garden site
- vegetable gardening
- flower arranging
- cooking with edible flowers
- butterfly gardening
- weeding, watering, pruning

- taking the garden down
- planting a tree
- flower gardening
- flower drying
- growing herbs
- growing birdseed plants

- planting a pizza garden
- growing fruits
- making a garden salad
- harvesting herbs
- building a compost pile
- taking cuttings of annuals

Craft and Nature-Related Programs

- making sachets
- flower pressing
- using dried flowers in arrangements
- decorating pumpkins
- pressed-flower bookmark
- taking nature walks
- building birdhouses
- collecting butterflies
- cutting brush in the woods
- chopping wood

- making a pine-cone feeder
- mixing a spice boil
- mixing birdseed from birdseed plants
- mixing herbs for potpourri
- making wreaths with dried materials
- building nature trails
- making an insect display
- field trips to greenhouses and nature centers
- drying fruits

Common construction materials can be used for container gardens. Here flue tiles and sewer tiles filled with soil demonstrate a simple, small space, wheel-chair-accessible garden. Pictured in the background is a wheelchair-accessible garden built with vertically set landscape timbers. Note the continuous barrier-free paving throughout the garden space. (Chicago Botanic Garden)

Components of Horticultural Therapy Programs at Public Gardens

Horticultural therapy programs at public gardens in the United States take a variety of forms due to several factors: the founding orientation of a specific horticultural therapy program, the public garden's mission and its relationship with its community, and the community response to horticultural therapy programs. This section will explore some of the diversity of programs currently being offered by public gardens in the United States.

OVERVIEW OF THE RANGE OF HORTICULTURAL THERAPY SERVICES AT PUBLIC GARDENS

Several U.S. public gardens have developed extensive horticultural therapy programs that have acquired a diversity of elements and services. Services have developed as a result of theses programs finding a niche in the community, by meeting a need, or by cultivating a market for the services they render.

Contracts, Consultation, Design, and Presentations

Contract Programs

Some public gardens contract to provide a series of horticultural therapy programs to health care and human service facilities. Fees generally are assessed based on costs incurred. Most public gardens are not-for-profit entities that seek to break even with fees covering expenses. In addition, revenue can also be raised from fees for services or support by grants or endowment funds.

Contract programs are a simple, market-oriented way to offer horticultural therapy services to a community. Under the auspices of a public garden, a horticultural therapy service can create a market for itself. The support of the public garden provides credibility, name recognition, and additional ancillary support including plant production, information resources, equipment usage, and special resources and materials like compost, unusual plant materials, and access to experts.

A standard contract form is presented in Figure 13.1. By its nature, a contract lends legitimacy to an agreement between parties. With signatures and terms made clear, a contract spells out what each side will provide to fulfill the terms of the arrangement.

Consultation Services

Horticultural therapy consulting services under the auspices of public gardens varies widely and may include

- information services about setting up horticultural therapy programs,
- site visit consultations to facilities considering the establishment of therapeutic gardens,
- phone consultations about programmatic components of a horticultural therapy project,
- consultations in the form of one-day in-services focusing on a particular application of horticultural therapy to a specific group or special population, and
- specialized information about the specific use of certain plants.

Public garden horticultural therapy consulting services generally are provided on a fee basis with sliding fees typical. Fees for such services

FIGURE 13.1. Sample Contract Agreement for Horticultural Therapy

We (your facility name)_____ agree to participate in the year-round horticultural therapy program sponsored by the New City Public Garden as outlined in the program proposal. We understand that payment will proceed in quarterly installments of $--------each. The following personnel from your facility have been designated to work with the program:

Name:————————————Title:————————————

Name:————————————Title:————————————

Name:————————————Title:————————————

Name:————————————Title:————————————

. . . .

Please sign below

Facility Representatives:

Facility Administrator:————————————Date:————————

Activity Director:————————————————Date:————————

New City Representatives:

————————————————————————Date:————————
New City Public Garden Director

————————————————————————Date:————————
New City Horticultural Therapy Director

provide additional revenues, which in turn support the existence of a horticultural therapy program.

Design

A horticultural therapist has expertise in developing therapeutic gardens that provide interactive experiences and daily activities to stimulate the senses and challenge the ability level of a diverse client group.

With the input of a horticultural therapist, gardens can be built to be accessible, so that containers are the proper height, the surfaces have sufficient rigidity, paths are wide enough, and plant materials are properly selected. In this manner, public gardens not only encourage horticulture and gardening, but also encourage the establishment of gardens that can be effectively used and enjoyed by a wide variety of people. Design services are an excellent way for a public garden to expand its reach and broaden its contact with its community.

Kay Radnor, a Chicago Botanic Garden volunteer in the Enabling Garden, is shown removing old flower stalks from an annual fountain grass. The garden bed in front of her is built in a raised fashion, so that the gardener does not have to reach all the way to the ground. Note also the two- by ten-inch board set on top of the bed. This makes an ideal place for sitting. (Chicago Botanic Garden)

Public Presentations

Horticultural therapy presentations are made under the auspices of a wide variety of public gardens and further the mission of a public garden to promote horticulture. Horticultural therapists address such diverse groups as university classes, allied health professional conferences, teachers groups, garden clubs, municipal organizations, corporate employee enrichment programs, and specialized workshops. Presentations can raise revenue for a horticultural therapy program and raise the profile of a public garden in its community.

Program Models

Two types of programmatic models may be used by horticultural therapy programs at public gardens:

1. Outreach versus on-site program
2. Direct service versus staff training

Outreach Versus On-Site Program

Horticultural therapy programs are carried out not only under the direct sponsorship of the public garden, but also under the public garden's umbrella per se. With this model, horticultural therapy programs take place on the grounds of the public garden itself. With this orientation, both short- and long-term programs can take place. Short-term programs can be as brief as a one-time event, or a segment of programs conducted once a week over a series of weeks. Long-term programs can run from six months to a year or more and become multiple year contracts. A number of gardens provide this kind of on-site horticultural therapy programming.

Over the years, such a program orientation can lead to the development of a specialized garden/facility for horticultural therapy programs. Several U.S. public gardens have developed or are currently developing on-site horticultural therapy gardens. Such a garden focuses a public garden's horticultural therapy programs in that area and creates a place where the programs are housed. In addition, such a model also serves as a public demonstration of horticultural therapy techniques and methods. This point underlines the role a public garden can play in fostering horticultural therapy programs throughout a community. These specialized therapeutic gardens become both a place to hold horticultural therapy programs and a place for agencies, health care facilities, and interested members of the general public to learn about how to build or how to use a therapeutic garden.

The alternative model to the on-site HT program is to provide outreach programs to interested groups and organizations at locations in the community. Such a program is also sponsored by the public garden, but its implementation is away from the public garden itself and located at the site where the contract groups and populations are serviced, where they live, where they go to school, etc.

This outreach approach can be very powerful because the public garden's program goes directly to the site. Being in-house and on location, such an HT program often has a broader impact on staff and clients beyond the horticultural therapy group working with the public garden. A program established at a contract site can be utilized by multiple groups and individuals.

Direct Service Versus Staff Training

The other model of program orientation is whether to provide direct service horticultural therapy programs or staff training in horticultural therapy programming. These are two very different approaches to providing HT.

With a direct-service horticultural therapy program, the aim is to provide the contract site and its clients with a therapeutic experience of horticulture. Activities run the gamut from an in-depth program focusing on vegetable gardening from A to Z, to having a couple of flower-arranging sessions. The focus is on the experience of the client with the medium. Generally, a program of this sort relies heavily on the expertise of staff from the public garden. When the involvement of the public garden ends, typically the program also comes to an end. Without the input and support from the public garden staff, the program has difficulty continuing. The advantage of this approach for a contract facility is to access the benefits of a horticulture therapy program without incurring the costs associated with hiring staff and purchasing materials.

The alternative model to the direct-service approach is to provide horticultural therapy staff training programs. HT activities are carried out with contract site clients and most importantly, contract site staff. The goal is to teach methods of using horticulture and gardening with clients. With this approach, staff become trained in gardening techniques so that when the public garden's involvement comes to an end, the facility's staff are able to continue to carry out the program independently.

The staff training approach is in line with broad, current public garden educational program efforts to train teachers, putting skills in the hands of staff who can then carry on certain activities after contract completion.

OVERVIEW OF CLIENT GROUPS AND AGENCIES SERVED BY PUBLIC GARDEN HORTICULTURAL THERAPY PROGRAMS

The client groups and agencies that public garden horticultural therapy programs serve are similar to those served directly by horticultural therapists around the country without adding additional staff. The public garden staff can expose the contract facility to the practical application of horticultural therapy with its clients and assess its impact and effectiveness to determine if it would like to incorporate horticultural therapy into an overall treatment or activity program.

The types of agencies with which public garden horticultural therapy programs contract are presented in Table 13.2.

TABLE 13.2. Types of Agencies with Which Public Garden Horticultural Therapy Programs Contract

• Psychiatric hospitals	• Correctional facilities
• Training centers for people with developmental disabilities	• School systems
• Residences for older adults and senior care facilities	• Municipal services for people with disabilities
• Public housing agencies	• Shelters for battered women
• Children's hospitals	• Residences for abused children and wards of the state
• Substance abuse programs	• Community mental health centers
• Veterans Administration hospitals	• Programs for at-risk youth
• Special recreation/park district administrations	• Early intervention programs
• Intergenerational programs	• School and training programs for people with visual and hearing impairments
• Cancer treatment facilities	
• Vocational training programs	• Hospice care programs
• Rehabilitation hospitals	

CONTENT OF PUBLIC GARDEN HORTICULTURAL THERAPY PROGRAMS

So what then is contained in terms of program content in a public garden horticultural therapy program? A public garden horticultural therapy program focuses on using plants and nature-related activities to address treatment and activity goals for clients. However, since public garden horticultural therapy staff are typically engaged as outside contract providers, such goal setting and tracking is generally the responsibility of the contract facility staff. Horticultural therapy includes projects that broaden the life experiences of clients. Some of these projects are listed in the charts at the back of this chapter.

HORTICULTURAL THERAPY STAFFING PATTERNS AT PUBLIC GARDENS

The majority of public gardens that have horticultural therapy programs utilize one staff person to carry out a wide variety of functions, including the following:

- Program developer: In addition to establishing the overall orientation, the horticultural therapist is charged with selecting, developing, organizing therapeutic content, and writing all program materials.
- Garden curator: The staff horticultural therapist is responsible for design, care, and overall maintenance of therapeutic gardens.
- Project coordinator: The horticultural therapist carries out all the HT programs with specialized groups, in-service training programs with staff or direct-service programs aimed at clients from human service agencies.
- Materials and supplies chief: The horticultural therapist gathers and organizes all specialized materials to be used in programs, including everything from pumpkins, to soil, to magic markers, etc.
- Marketer: Because public garden horticultural therapy programs are offered as a for-fee service contracts, the more contracts there are signed, the more solidly the horticultural therapy program will take root at a public garden. This job of lining up new sites is also often the staff horticultural therapist's.
- Volunteer coordinator: Volunteers are utilized to provide additional hands with all the tasks involved. Horticultural therapists oversee the scheduling of HT volunteers.

- Internship supervisor: Most horticultural therapy public garden programs offer horticultural therapy internships, providing students with a supervised six-month training in the use of horticulture in therapeutic settings.
- Workshop organizer: In addition to providing horticultural therapy programs, many public gardens also offer training workshops as a way to provide short-term exposure to horticultural therapy methods and techniques.

As of 1996, the horticultural therapy staff norm at public gardens in the United States is one staff person. Two gardens employ more than one person. Some gardens engage a horticultural therapist on an independent contract basis to provide services. At other gardens, horticultural therapy staff work on a part-time basis. Horticultural therapy programs are also

Patsy Magner and Kay Radnor, Chicago Botanic Garden volunteers, prepare to hang fresh herbs for drying. Herbs grown in the summer garden are dried and used in fall indoor activities. (Chicago Botanic Garden)

offered through public garden education departments where programs are run by "education specialists."

THREE CASE STUDIES OF PUBLIC GARDEN
HORTICULTURAL THERAPY PROGRAMS

The Holden Arboretum, Mentor, Ohio

The Holden Arboretum is located on over 3,100 acres and serves the greater Cleveland area. The arboretum was envisioned by Albert Holden, a wealthy mining engineer who died in 1913. His estate was left in trust to benefit the creation of an arboretum. This arboretum began to take form in 1931 and has grown vigorously from an initial acreage of just 100 acres.

The horticultural therapy program began in 1954 with strong support from an early supporter of the arboretum, Mrs. Warren Corning. She had been inspired by viewing such programs in England where they were widely practiced. Together with Lewis Lipp, the plant propagator from the arboretum, Mrs. Corning began to bring men and women from nearby retirement residences and a rehabilitation hospital to the arboretum for workshops.

Interest in the program grew, and the arboretum received increased requests for the plant activity programs. These programs had on-site orientations with groups coming from the surrounding area to the Holden Arboretum. Initially, the classes were organized in five-week blocks in both the spring and fall. A wide variety of topics were taught, including sowing and sprouting of seeds, plant propagation activities, and tree and shrub identification. These hands-on sessions were interspersed with short presentations by knowledgeable arboretum staff.

Grant support helped the program off to a substantial start. Funds were used to purchase program materials and cover staff and transportation costs. Throughout the 1960s, the arboretum program generated enthusiastic response that culminated in the formation of a horticultural therapy program in the education department in 1970. The mission statement of the Holden horticultural therapy program is to enable individuals with special needs to benefit from therapy through horticulture by engaging professionals in educational programs, providing therapeutic programming, and instigating research. Grant awards in 1973 and 1976 were used to develop horticultural therapy programs, expand services, and broaden the scope of horticultural therapy programs. A registered horticultural therapist was hired, numerous horticultural therapy interns were trained, and horticultural therapy programs have been developed both on- and off-site. In the 1980s, the Holden horticultural therapy program developed traveling workshops that brought horticulture and garden training programs to contract sites four times a year.

Holden arboretum serves a three-county area with contractual program arrangements with health care facilities throughout the greater Cleveland metropolitan area. Some programs have a direct service orientation, while others train contract site staff to carry out programs after contract completion. Horticultural therapy staff provide consulting services to organizations in Ohio interested in developing therapeutic garden spaces and horticultural therapy programs. Community services include a radio program, a monthly senior horticulture group, participation in display events such as the Abilities Expo and Senior Day. The establishment of "satellite" horticultural therapy programs extends the reach of the arboretum and broadens its impact. Professional training is supported by means of a newsletter, workshops, in-service trainings, and student internships. The arboretum horticultural therapy program extends the arboretum user group to include a broader base of the local community, demonstrates the arboretum's interest in the community, provides a positive horticultural impact, and exhibits accessible gardening methods to the public.

The Minnesota Landscape Arboretum, Chanhassen*

The Minnesota Landscape Arboretum was established in 1958. It is an operating unit of the Department of Horticultural Science and Landscape Architecture of the University of Minnesota. The mission of the arboretum is to provide a national resource for horticultural and environmental information, research, and public education; to develop and evaluate plants and horticultural practices for cold climates; and to inspire and delight all visitors with quality plants in well-designed and maintained displays, collections, and model landscapes.

The Minnesota Landscape Arboretum has a number of people-oriented programs:

- The Learning Center program annually serves over 27,000 school-age children with educational tours, gardening programs, summer day camps, and the traveling Plantmobile.
- The adult education program offers year-round classes in horticulture, flower design, herbs, landscaping, pruning, and related topics.

---

*The information regarding the Minnesota Landscape Arboretum was contributed by Jean M. Larson, Coordinator of Therapeutic Horticulture.

- The Anderson Horticultural Library, a noncirculating library, houses over 10,000 volumes and various other publications on horticulture, botany, landscaping, gardening, and natural history and children's books.
- The Horticultural Research Center provides extensive information in fruit breeding, winter hardy propagation, and tree and shrub development.
- The Therapeutic Horticulture Services are available to persons with varying abilities and professionals in social service fields. It offers integrated programs for the general public through classes, symposia, direct services, and resource information.

Because of the healing, recreation, and integration properties of plants, the Minnesota Landscape Arboretum incorporated a therapeutic horticulture program into its educational agenda in 1992. A coordinator was hired in 1992 to develop therapeutic horticulture. The Minnesota Landscape Arboretum's program offers training, direct program delivery, resource information, and research projects that identify and support the beneficial effects of people and plant interaction. The program teaches social service professionals, community recreation organizers, teachers, horticulturists, and others to develop activities that use horticulture as a therapeutic tool.

Training emphasizes strategies and techniques of therapeutic horticulture primarily through design and implementation of HT in collaboration with a service agency, either on-site at the arboretum or at the community facility. The training is hands-on, practical, and individualized. Other educational opportunities include a spring conference, on-site classes, year-round workshops, and short courses for community agencies and organizations.

An agency may request activities in therapeutic horticulture on site at the arboretum or at a community center. Examples of therapeutic horticulture programming include summer intergenerational gardening for children and elders; recreational gardening for Native American youth; therapeutic gardening for adults with developmental disabilities; year-round horticulture education for detainees at a county correctional facility; recreational and educational gardening for children and women healing from domestic abuse; and outdoor education and gardening for adults with mental illness and developmental disabilities.

Resource information about therapeutic horticulture is available from the Anderson Horticultural Library and the office of the therapeutic horticulture coordinator. Books, articles, tools/equipment, and computer-based bibliographies are available.

The Clotilde Irvine Sensory Garden and Therapeutic Horticulture Program Center is a fully accessible garden and showcases a variety of accessible design, equipment, and containers to arboretum visitors. It incorporates plants and other structures that stimulate the senses and includes easy, low-maintenance plant selections. The center is a place where interactive, hands-on training and programming occur. Integrative programs for all are held at the program center throughout the summer weekends (e.g., "Sense-Sational Sundays"). The arboretum believes the Clotilde Irvine Sensory Garden and Therapeutic Horticulture Program Center, along with the therapeutic horticulture program itself, serve as a national model of inclusiveness and accessibility at public gardens.

Chicago Botanic Garden

The Chicago Botanic Garden opened in 1965. It is owned by the Cook County Forest Preserve District and managed by the Chicago Horticultural Society, an organization in existence for over 100 years. The mission of the Chicago Botanic Garden is to stimulate and develop interest, appreciation and understanding related to gardening, horticulture, botany, and the conservation of natural resources. A variety of gardens, plant collections, education, and research programs of excellence have been developed while recognizing a need to provide a continuing aesthetic landscape experience.

The establishment of two programs contributed to the establishment of the Department of Horticultural Therapy:

1. In 1974, the Founders Fund of the Garden Clubs of America provided funding to build the Learning Garden, later renamed the Enabling Garden for People with Disabilities. This garden serves as a demonstration of specialized garden structures and surfaces that allow someone with a physical disability to garden. Some of the structures include raised beds and planters, gardens built in a vertical manner with soil encased in wooden upright structures, and the utilization of accessible hanging planters connected to pulleys and winches. This garden was one of the first of its kind in the United States and has inspired a number of similar therapeutic gardens throughout the United States.

2. The Chicago Botanic Garden established the Department of Horticultural Therapy in 1977. From an early on-site programmatic orientation of individual classes, workshops, and special trainings focusing on a wide variety of horticultural topics, the Horticultural Therapy Department developed with grant support a pilot off-site component in 1978. Initially, funding was provided to establish "satellite" therapeutic horticulture at four health care and human service facilities. This program was well received and expanded with addi-

tional grant support in 1981 to provide services to ten nursing homes. A series of four horticultural therapy manuals were published by the Chicago Botanic Garden and the first horticultural therapy interns were placed. In 1982, the HT Department was expanded to the regional health care network and nine facilities contracted with the botanic garden to provide year-round horticultural therapy training and activities.

Currently, the Chicago Botanic Garden's Horticultural Therapy Services is staffed by two horticultural therapists and a student intern. Salaries are paid from the botanic garden's operating budget. Program costs other than salary (e.g., materials, operating costs) are offset by contractual revenues generated from year-long contracts with up to twelve contract sites. The current contract fee is $4,500 per year for sites in Cook County and $5,000 outside the county.

The central aim of the Chicago Botanic Garden's HT Services is to provide both horticultural training and training in the specialized adaptation of horticulture in health care settings to staff from contract facilities.

This photo shows a place in the Enabling Garden where a group can sit around a table and work together. Working on joint tasks together is a wonderful way to bring people in contact with one another. (Chicago Botanic Garden)

This active, hands-on training takes place by means of a schedule of forty weekly off-site therapeutic horticulture program sessions. Chicago Botanic Garden staff provide all materials necessary to conduct these sessions and they staff the program on an alternating weekly basis. In the intervening weeks, staff from the contract facility carry out a lesson plan Chicago Botanic Garden staff have left behind. Facility staff learn horticultural therapy techniques by observing Chicago Botanic Garden staff work with their clients. In the intervening week, facility staff have an opportunity to try their hand at the horticultural therapy programs. With a program of forty weekly sessions, the projects take place throughout each of the four seasons and reflect the type of activities described in the earlier section, Content of Botanic Garden Horticultural Therapy programs.

Another training opportunity is the annual spring and fall gardening workshops at the Chicago Botanic Garden. These full-day training sessions are open to contract staff, staff from other health care and human service agencies, students, and interested members of the general public. These workshops provide an exposure to concepts and practices used in horticultural therapy, as well as horticultural information pertaining to specific topics (for example, water gardening, drying flowers, basics of propagation, and integrated pest management).

THE FUTURE OF PUBLIC GARDEN HORTICULTURAL THERAPY PROGRAMS

The question that arises from this examination of horticultural therapy programs in public garden settings asks, is this phenomena peculiar to a few gardens spread throughout the United States or is it a broader trend? This chapter argues that American public gardens realize that the greater the number of groups served, including the specialized groups served by horticultural therapy programs, the broader the impact the public garden has on its community.

Reexamination of the Role Public Gardens Can Play to Promote the Development of Horticultural Therapy

It appears that certain gardens have developed horticultural therapy because the funding support was present at a critical point in the founding of the program. Once these pilots have gotten up and running, local awareness has grown about these special programs and they have been valued (both monetarily and in terms of the demand for services). The marketing

This shot shows how the sitting ledge can be used to make it easier to harvest vegetables from a raised bed. Such an approach also presents an opportunity to exercise the waist and trunk by turning toward the task in the bed from a seated position. (Chicago Botanic Garden)

of horticultural therapy to the social service network is a hallmark of successful public garden-based horticultural therapy programs. This market success has contributed to the stability and longevity of public garden-based horticultural therapy.

But public gardens are not social service agencies. So why have they gotten involved in horticultural therapy? Probably the easiest explanation for this is in the effort on the part of all public museums to serve and increase their audience communities. A public garden with a diverse array of services that reaches out to all segments of the community becomes enlivened and *peopled* by these programs. Such a garden becomes a much richer place and in turn becomes more meaningful to its audience, its community. Particularly for public gardens that receive operating funds from municipalities and taxing bodies, a diversity of services, including horticultural therapy is an explicit way to provide activities and services to its direct supporters.

A panoramic shot of the Enabling Garden and some of its features. (Chicago Botanic Garden)

In broad terms, public garden horticultural therapy can be understood in light of the growing awareness of the human dimension involved in horticulture, and in this case with horticultural institutions. What better group to take on a nature walk through a public garden forest but a group of children from a psychiatric hospital? Is bringing an amaryllis planting activity to an Alzheimer's unit not within the realm of what a contemporary American public garden is? Clearly such programs expand and grow public garden institutions. So it is understandable that new gardens are highly attuned to the "people-program," or humanistic aspect of a public garden. It is in this regard that a number of public gardens have recently established such programs, or gardens and others are investigating such programs.

Horticultural therapy at public gardens can provide a vehicle for a public garden to share its expertise with its community. Horticultural therapy training, workshops, symposia, and information resources serve as a community platform from which a public garden can declare its expertise and usefulness to the community.

ORGANIZATIONS

American Association of Arboreta and Botanic Gardens, 786 Church Rd., Wayne, PA 19087 (610) 688-1120.

American Community Gardening Association, 325 Walnut St., Philadelphia, PA 19103 (215) 625-8280.

National Gardening Association, 180 Flynn Ave., Burlington, VT 05401 (802) 863-1308.

SOURCES FOR ADDITIONAL INFORMATION

Bottle Biology. 1993. Paul H. Williams. Kendall/Hunt Publishing Co., 4050 Westmark Dr., PO 1840, Dubuque, IA 52004-1840 (800) 346-2377.

The Enabling Garden: Creating Barrier Free Gardens. 1994. Gene Rothert. Taylor Publishing Co., 1550 W. Mockingbird Ln., Dallas, TX 75235.

Growing with Gardening, A Twelve-Month Guide for Therapy, Recreation and Education. 1989. Bibby Moore. University of North Carolina Press, P.O. Box 2288, Chapel Hill, NC 27515-2288.

Growth Point Magazine. Horticultural Therapy, Goulds Ground, Vallis Way, Frome, Somerset, BA113DW, England.

Horticultural Therapy at a Psychiatric Hospital. 1981 (out of print). Eugene Rothert and James Daubert. Chicago Horticultural Society, P.O. Box 400, Glencoe, IL 60022 (708) 835-8250.

Horticulture for the Disabled and Disadvantaged. 1978. Damon R. Olszowy. Charles C Thomas, Publisher. Bannerstone House, 301-327 E. Lawrence Ave., Springfield, IL.

Horticulture Therapy for the Mentally Handicapped, Chicago Horticultural Society.

Horticulture Therapy for Nursing Homes, Senior Centers and Retirement Living, Chicago Horticultural Society.

Horticulture Therapy at a Physical Rehabilitation Facility, Chicago Horticultural Society.

People-Plant Council News. Virginia Polytechnic Institute and State University, Blacksburg, VA 24061-0327 (703) 231-6254.

Plants for People: The Psychological Effects of Plants. 1992. Judith Keane. SRB 92-04 Special Ref. Briefs. Reference Office, Rm. 111, National Agricultural Library, U.S.D.A., 10301 Baltimore Blvd., Beltsville, MD 20705- 2351.

Chapter 14

Community Gardening: Design, Techniques, and Tools

Patricia Schrieber

INTRODUCTION

Beginning in the 1970s, community-based greening efforts emerged across the country in response to the deteriorating quality of life in inner-city neighborhoods. Shifts in industry and population since World War II had caused an increase in the number of abandoned, decaying structures and vacant lots in inner cities with a corresponding decrease in the tax base and social services available to the people who lived there. Community leaders began working with city agencies to seal buildings, schedule demolitions, and clean trash-strewn vacant lots, which they still do today, twenty-five to thirty years later. City dwellers with rural roots and an urge to garden began to reuse vacant land for community gardens, illustrating their commitment to cultivating a pride of place.

At the same time, in cities such as Boston, New York, Philadelphia, Chicago, San Francisco, Los Angeles, Denver, and Seattle, community greening programs began taking shape under the guidance of an array of sponsoring city agencies, nonprofit organizations, and agricultural exten-

The author wishes to thank the following colleagues for providing information used in the preparation of this chapter: Michael Adrio, Executive Director, *Gateway to Gardening*, P.O. Box 299, St. Louis, MO 63166; Kristi Appelhans, *Idaho Falls Community Garden Association,* 6643 Limousin Ave., Idaho Falls, ID 83404; H.R. Draper, *Pennsylvania State University's Urban Gardening Program,* 4601 Market St., Philadelphia, PA 19139; Allison Mowry, Outreach Horticulturist, *BC Green, Leila Arboretum Society, Inc.,* 928 West Michigan Ave., Battle Creek, MI 49017; Lorka Muñoz, Director, *Grow with Your Neighbors, Five Rivers Metro-Park, Dayton Montgomery County,* 1301 E. Siebenthaler Ave., Dayton, OH 45414.

377

sion programs. The common denominator shared by all community green-ing organizations has been their cooperation with community residents working to keep resources and positive energies flowing in their neighbor-hoods. Funding for community greening programs has been provided by city, state and federal sources as well as foundations, corporations, and individual donors. The track record set by numerous community greening organizations over the past two decades illustrates horticulture's excellent healing potential for urban communities.

This chapter illustrates the beneficial impact of community greening to urban communities, offering a different kind of therapeutic experience than more traditional horticultural therapy models. Greening projects can change communities and the individuals who live in them, often bringing community members together for the first time. The reader will gain knowledge about the components of community greening programs, the kinds of groups served by community greening organizations, and how partnerships with other organizations support the progress of community greening programs. Case presentations will offer examples of community greening projects from different cities, and the challenges groups faced in getting them going.

COMPONENTS OF A COMMUNITY GARDENING PROGRAM

Gardening Opportunities

People living in urban areas who participate in community gardening programs can find a variety of opportunities to support their efforts in (1) creating community gardens, (2) caring for street trees, (3) reclaiming parks, (4) managing open space, (5) training for community leaders, (6) cultivating youth, and (7) sustaining interest through events.

Creating Community Gardens

The success of a community garden rests on the energies of the core group of residents who see the possibilities for neighborhood beautifica-tion in a vacant lot filled with trash. The neighbors may long for a garden spot with benches, flowers, and several small trees. They may dream of growing the vegetables and herbs they remember from their grandparents' gardens long ago. Some may envision a nurturing, safe place for neighbor-hood children, where youngsters can taste, touch, and learn about the power of plants.

The gardens with the greatest longevity are typically run by an organized group of cooperative people. They must get permission to use the lot; locate sources for fencing, soil amendments, plants, and seeds; and divide the responsibilities for getting the jobs done. If the group receives funding or materials from a community greening organization, the time line for the progress of the project may be shorter than for those groups that must raise funds themselves. Whether a group works from a formal plan or adds elements as time goes by, the garden acquires unique qualities flavored by the preferences and personalities of the gardeners who collaborate on its development.

Caring for Street Trees

Community groups have learned how easy it is to shift the environmental balance in their neighborhoods by planting trees and improving the health of existing trees along their streets. In cities like San Francisco, New York, and Philadelphia, community members are passionately inter-

Community gardeners provide a colorful front border as a welcome for visitors and friends to Mr. McGregor's Garden, sponsored by the Grow with Your Neighbors Community Gardening Outreach Program of Five Rivers Metro-Parks, Dayton, Ohio.

ested in maintaining a healthy urban forest. They become familiar with the government agencies and nonprofit organizations able to respond with funding, training, and technical assistance. They organize their neighbors to work together to plant and maintain new trees. As the trees mature, these tree advocates take on some of the more challenging aspects of long-term maintenance, raising funds for pruning, removing dead trees, and planting new trees.

Reclaiming Parks

Economic conditions in the nineteenth century crowded workers and their families into neighborhoods adjoining manufacturing centers. The resulting communities, where brick, concrete, and asphalt dominate, offered little or no room for open space. Social reforms of the late nineteenth century created large parcels of city parkland and added smaller neighborhood squares to enhance the more populated communities. Diminishing city budgets have caused the deterioration of these green spaces, motivating community residents to join forces to save these precious urban treasures. Community interest in cities like Portland, Chicago, Philadelphia, and New York have given birth to park groups that vigorously campaign for improvements. These groups perform multiple tasks: lobby for government funding; organize fundraising events; and inventory maintenance problems such as eroding slopes, overgrown shrubs, broken benches, and trees in decline; create master plans; and commit to the physical labor needed to clear debris and replace dead or damaged plantings.

Managing Open Space

A number of organizations with strong community development agendas have emerged in inner cities over the past two decades. These organizations are involved in a wide range of activities focused on community rebuilding, attracting resources for housing rehabilitation, new housing construction, and revitalization of the business base of the community. More recently, these organizations have begun to look beyond housing and economic development to the broader community environment. Adding new housing units to the existing stock is only part of the solution when the remaining vacant land is still filled with debris. A community organization can identify degraded places most in need of revitalization within the neighborhood through a vacant land inventory. Priorities can be set based on available resources. Low and moderate cost options can be studied. For example, one short-term, low-cost solution for a large tract of vacant land

Sun Circle Community Gardeners lovingly care for the garden created from a large, trash-filled vacant lot, with the help of the Pennsylvania Horticultural Society's Philadelphia Green Program.

could be a perimeter tree planting where frequent mowing of weeds can result in a soft-surfaced playing field. Long-range planning and fundraising could later turn this space into a formal recreation area with ball fields, playgrounds, and staffing to support year-round recreation programs.

Training for Community Leaders

Community greening organizations often bring community leaders together to share success stories about how they recruit and maintain volunteer support for their projects. Through participation in training sessions, leaders acquire a variety of skills to sustain gardens and other greening projects when outside resources become more limited. Leaders who coordinate successful greening projects are often energized to take on other community issues. Training provides opportunities for new leaders

Two generations of Norris Square community gardeners work in the Philadelphia garden known as Las Parcelas, where a brightly colored little house, or
"la casita," helps teach young people more about their Puerto Rican heritage.

to get involved in planning for revitalized communities, giving them a
greater stake in the decision-making process.

Cultivating Youth

Youth can broaden their understanding of the natural world by participating in classroom investigations, gardening activities, and field trips to
community gardens and parks. Yet giving youth horticultural experiences
within the school curriculum is only part of the picture. Community greening organizations are making a greater effort to tap the positive energies of
youth and to cultivate in them a sense of community stewardship. Young
people help plant community gardens, parks, recreation centers, and other
community open spaces. They join after-school and summer programs that
sponsor neighborhood-wide environmental programs. They learn how to
care for gardens and street trees within their neighborhoods and teach
what they have learned to their neighbors. Through involvement in these

community-based activities, a young person can acquire a deeper understanding of the urban environment, a vision of his or her own community's place within that environment, and an opportunity to explore alternatives more constructive than graffiti or vandalism.

Sustaining Interest Through Events

Community greening succeeds in large part because of the ongoing social interaction of the participants. While community groups can sustain interest by holding periodic meetings and work days, the bonds among members are strengthened by hosting potluck dinners, bake sales, plant sales, and raffles. They may participate in city-wide activities like a garden contest or harvest festival sponsored by a city agency or nonprofit organization. Contest winners receive well-earned recognition for their community revitalization efforts. Groups that participate in harvest festivals celebrate the end of a bountiful gardening season and proudly exhibit the fruits of their labors. These city-wide activities also carry the cross-fertilization of ideas and experiences beyond neighborhood boundaries and can strengthen the formal and informal network of community greening leaders.

Groups Served by Community Gardening Organizations

Community gardening programs generally provide support to (1) community and block organizations, (2) target communities, and (3) special audiences.

Community and Block Organizations

People living in urban areas benefit from membership in neighborhood associations. Whether a group is composed of residents from one city block or a coalition of neighboring blocks, members are brought together by the need to improve conditions or to fight against the impact of an undesirable project like a highway that would divide a community. The strength of a neighborhood group depends on the level of experience and commitment shared by its members. Some groups have by-laws, hold regular meetings, develop action plans, and maintain treasuries for handling problem situations. Others exist on a more informal basis, with members showing a willingness to act only when a particular threat warrants it. In all cases, a strong circle of volunteers is essential to carry out efforts on behalf of the community. Otherwise, an individual with a vision becomes overburdened by the project if he or she is unable to attract or accept the assistance of neighbors ready to work for the good of the community.

Community groups may look to community greening organizations as sources for information and inspiration. A community greening program may provide plants and soil, help set up fencing for a new project or offer rototilling at the beginning of a growing season. The organization may conduct training sessions that clarify the basic steps needed to establish a successful greening project and cover issues related to land ownership and permission to use the land, gardening skills, and fund-raising techniques. When roles and expectations are defined at the outset, there is a better chance that all parties can move ahead knowing what they must do to ensure the success of the greening projects.

Target Communities

Community greening organizations in Philadelphia and Battle Creek are among those that have concentrated greening initiatives in specific geographic areas in cooperation with residents, community organizations, and other technical assistance agencies. When resources are focused

A Leila Arboretum master gardener joins residents of Stratford Park Town Homes in Battle Creek, Michigan, in a planting sponsored by BC Green.

within a concentrated area, positive results are almost immediately visible. A community-wide, block-by-block revitalization effort can promote community awareness, provide training for block captains and tenant councils, and introduce younger leadership into the cadre of community stewards. Neighborhood successes that capitalize on improvements through greening often attract private and public monies for new housing, housing rehabilitation, new streets, and sidewalks. Projects targeted within public housing communities have the power to change assumptions about public housing as a source of conflict and crime. Community activists residing in public housing complexes show the same dedication and desire for a better quality of life as in any other community neighborhood. Community greening becomes an integral element in the synergism created with concurrent community development efforts, such as housing, social service programs, and economic development.

Special Audiences

Involvement in community gardening provides significant physical and social benefits for older adults and people with disabilities. Older people are among the most enthusiastic, vigorous, and adaptable participants in the community greening movement. While younger people are frequently caught up with family and work responsibilities, community elders usually maintain the standards of community well-being. Though the majority of gardens are established without any special amenities for seniors, community greening programs sometimes work with senior centers or community gardeners to set up raised beds and paved walkways to accommodate physical limitations.

Community gardens are often started on a shoestring budget, with every attempt being made to find free or inexpensive materials. When people with disabilities are involved in a garden project, the community group must attract additional resources since construction costs for paved pathways and wheelchair-accessible raised beds are relatively high. Groups must look for creative ways to raise funds and garner in-kind services from designers and construction crews to develop accessible gardens.

Partnerships with Other Organizations

The effectiveness of a community gardening program is frequently enhanced by the partnerships created with (1) city and other government agencies, (2) foundations, corporations, and private donors, and (3) universities and colleges.

The Idaho Falls Community Gardening Association was founded on the principle of accessible gardening for all community members.

City and Other Government Agencies

Community greening organizations depend on their ability to create partnerships with government agencies because of complex issues such as obtaining permission to use publicly owned land, adhering to guidelines for using the land, and finding public monies to support projects. Those community greening organizations that are well connected to government networks are often in the right place when word gets out about available funding. At the same time that greening organizations administer program activities, they can leave the lobbying efforts to those community leaders who are politically aware and know what to do and whom to call when major obstacles threaten to derail their projects.

Foundations, Corporations, and Private Donors

While a community greening organization may rely on government sources of funding, frequent cutbacks and current fiscal realities motivate program administrators to diversify funding and discover the extent of available assistance, both locally and nationally, from foundations, corporations, garden clubs, civic organizations, and individual donors. Program administrators for community greening organizations must prepare letters, proposals, and substantiating documents highlighting the past successes and future plans for their organizations. Fortunately, a number of funding organizations have set their funding priorities on improving the quality of life in urban communities and are very interested in supporting community greening efforts.

Universities and Colleges

Local universities and colleges can support community greening initiatives in several ways. While community greening organizations possess anecdotal information, they often lack the data and documentation required by funders to substantiate the impact that community greening has on the quality of life in environmentally challenged communities. Academic research teams can gather information and analyze results of studies that verify the benefits of community greening. Community greening initiatives can provide rich learning experiences for students participating on study teams. These students can acquire valuable practical training by working cooperatively with community organizations. Students who learn to listen acutely while community members express their ideas will be better equipped as professionals involved in community revitalization projects in later years.

CASE STUDIES

Stories of community gardening efforts from Dayton, Philadelphia, Battle Creek, Idaho Falls, and St. Louis are offered as examples of projects developed in collaboration with a broad range of participants from (1) community and block organizations, (2) target communities, and (3) special audiences.

Community and Block Organizations

Riverdale Neighborhood Association, Dayton, Ohio

The Riverdale Neighborhood Association approached Grow with Your Neighbors Community Gardening Outreach Program of the Five Rivers MetroParks at a time when a great deal of negative publicity was being broadcast about their community. The group wanted to develop a garden at a major intersection as a way to publicize the good news about their neighborhood.

As people worked together to articulate their needs, the garden became a centerpiece that highlighted the positive energies existing within the community. Within four years, Mr. McGregor's Garden became a neighborhood showcase. A 110-foot-long border of lavender and yarrow frames the front of the garden. A rose-covered arbor welcomes the visitor. Formal beds in the center are filled with herbs and bulbs. Twenty dedicated gardeners care for the garden along with forty children from a local day care center who have adopted the perennial border along the street as well as a couple of plots within the garden. One man, living across from the garden, gardened there and designated himself the garden watchman. When a rash of thefts almost caused this neighbor to stop gardening, children from the day care made cards saying how much they appreciated what he had been doing. They helped revive his enthusiasm and he became involved once again.

Mr. McGregor's Garden has provided an opportunity for people to make significant changes in their lives. The original goal of giving hope and restoring pride to the community has been fulfilled. This project, like other community greening initiatives, proved to be a doorway through which people could participate more intimately in their community, helping them handle challenging issues in a tangible way. From the conceptual stages of Mr. McGregor's Garden and onward, the Riverdale Neighborhood Association was headed by a very energetic woman who was accustomed to being very vocal and doing things by herself. Her dominant leadership style helped the garden to grow and flourish as she motivated residents to become involved in the construction of the garden. Once the most challenging aspects of establishing the garden were completed, this strong leader began to turn the garden coordination over to a new garden leader with a less aggressive management style, who shared her passion for the garden. She moved on to lead other neighborhood improvement projects, no longer the neighborhood association president. She learned from her garden experiences that in order for a community project to be truly sustainable, a leader not only delegates responsibility but inspires others to become leaders as well. And so, the natural process of renewal continues.

Sun Circle Garden, Philadelphia, Pennsylvania

A coalition of residents from several neighboring blocks created the Sun Circle Garden in the mid 1980s with help from the Pennsylvania Horticultural Society's Philadelphia Green Program. The group's history with Philadelphia Green dated back to the time they planted street trees together. After five vacant houses were demolished at a busy corner, group members organized frequent cleanup days while waiting three years before resources were available from Philadelphia Green.

Leaders of the garden attended planning workshops run by the Philadelphia Green staff. They made a low-maintenance plan for a parklike setting, with trees in island beds, and a colorful flower bed in the center, and added vegetable beds to the plan as a way to encourage a higher level of maintenance. The gardeners reported that they found it much easier to care for the trees, shrubs, and flowers while they were out in the garden tending their vegetables.

The Sun Circle gardeners have been frequent winners in the City Gardens Contest and the Harvest Show, events sponsored by the Pennsylvania Horticultural Society. They have demonstrated an impressive ability to be resourceful and to use advice from staff and contest judges to continually make improvements in the garden. They arranged for the painting of a mural on an adjacent house wall. The gardeners are delighted with their success in transforming the Sun Circle Garden into a valuable asset for their North Philadelphia community.

Target Communities

Norris Square Greene Countrie Towne, Philadelphia, Pennsylvania

When the Pennsylvania Horticultural Society's Philadelphia Green Program selected the Norris Square neighborhood to be developed into a

Greene Countrie Towne in the late 1980s, community leaders were focused on the opportunities afforded by such a designation. Each community named as a Greene Countrie Towne was offered a concentration of greening projects that would transform vacant lots into gardens and add color and shade of trees to stark streets. However, a high crime rate, caused by drug dealing on local street corners and vacant lots, had to be reversed before work on their Greene Countrie Towne could get underway. Three years of efforts on the part of the community residents, with assistance from police and drug enforcement agencies, finally forced the drug dealers to move out of the neighborhood. As the drug dealers loosened their hold on the community, work began on the greening of vacant lots and streets in Norris Square.

All the time the community focused its attention on forcing the drug dealers to leave the area, the leaders kept the image of Norris Square as the first Hispanic Greene Countrie Towne firmly planted in their minds. They intended to use the gardens as a way to honor their Puerto Rican culture. They now celebrate their blended Native, African, and European heritage, illustrated by a colorful mural in the garden named *Raices*, meaning "roots." The stone circle in the El Batey Garden reminds them of the place held sacred by the native people of Puerto Rico.

The major community organization involved in this greening initiative has been the Norris Square Neighborhood Project (NSNP), operating since the mid 1970s as an urban environmental education center. NSNP offers environmental programs that generate a sense of hope and pride in the youth of that community. In classroom and after-school activities, the young people learn to appreciate the people and places within their immediate neighborhood. They work in the neighborhood gardens and they participate in the care of the community park, sing the songs of Puerto Rico at community festivals, and learn from their elders, who teach them how important it is to work in service to their community.

Several other organizations have supported development of the Norris Square Greene Countrie Towne. The Pennsylvania State University's Urban Gardening Program (PSUG) provides workshops and site visits to Philadelphia community gardeners in food and nutrition under the guidelines of the U.S. Department of Agriculture. PSUG established a Norris Square demonstration garden to teach good gardening practices. The artists from the Philadelphia Anti-Graffiti Network (PAGN) created dramatic murals in partnership with Norris Square. One mural depicts youth and elders in the act of rebuilding their community, while another illustrates the rich heritage of Puerto Rican culture.

BC Green, Battle Creek, Michigan

BC Green, sponsored by the Leila Arboretum, has worked in Battle Creek's low- and moderate-income communities since 1993. Stratford Park Town Homes, a low-income housing development, is one of BC Green's early success stories. Stratford Park had become a community where crime rates were high and a strong police presence was a constant source of frustration to the residents. Several years ago, Stratford Park's new manager developed revised policies and regulations that clearly stated that every household would be required to provide a minimum number of hours working on community improvement projects. The manager enlisted BC Green's help as a partner in support of community-wide revitalization efforts.

Community residents started a vegetable garden and planted flower beds around the complex with the help of BC Green staff. Several tenants attended the arboretum's master gardener training program, and used their knowledge to plan and plant perennial gardens and street trees in the community. After taking the master gardener training, one neighbor put his newly acquired skills to work as an enthusiastic member of the tree selection committee. Other residents worked with arboretum staff to create a plan for plantings along a series of blank walls at one end of the neighborhood.

Stratford Park's community greening efforts brought about a significant reduction in crime when the residents began to work together. Their efforts showed that people who lived in Stratford Park really cared. The community as a whole felt a sense of appreciation for all the hard work, and people became much better able to nurture the environment in which they lived.

Special Audiences

Idaho Falls Community Garden Association, Idaho

Idaho Falls, a comparatively small city of 50,000, has few of the day-to-day challenges faced by larger urban areas, yet the concept of community gardening, a fairly recent phenomenon in Idaho Falls, provides a way to encourage productive interaction among people, from different socioeconomic levels, who may have dissimilar goals and interests but share a living place.

The idea for the Idaho Falls Community Garden Association was first inspired by a garden created at a sheltered workshop for clients with multiple handicaps. A volunteer, who was familiar with the workshop, offered to set up a gardening program there as a way to fulfill a community service requirement for the master gardening training she had just completed. She built two thirty-foot-long wheelchair-accessible raised beds with the help of other volunteers. At the end of two years, officers of the nonprofit organization that administered the workshop met with the master gardener to discuss how well the clients had benefitted from the gardening activities. She suggested that the garden be expanded as a way to improve relations with the surrounding community, which was not always hospitable to workshop clients and their special needs.

The nonprofit organization gave permission for a community garden to be established on an adjacent piece of land. A planning group was set up to engage the support of a cross section of community participants, including the cooperative extension service and its master gardeners; senior citizens; social service agencies; day care facilities; businesses; governmental agencies; fraternal organizations; scouts; and individuals. The city and county cooperated by grading the site; hauling materials such as soil, woodchips, and leaves; and laying water pipes. The garden start-up was delayed by extremely wet weather, but within a year after planning began, it was in full swing.

The garden is still young, and situations continue to test the strength of the group. The community garden association hopes that with enough enthusiasm and inspiration from the success of the first garden, other neighborhoods will be motivated to create community gardens accessible to all members of the community, no matter what their special needs may be.

Meade Elementary School, Philadelphia, Pennsylvania

The Beech Corporation, a local community development corporation in North Philadelphia, asked the Pennsylvania Horticultural Society's Philadelphia Green to establish a school gardening program for the Meade Elementary School in 1992. Parents, teachers, and students participated in planning sessions to create designs for a series of garden spaces. Their collaboration resulted in a multifaceted teaching garden and a school tree lab installed at the end of the school year. Students designed tiles that were incorporated into the garden paths and worked with an artist to create a wall mosaic depicting natural and built elements of the urban environment. Classes use the teaching garden for outdoor lessons on various subjects. Students tend small trees in the tree lab until the trees are large enough to plant at various sites within the surrounding community. Teachers lead the students in classroom gardening activities, growing plants under lights and on windowsills to later be moved out to the gardens. Students learn how to care for flowers, herbs, and trees growing in the gardens. They also grow entries for the city-wide Junior Flower Show where they receive numerous awards each year.

Philadelphia Green staff provided frequent teacher training sessions over the first two years of the school gardening program. As the faculty acquired more skills, they were ready to manage the program themselves, with a minimum of assistance from Philadelphia Green. They raised funds to establish a parenting-skills project using gardening as a focus for a series of parent-child projects. Participating parents later helped lead introductory workshops for teachers new to the gardening program. The faculty maintains a bulletin board in the school lobby, highlighting seasonal gardening activities. Teachers, parents, and students are proud of their school gardening program. Their cooperative efforts stand out as positive contributions for a community rebuilding itself.

Bell Garden, St. Louis, Missouri

Bell Garden was one of the first food gardens started by St. Louis' Gateway to Gardening (GTG), which has supported neighborhood revitalization through greening since 1984. Bell Garden had fallen into decline after six years in operation because many people had moved from the neighborhood. When GTG found itself spending a great deal of money maintaining the large site without much community support, the program director began looking for a reasonable solution to the situation. A highly motivated GTG volunteer, trained as a master gardener, suggested turning the garden into a training site for community gardeners from all over the city. She and a small group of GTG volunteers joined together to reclaim the garden that was conveniently located in the center of St. Louis. They sent flyers to other community groups to see who would be interested in helping to bring the garden back to life. A small corps of hard workers attacked the shoulder-high grass, and with funding provided by GTG, set up demonstration areas for square-foot gardening, composting, and children's garden activities. They built a picnic shelter and a lath house for winter protection of young trees and shrubs. Every Saturday, from March to October, at least one volunteer is on hand at the garden to offer guidance to visitors and coordinate GTG's distribution center for seeds and seedlings.

Now that the garden lives again, the original leader is looking to reduce her involvement. The Bell Garden committee and other members of the city-wide GTG Community Garden Committee are beginning to share leadership roles. They are establishing tasks and becoming more involved in planning and programming for the garden. Not only has Bell Garden been restored as a working garden, but its role as community training center has broadened, attracting people from different walks of life, who are crossing generational, racial, geographic, and economic boundaries.

A Master Gardener, who can be found working in the Bell Garden on a daily basis, prepares to greet volunteers on a workday sponsored by the community gardeners and Gateway to Gardening, St. Louis, Missouri.

Carousel House, Philadelphia, Pennsylvania

People with disabilities travel to Carousel House from all parts of the city to participate in a wide range of recreational activities, including gardening. Carousel House is the only Department of Recreation facility in the city of Philadelphia specifically designed to address the needs of disabled people. Although individuals go to Carousel House from many different neighborhoods, the staff works to create a sense of community for all participants.

The garden holds a mix of opportunities, with room for both individual gardening and communal areas shared by different groups that bring their clients to the garden. Penn State's Urban Gardening Program provides support with demonstration plots and the latest tools and techniques to assist the disabled gardener. Staff members guide each individual in a direction that is appropriate to his or her own needs. People involved in the step-by-step process of gardening learn how to use the energies and resources available to them. They see that it is not necessary to do everything at once. Simple successes keep them interested. Each participant comes away with a better understanding of gardening and the incredible power of the plant world. Each person can move past frustration into an awareness of her or his own sense of self. Working in the garden at Carousel House gives each individual the feeling that he or she belongs to this special community of gardeners.

MOVING INTO THE FUTURE WITH COMMUNITY GREENING

Issues and Opportunities

Nurturing Communities Through Greening

While all the problems currently facing urban communities cannot be solved by planting gardens on vacant lots or trees along city streets, people involved in community greening activities foster a spirit of hope and demonstrate an ability to nurture, two elements integral to community revitalization. People have helped change their neighbors' perceptions about the immediate community because of their efforts to improve their surroundings through greening. Vacant lots piled high with garbage have been replaced by gardens that are centers for community activity. City streets lined with trees and colorful containers have created a different impression for visitors and have given a boost to neighborhood morale. Those who grow food in community gardens help themselves, their families, and others within the community with whom they share their garden bounty. People are less stressed when they work in gardens. Young people who join greening activities acquire new skills, some of which have lead to employment. Their involvement in greening projects has brought increased self-esteem because of their accomplishments. Community gardeners have honed leadership skills, and some have become committed to civic-minded activities even if this was not their original intention.

Vision for a New Urban Community

Picture a new urban community in the twenty-first century. Rather than being limited by the twentieth-century model, where blocks of houses faced each other over paved streets, houses in this new urban community are arranged in clusters, surrounded by open land, a community garden at the center, with tree-lined paved paths radiating from it, which neighbors use for bikes, wheelchairs and walking, far from street traffic. A banner over the entrance to the garden acclaims the surrounding neighborhood an award-winning community oasis in the city's community greening competition. The community bulletin board at the garden gate announces the topics for the monthly gardening workshops hosted by the gardeners for the rest of the community.

It is a Saturday morning in spring, and community residents are gathering at the garden, carrying the tools they will use to maintain the common areas of the garden and open spaces as they always do on the second and fourth Saturday of each month. Several neighbors are reading the special notice on the bulletin board offering the services of teens from the after-school program sponsored by the recreation center several streets away. The teenagers are looking for assignments to fulfill their community service requirement for the urban environmental studies course conducted by the recreation center in conjunction with the local high school. The two people, who are this year's garden coordinators, hold an impromptu meeting and ask the gardeners to suggest some projects the teens could work on. Repairing and painting benches, mulching the trees along the paths and out along the streets, building the new raised beds, and repainting the toolshed are a few chores that immediately come to mind. The group looks forward to the teens' help again this year because the service projects have been very successful since the program started ten years ago. In fact, several of the neighborhood's most dedicated gardeners came through the program and have continued on as mentors.

The gardeners hear the news that community members are invited to attend a special meeting called by the City Planning Commission and the Office of Housing and Community Development being held at the community center to discuss plans for the development of another new urban community in the district. City planners and housing coordinators want to hear reactions to the drawings they developed from all the suggestions made at the last community meeting. Everyone shares a jubilant moment, acknowledging all the work that went into the collaborative planning process.

Community Greening Models Support the Vision

The basic building blocks needed to turn this vision into reality are already in position. Given the current trends in urban areas, the amount of urban vacant land will increase over the next several decades. Community organizations can turn this additional open land into valuable neighborhood assets, using the many successful community greening models created over the past several decades as sources of inspiration for rejuvenating the urban environment in future years. They can build on the partnerships created among community residents of all ages, as well as with public and private institutions to keep the community revitalization momentum going. They will succeed as long as they nurture the tradition of strong community participation so that others will continue to help restore, protect, and enhance the communities in which they live, building on the efforts of the generations that came before them.

ORGANIZATIONS

American Community Gardening Association, c/o The Pennsylvania Horticultural Society, 100 N. 20th Street, Philadelphia PA 19103.

American Forests, Urban and Community Forestry, P.O. Box 2000, Washington, DC 20013.

American Institute of Architects, Philadelphia Chapter's Community Design Collaborative, 17th & Sansom Streets, Philadelphia, PA 19103.

National Gardening Association, 180 Flynn Avenue, Burlington, VT 05401.

BIBLIOGRAPHY

American Community Gardening Association. *Community Gardening Review.* Vols. 1-5, Philadelphia, PA: American Community Gardening Association, 1991-95.

American Community Gardening Association. *National Community Gardening Survey.* Philadelphia: American Community Gardening Association, 1992.

Brooklyn Botanic Garden. *Community Gardening.* Brooklyn, NY: Brooklyn Botanic Garden Publications, 1979.

DiSalvo-Ryan, DyAnne. *City Green.* New York: Morrow Junior Books, 1994.

Francis, Mark, Lisa Cashdan, and Lynn Parson. *Community Open Spaces.* Covelo, CA: Island Press, 1984.

Huff, Barbara A. *Greening the City Streets: The Story of Community Gardens.* New York: Clarion Books, 1990.

Hynes, H. Patricia. *A Patch of Eden: America's Inner-City Gardeners.* White River Junction, VT: Chelsea Green Publishing Company, 1996.

Jobb, Jamie. *The Complete Book of Community Gardening.* New York: Morrow Quill Paperback, 1979.

Naimark, Susan (ed.). *A Handbook of Community Gardening: Boston Urban Gardeners.* New York: Charles Scribner and Sons, 1982.

The Pennsylvania Horticultural Society. *Urban Vacant Land: Issues and Recommendations.* Philadelphia: The Pennsylvania Horticultural Society, 1995.

Rodale Press. *The Community Land Trust Handbook.* Emmaus, PA: Rodale Press.

Sommers Larry. *The Community Gardening Book.* Burlington VT: National Gardening Association, 1984.

Trust for Public Land. *Participatory Design, Development and Maintenance.* New York: Trust for Public Land.

PART FOUR:
SKILLS FOR HORTICULTURAL THERAPY PRACTICE

Chapter 15

Consultation Services
for Horticultural Therapy Practice

Melanie Trelaine

INTRODUCTION

In the health care field today, new opportunities for providing horticultural therapy services as a consultant or independent contractor afford flexibility and innovation in program possibilities. Contracting and providing consultation services offers hospitals and other large organizations the skills and services of a horticultural therapist where budgets cannot provide a full-time staff position. Although the professional field of horticultural therapy has existed since the late 1940s, many health care professionals are not aware of the growing number of individuals working in the field; therefore, as a consultant, individuals can develop and demonstrate the benefits of the partnership of people and plants.

The purpose of this chapter is to provide an outline to develop, market, and implement a horticultural therapy program for a variety of health care facilities as a consultant. This chapter will explain contracts and setting up your business organization, and will provide pointers to help you negotiate a contract, develop your services, and effectively market them to the health care field. Practical advice about purchasing supplies and estimating fees for services and other helpful information is provided to get you started.

WHY CONSULTANTS ARE NEEDED

A new field such as horticultural therapy (HT) takes time to demonstrate its benefits to the established professionals working in allied fields. Often those benefits have been brought to the attention of administrators through contact with professional organizations and literature. The desire to bring a qualified HT program to those facilities often initiates the need for HT consultant services. Many times it is the HT professional who contacts the administrators to educate them about the benefits of HT and offers to assist in the development of a program at the facility.

To ensure that the HT program meets all the necessary requirements for documentation, evaluation, and assessment, an agreement or contact is drawn up between the facility and an HT consultant. Specific work to be performed as well as a clearly defined responsibilities of the facility and the contractor is included in the written agreement. Further items to be included in this agreement would be a definition of a reasonable time frame in which to accomplish the goals and defined procedures to develop criteria for evaluation and assessment of the program.

Consulting offers flexibility, innovation, and entrepreneurial possibilities for many people contracting their horticultural therapy services. In areas where employment as a horticultural therapist is limited, consultants are able to provide services to a variety of health care providers. Horticultural therapy consultants are skilled in developing specific programs for a particular facility or group. Their job begins with a comprehensive assessment, taking in all the factors involved with initiating an HT program. Then they work with the facility's staff to develop and implement the program, assess the outcomes, and further adjust the program to fit the needs, budget, and expectations of the facility. A consultant may have specific goals outlined in a contract or memorandum of understanding between the facility and the consultant business organization. Many facilities hire a consultant to help them assess and develop programs, or they may want the HT to provide programs on a regular basis. With budget concerns, they may prefer to pay specific fees for services rather than employ an HT as full-time staff.

DEVELOPING A CONSULTATION SERVICE

Setting Up Your Business Organization

Consultants usually set up a business organization when they begin consulting. Before you begin to contact other business organizations, you should first start with this important step. Not only will you present yourself as a professional in the field, but there may be liability and tax protection issues that should be considered before you begin. Various types used by practicing professionals appear in Table 15.1.

Horticultural therapists, working as consultants, predominantly use sole proprietorships and corporations. Before setting up a business to perform HT consulting or contracting work, thoroughly investigate the laws and

TABLE 15.1. Various Types of Business Organizations Set Up by Practicing Professionals

	Sole Proprietorship	Partnership	Corporation
Advantages	• Low startup costs • Greater freedom from regulation • Owner in direct control • Minimum working capital requirements • Tax advantages to small owner	• Ease of formation • Low startup costs • Additional sources of venture capital • Broader management • Limited outside regulation	• Limited liability • Specialized management • Ownership is transferable • Continuous existence • Legal entity • Easier to raise capital • Unity of action having centralized authority in board of directors
Disadvantages	• Unlimited liability • Lack of continuity • More difficult to raise capital	• Unlimited personal liability • Lack of continuity • Divided authority • Difficulty in raising additional capital • Hard to find suitable partners	• Closely regulated • Most expensive to organize • Charter restrictions • Extensive record keeping necessary • Double taxation except when organized as an "S Corporation" • Difficult to liquidate investment

legal ramifications as well as taxation requirements for your state and local municipalities.

Several organizations can be of assistance:

• U.S. Small Business Administration and SBA District offices
• Small Business Development Centers (SBDCs)
• Service Corps of Retired Executives (SCORE)
• Small Business Institutes (SBIs)

Consult your telephone directory under U.S. Government for your local SBA office or call the Small Business Answer Desk at 1-800-UASK-SBA for information. You also can request a free copy of *The Small Business Directory* listing of business development publications and videotapes from your local SBA office or the Answer Desk.

Other sources for help are state economic development agencies, chambers of commerce, local colleges, libraries, and manufacturers and suppliers of small business technologies and products.

Review Your Liabilities and Obligations

Rapidly changing regulations in any aspect of the health care profession today increasingly places greater responsibility on client outcomes, particularly those involving therapy and behavior. Consultants should seek advice from their insurance providers or obtain legal counsel to define particular areas of liability (professional or general liability) for any contract work being performed.

Administration

How payment for services is received will be determined by the facility and should be outlined in the contact agreement. Many facilities require the HT to provide a bill for services to be submitted weekly or monthly to the finance office. Most contractors can expect payment to be received within seven to fourteen days after receipt of the bill for services. Some facilities will have a check ready at the time of the HT program and others will pay for services in cash.

Taxes

Tax information and forms regarding self-employment tax, social security benefits, and other related fees can be obtained from the Internal Revenue Service Center or the state department of taxation listed in your local telephone directory. The HT contractor should have an established method of payment of any applicable taxes as required by law and determined by the type of business organization.

Insurances

With today's professional liability risks, insurance is an important and expensive part of private contracting. Facilities may require the HT to have at least $1 to 3 million professional liability insurance. Seek professional advice about protecting your business organization from possible claims and understand the liabilities risk before engaging in a contract agreement. Information about personal health insurance can be obtained by contacting your local insurance representative.

MARKETING YOUR SERVICES

Starting an HT contract program requires research and specific expected outcome projections. Learn as much as you an about the facility in which you have chosen to start a program before making contact with them. The business department at the local library is an excellent resource for business journals that publish lists of companies (by type) along with the names of directors, size, and financial data. This information will assist you in determining potential organizations by using the information in directories, contacting business, and community organizations, as well as thru professional networking. After your business organization is established, you can send letters, flyers, or brochures personally to directors and administrators of potential facilities outlining your services and credentials. HT contractors often follow up those initial letters, brochures, or flyers with a personal phone call to the directors/administrators to answer questions and discuss the prospective program further.

Community exposure to your horticultural therapy services can come through in-service presentations, television, radio, newspapers, and magazines, as well as health care trade shows, small business networks, and volunteer service organizations. Many local stations have gardening or related programs on the air and usually will host an interview as a public service or as a special interest program. As the HT professional gains experience and exposure, word of mouth is one of the best methods of communicating that new programs are started. A database of organizations and contacts created from the HT's research and information is also beneficial for further consideration or future contacts.

Donations

Garden Clubs and horticultural organizations, plant-related businesses, hotels, shopping centers, and other businesses are happy to donate unwanted plant materials. Determine how much time you are willing to spend picking up donations and what records may be necessary to keep for any donations received. Consider where you will keep the donated plants and materials until they will be used in your program. To begin receiving donations, flyers mailed out to prospective donors announcing your needs usually work well.

TABLE 15.2. The Research Process

	Hospital/Hospice	Rehab Ctr/Training	Skilled Nursing	School/University	Community Garden/Park	Adult Day Program	Correctional/Substance Abuse	Retirement/Senior Center
Sensory stimulation	X	X	X		X	X	X	X
Floral design	X	X	X	X	X	X	X	X
Accessible garden design	X	X	X	X	X	X	X	X
Horticultural crafts	X	X	X	X	X	X	X	X
Art and design	X	X	X	X	X	X	X	X
Adaptable tools/techniques	X	X	X	X	X	X	X	X
Landscape design	X	X	X	X	X		X	X
Community gardening		X			X	X	X	
Vocational/job coaching		X		X	X	X	X	
Cooking/nutrition	X	X	X	X	X	X	X	X
Staff training/workshops	X	X	X	X	X	X	X	X
Educational seminars	X	X		X	X		X	X
Greenhouse operations		X		X	X		X	X
Physical therapy	X	X	X	X		X	X	X
Recreational therapy	X	X	X	X		X	X	X
Occupational therapy	X	X	X	X		X	X	X
Speech therapy	X	X	X	X		X	X	X
Plant physiology	X	X	X	X	X	X	X	X
Cognitive stimulation	X	X	X		X	X	X	X

RECRUITING CLIENTS

Once the business organization is formally and legally established, horticultural therapists begin to contact local health care organizations about starting HT programs. Research all you can about the facility and its mission. Tailor your marketing strategy to fit their current needs for services to patients. Find out what criteria they usually require for securing work as a contractor. Focus on the work you will be performing rather than securing a long-term contact with the facility. Administrators may not be familiar with horticultural therapy and may not be interested in a lengthy formal contact. You might offer a demo session to show how the program works and how a regular program at the facility would benefit their patients. Many medical facilities are undergoing staff and budget shortages; therefore, potential HT work must stress assistance with those critical areas of need in order to gain the ear of the administrator. From your researched information as well as interviews with the facility staff, you should be able to develop an initial outline of what your program can do to help the facility.

FORMULATING CONTRACTS
AND DEVELOPING BUDGETS

There are three main types of contracts for consulting services:

Professional Services Contract

A professional services contract is the most formal and is appropriate for large organizations and when there is a broad scope of work to be performed. Some organizations may be hesitant to sign a complex agreement with the HT if they are unfamiliar with horticultural therapy or their business organization requires signatures of the board and/or legal department.

Memorandum or Letter of Understanding

Less formal is the memorandum or letter of understanding, which is usually brief and outlines the work to be performed, goals to be met, and fee to be paid. Most organizations are more comfortable with this type of agreement due to simple language and signing authority of staff.

Verbal Contract

Verbal contracts are often used by HTs where each party has an understanding of the work to be performed and goals to be accomplished. This type of contract might be acceptable in specific circumstances where a facility has just been introduced to horticultural therapy and is hesitant to enter into a formal agreement initially. After a brief introductory period, a written contract would be developed. A verbal agreement is not recommended by professionals as a general rule because it offers no protection to the horticultural therapist in the case of a dispute or liability.

HORTICULTURAL THERAPY SERVICES

Assessment, Needs, and Feasibility of Starting a Program

Any medical group or agency in the health care industry needs accurate information and assessment of a patient's participation and benefit thru the HT contract program. Assessments may be in the form of clinical end-points such as routine patient charting, client functioning, program outcomes, facility logistics, staff meetings, or there may be none at all. Each facility has different needs as to what they need to assess to provide the needed information on your program outcome. An assessment would outline the goals and objectives of the program as well as client goals and objectives. If an assessment is desired, the development of its form and content should begin with the initial contact and may take several revisions before an adequate assessment tool is ready to use.

Assessment begins with definition of the population group to determine the functioning level of clients as well as definition of specific program goals. Following this, a schedule or calendar of program dates should be developed and staff involvement and support outlined. The HT should find out what documentation or evaluation is needed by the facility to chart client progress and goal achievements. Following a facility tour, the HT consultant should familiarize him or herself with the logistics of the physical buildings and determine specific challenges to resolve (e.g., transportation of materials into areas of the hospital, use of carts, access to locked areas). The HT should have a clear outline of all client needs/goals in addition to the facility assessment prior to the start of the program.

NEGOTIATING THE CONTRACT OR AGREEMENT

Compensation for HT Services

Probably the most difficult part of getting started in contracting is determining what to charge for your services. Assessment of the facility, clients, and budgets would help the HT determine an appropriate fee. Based on the frequency of the program, number of clients, and types of programs, HT professionals generally charge a fee per hour. Some facilities charge per client, which helps them keep within budgets. Program materials may or may not be figured in this fee. The HT professional must determine all the program costs (i.e., preparation time, transportation, supplies, taxes, insurance) in order to figure a profit margin. For example, driving thirty miles to deliver one program per week with five clients may not be as profitable as driving five miles to a nearby hospital and providing three back-to-back programs for fifteen clients.

Discuss Budget for Program

An important issue of any contact is always the cost. The HT program budget should be defined and clearly stated in the contract. The HT professional must determine how many clients are to be involved with the program, who will be responsible for purchasing materials, and how they will be paid. Payment is either up front or by reimbursement. Similar to the assessment, each facility has a specific way to pay for contract services, such as cash at time of program, check mailed monthly, or through billing. The HT professional should prepare a program outline for the term of the contract and provide a budget and outline of specific materials that need to be purchased and/or provided by the facility. If resources are limited for the facility, offer alternate methods for procuring materials such as donations from garden clubs, garden centers, and related businesses, as well as from wholesale nurseries. Incorporation of community resources and volunteers will benefit the facility as well as ensure success for your program. In some cases, the use of volunteers may not be allowed by the facility, so it is a good idea to ask when you are negotiating the contract.

Estimating Supplies

Depending on the number of clients and how often you provide HT programs, you will need to estimate cost and quantity of materials. Initially, this can be difficult, but as you gain more experience, you will become proficient at it. You can experiment by measuring cups of soil out into a smaller bags, for example, dividing larger flats of plants into smaller and easier to carry groups, as well as precutting fabrics for potpourri. Keeping a record of your project materials will help you not only estimate future projects more easily, but will give you the information you need to develop an accurate budget for materials for future contracts.

Suggested Material and Supplies to Get You Started

1. Seeds for flowers and vegetables
2. Soils such as perlite, vermiculite, peat moss, and speciality mixes for cactus, orchids, African violets, etc.
3. Dried plant materials for nature crafts and potpourri projects
4. Variety of aromatic herbs (mints and scented geraniums are excellent)
5. Fresh flowers for pressing and crafts
6. Books and colorful pictures and posters of plants, particularly flowers
7. Pots, clay and plastic, in a variety of sizes and colors
8. Small hand tools (foam-grip tools for outdoor gardening and long-handled tools for beds)
9. Plastic cups, all sizes and colors, to make excellent pots for annuals indoors decorated with ribbon, and clear cups to make mini-greenhouses for windowsills and propagation
10. Moist towelettes or washcloths for cleanup
11. Ribbons and decorative stickers
12. Clear or decorated cellophane bags for wrapping flowers and potting cups
13. Cart for transporting materials to program site
14. Scissors and pruners, and plastic spoons and forks make good indoor shovels
15. Tablecloths, cloth and/or latex gloves, and apron and protective covering for clothes
16. Trays to keep tables and floor area clean—assists visually impaired locate objects
17. Cuttings for propagation programs
18. A variety of houseplants, including cactus and succulents

Evaluate Client Needs and Provide Assessment Tools

The success of an HT program lies in the ability to demonstrate noticeable improvements in client functions thru contact with your program. It is helpful to know as much as you can about the goals of clients when you begin. Access to charts, progress notes, and care plans is helpful to the horticultural therapist because he or she will be able to design the program to insure successful participation and overcome client disabilities (i.e., grasping a cup, inability to see or hear, or a short attention span). Discussing program outcomes with staff and others who come in contact with your program will help you get to know your clients faster and help them be successful in activity participation. HT programs can assist allied professionals such as physical therapists, occupational therapists, and speech therapists in achieving client goals. Clients achieve goals through active involvement in the program coupled with staff support of client plan-with-care outcomes.

IMPLEMENTING THE SERVICE

Start small, start simple. Introduce clients to horticultural therapy through the use of stimulating and goal-oriented programs. For example, regular scheduling provides anticipation and opportunities to experience this type of therapy. Getting to know participants in the program will provide insight about their needs, interests, skills, and interpersonal relationships. Programs can be designed around those areas of interest and at the same time meet the goals of the client and facility. For example, a cooperative horticultural therapy program may include a PT helping a client improve his or her ability to grasp by holding a pot while filling it with soil, while at the same time a speech therapist may coach a client with aphasia to sound the names of the flowers on the table. Groups can be formed around the HT program and stimulate other ideas for events at the facility, such as a holiday workshop making potpourri ornaments.

Purchasing

Once you have established your budget and have a schedule for your first HT programs, you will begin to purchase the necessary supplies needed to provide HT programs. A schedule of programs and supplies should be worked out three to six months in advance to ensure availability as well as provide adequate budgetary resources for purchasing. Your contact should specifically outline who is responsible for the supplies for programming. If the facility is to purchase supplies, you will need to follow up to ensure that the supplies are delivered when you need them for the program. Shop ahead, sometimes things are available abundantly, but when you actually need something for the HT program, it may be either unavailable or in short supply. Purchasing in advance may be a budget concern for both you and the facility, so be sure to discuss this with the program coordinator.

Storage of Supplies

Many facilities have a limited amount of space to store soils, pots, etc., for your program. Often it is locked, and access to the key may be difficult if personnel are not present on the day of your program. Maintaining control over your supplies is important to ensure you can set up and organize prior to the start of your program.

Sanitation and Infection Control

Use of trays or towels in the HT program is a concern if those are to be rewashed and returned for patient use. Consider using designated towels or other supplies only for the HT program. Wash and return them to the HT supply cabinet. Soil components such as perlite may pose a potential health risk, therefore clients should wear protective covers and avoid breathing dust. Sterilized soils are recommended for use in hospital or sanitary environments as some soils and mosses carry bacterial spores. Discuss these issues with the staff person involved with infection control for suggestions and precautions.

Transportation

Consider how you will pick up and deliver the materials for your program. Sometimes, if you are doing a plant propagation program and need to bring in large plants to be cut, you may find it difficult to transport or carry them into the facility. A small, collapsible cart is a wise purchase, especially when your program is on the tenth floor of the hospital wing! Many facilities will designate a cart for your use if you ask.

Starting Up the Program

On your first day of the program, have the facility staff introduce you to patients and other therapeutic staff members. Before you begin, briefly outline your program and explain what you will be doing with the group. Depending on the number in the group, you will need assistance to bring the group together at the set time. This often can be a problem when the facility is short staffed. Try to have this prearranged with staff to eliminate delays in getting your program started. Those in wheelchairs or walking with assistance devices will take longer to get into the area where your program is to take place. As a consultant, you are being paid to provide the horticultural therapy program, not provide transportation for their patients. Staff support is very important and should be a key item outlined in your agreement!

Your first program should be a sensory stimulating, introductory, and/or easy project that can be done by everyone in the group. Always provide programs that are positive and can be completed successfully by each member of the group within the program period. Avoid activities that require several sessions to complete in the initial period of the program. Everyone should be able to take their completed project back to their room at the end of the program. This is further reinforcement of their success. The completed HT project also provides a way for staff and visitors to engage the patient in conversation about the activity. This enhances their social skills and builds self-esteem.

As your program develops, and as your relationship with the patients and staff is established, you can begin projects that span several days or weeks. Focus on success and visible results during the initial program period. Your effectiveness as a group leader and completion of goals set by the facility staff will be observed and evaluated. It is recommended that you have periodic meetings with the administrator or key staff to discuss the program and areas for improvement or support.

Program Documentation, Evaluation, and Billing for Services

Prompt return of assessment or attendance forms to the staff is very important to avoid delay in their record keeping. Submit program billing requests in a timely manner to the designated department/person to avoid delays in payment for services.

CASE STUDIES OF HORTICULTURAL THERAPY PROGRAM CONTRACTORS

Judy Allen, HTT

Judy lives in Jackson, Wyoming, and has been consulting for over ten years. She organized her business as a sole proprietorship and began consulting because there were no other horticultural therapy employment opportunities in the area. She created HT programs at three sites: residential treatment programs for physically challenged individuals, people who have been sexually abused, and people with learning disabilities in a summer day care facility. Her networking began through a direct contact with the facility administrators. Horticultural therapy programs are delivered to these organizations one to two times per week with about one-quarter to one-half of the total time getting supplies and planning. This time is added to the client's bill. She uses a contract and charges about $12 per hour. Sadly, she has encountered many potential programs that never became reality because funding from facility budgets could not afford the HT program.

Karen Belanger, HTR

Karen lives in Toms River, New Jersey, and began to consult because, like Judy Allen, there were no horticultural therapy positions available where she lived. She has been providing HT programs for over six years to between twenty and twenty-five organizations with clients who are children, adolescents, seniors, developmentally disabled, physically challenged, and Alzheimer's and stroke recovery patients. New programs are now started by word of mouth after initial marketing by mailing, phone calls, and networking with activity professionals proved time consuming and expensive. She does an initial facility assessment to ensure potential programs meet her goals as well as theirs. Most of the nursing home programs are on a monthly schedule while other programs are daily. Some of her payment is received through invoicing, but most facilities provide a check at the time of her visit. Special challenges are lack of realization and support for the full impact her HT program has on clients and continuing to provide creative, inexpensive, and interesting programs.

C.J. Blades

C.J., who lives in Pomona, California, began consulting with a sole proprietorship established over seven years ago because he saw an obvious need. Presently, he provides weekly programs in hospitals, day programs, and retirement centers to clients, young and old alike, who are physically challenged, aged, mentally disabled, or are suffering from Alzheimer's and stroke. He initiates new business through service referrals and cold-call letters. His assessments and evaluation use clinical endpoints, client functioning, program outcomes, and routine hospital charting. His monthly billing invoices do not include prep time and average $40 to $60 per hour plus materials. His challenge has been to keep horticultural therapy from being just another activity, and like Karen, he would like to see more appreciation for the impact HT makes on client well-being and increased life skills.

Melanie Trelaine

Melanie lives in Cincinnati, Ohio, and began consulting and providing contractual horticultural therapy programs seven years ago by forming a nonprofit corporation. Like the others, contracting began out of a need to provide services that were unavailable in the area. Initial contact with a variety of facilities began with letters and cold calls. Several grants were applied for and awarded, which proved to be a mixed blessing. Eventually, word of mouth started new programs until over twenty different organizations were involved. Evaluations and assessments were completed at the beginning and throughout the life of the program; some were monthly and others weekly. Payments were received through billing invoices, cash, or payment at time of program, with an average of $45 per hour, which included materials. Keeping program costs down; soliciting and picking up of donated materials to use in programs; travel time and weather conditions; and continually bringing attention to administrators, other professional staff, and the general public about the benefits of HT continually challenge the HT consultant.

Chapter 16

Applied Research
for Horticultural Therapy Practice

Patrick Neal Williams
Patricia Myroniuk Williams

INTRODUCTION

Research can be scary and foreboding when the language and process are not understood. The purpose of this chapter is to outline the research process from beginning to end. This will not be a complete cookbook recipe to conducting research, but it should demystify the research aura and make it more approachable for students, therapists, practitioners, and horticultural therapy facilities.

Horticultural therapy is not yet a household term. The need for applied research is central in becoming universally and unequivocally recognized as a therapy while gaining acceptance in the medical field. This chapter has been included because research is a significant resource for advancing the practice of horticultural therapy. The amount of research must increase as the profession grows to ensure a long future of success as a viable profession.

The learning objectives for this chapter are

1. to give the reader an understanding of the overall research process;
2. to dispel myths regarding research;
3. to motivate the reader to support horticultural therapy research; and
4. to encourage the reader to carry out horticultural therapy research.

COVERING THE REALM OF RESEARCH

Most students and researchers would be ecstatic to find a step-by-step procedure for conducting research. However, there is great difficulty in producing *a recipe for research* due to the variability between each research project. The research settings will change as well as the hypotheses and subject matter. Thus, a basic understanding of the research process and steps will come as close to a recipe as can be produced here.

This chapter will attempt to make the research concept and approach more friendly. If the research process can be understood, then hopefully individuals and facilities will want to support and conduct research. Some people have the idea that research must be performed by university scholars. This is far from the truth. The time is upon us to dispel these myths regarding research and to build a knowledge base for horticultural therapy. Therapists and practitioners need to become involved in research. They are the ones who work first hand with populations and understand the questions and goals best suited to produce publishable results. These individuals already have access to researchable populations and know the ins and outs of their facilities. For the universities doing research in this field, accessibility to facility populations can be more difficult and can take months to years to begin collecting data.

Facilities can supply practitioners with the support to perform research. A partnership between agencies that assist with procurement of funds and/or with government agencies, associations, clubs, and others can supply the knowledge and expertise to carry out this technical undertaking. The output will become substantial when the industry can collaborate freely with the research entities on such studies. This will make the research process easier for everyone involved, including therapists, practitioners, and students.

WHY DO RESEARCH?

Why should time, energy, and money be expended to complete research? Who uses the research findings? If the profession of horticultural therapy never completed any additional research, who would notice? The following comments will answer these questions.

The Benefits of Research

- Increase knowledge in horticultural therapy
- Improve treatment outcomes
- Enhance professional and personal advancement
- Build research base to promote horticultural therapy within the health care arena
- Explore new frontiers

Research has many benefits, from advancing the body of knowledge to improving social conditions of those we directly and indirectly contact. As this knowledge grows, that information is disseminated to others and everyone in the field has the opportunity for improvement. The profession becomes more cohesive. Individuals can then build upon what others are doing and avoid stagnation. The end result of this research is changing policy or structure to benefit those working in horticultural therapy programs.

Professional organizations gain advancement from research. One of the American Horticultural Therapy Association's (AHTA) long-term goals is to be widely accepted within the medical/health care community. Part of this goal will be accomplished through good, solid, applied research. This in turn will broaden the number of services available to individuals in health care settings and create more career opportunities for those practitioners in horticultural therapy.

Professional and personal advancement of an individual is enhanced through research. Some individuals use research to become better known in their field of expertise, to advance and earn promotions in their prospective field, to answer questions of interest, and/or to feel they are making worthwhile contributions to the overall body of knowledge. Students conduct research for some of those same reasons and to complete higher educational degrees. Ultimately, motivation is a key aspect driving all these individuals to conduct research. Successful research builds self-esteem, confidence, and can lead toward more research being completed.

As the body of research in horticultural therapy builds, more individuals and organizations within the medical/health care arena will become aware of the progress. Research will aide in establishing horticultural therapy as a true alternative therapy within the medical/health care system.

The exploration of unknown questions is the challenge and excitement of research. To be the first to explore a field, a procedure, or a niche has its own motivational components. The drive for pure knowledge or to be tops in one's field is but one component of finding new information. In horticultural therapy, exploration of unknown questions should help advance the profession and ensure acceptance throughout the medical arena.

The Not-So-Glorious Side of Research

The quest for knowledge is not always promotions and accolades. If the bumps in the road are known before traveling, one can prepare and often avoid problems. There are numerous items that can hamper the progress of research projects. Table 16.1 outlines some trouble spots experienced by those conducting research from outside the facility and also by practitioners from within a facility.

TABLE 16.1. Problems to Consider When Conducting Research

Problems Experienced by Outside Researchers	Problems Experienced by Practitioners Within Their Own Facilities
• Obtaining funds for research • Securing and maintaining participation from human subjects • Finding research sites • Maintaining and collecting data from long distances • Handling multiple sites • Taking years to complete research and having priorities change	• Soliciting funding from sources approved by facility • Securing and maintaining participation from human subjects • Convincing facility to support internal research • Sidestepping territorial issues within or between departments • Resolving discrepancies with job description responsibilities • Maintaining data consistency with staff turnover

APPLIED VERSUS BASIC RESEARCH

Advancement Through Applied Research

Applied research has its foundation in two areas. The first area is research sanctioned and supported by an organization for its benefit. The second area is research answering specific questions about real-life problems within the field.

In the first area, foundations, institutions or other groups can contact a researcher and ask to have a specific problem investigated. When the research is complete, the new information will be used to change structure or policy for the organization initiating the research. This in turn can have widespread effects on similar programs if the information is disseminated.

The second area is initiated by the researcher to look at one certain aspect (researchable hypothesis) of a process. After a testable hypothesis is formulated, researchers locate proper field conditions, set up the experiment, and conduct their research. The end results are often the same in that some aspect of the structure or policy may be changed. Through understanding the working arenas of our field, these applied studies over time will add knowledge and in themselves become a foundation for future researchers.

Basic Research: Building a Foundation

Basic research projects are initiated and completed to add knowledge in a field of study. This knowledge may or may not have direct application to therapeutic environments, but the study outcomes lead to better understanding the total picture. In a field as relatively young as horticultural therapy, research needs to be completed in both basic and applied research areas.

THE RESEARCH STUDY, START TO FINISH

The brainstorming process to arrive at a researchable question parallels the earlier discussion on why research is done. These same processes also produce researchable questions.

The Initial Idea

This idea can be from one of three areas. Either the idea repeats others' work to check the validity of past research, continues on where a previous research project left off, or is a completely new researchable idea. These ideas can add to the basic knowledge of a discipline or manifest from applied research topics. An initial idea may come from one person, a couple of people, or a whole research team. Imagination is the only limiting factor for initial ideas. Once the idea has been generated, it must be converted into a testable hypothesis.

At this point in the research study, many processes must be thought about concurrently (Table 16.2). Because so many of these different topics influence each other, it is impossible to separate them into a distinct sequential flowchart. The following topic areas are not necessarily in the order individuals might follow to complete a successful research project. One suggestion at this point in the process is to formalize the idea in a research proposal. Components of a research proposal can be found in Table 16.3.

Literature Review

One of the first tasks after selecting a research idea is to collect information to learn what others have done. This is accomplished by conducting a literature review.

TABLE 16.2. The Research Process

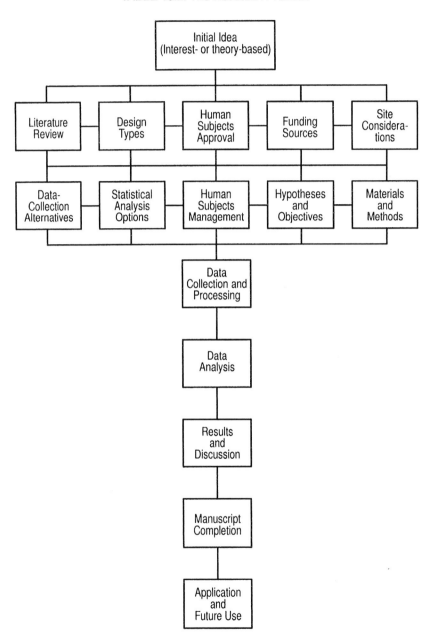

TABLE 16.3. Research Proposal Outline

Outline for the Research Proposal
• Title Page

• Introduction
    Purpose of Study
    Problem Statement
    Variables
    Background
    Study Significance
    Research Hypotheses, Objectives, Questions

• Literature Review
    Historical Overview
    Conceptual Framework
    Strengths and Weaknesses in Research Area

• Materials and Methods
    Subject Sampling
    Instrumentation
    Procedural Methods
    Research Design
    Ethical and Practical Considerations
    Data Collection Methods
    Results and Methods of Data Analysis

• Time Line
• Budget/Funding Consideration
• Literature Citations

In performing a literature review, the investigator hopes to find past studies, theories, and settings where similar research has been carried out. This can aid in completing the research design, pose new questions, help see the feasibility of the research, and act as an historical reference to the study. It can also provide information on whether a similar experiment has been completed beforehand.

Conducting a literature review may seem insurmountable if the investigator is not familiar with the use of libraries or databases. As with any of the steps in research, answers to questions and concerns are readily available. Librarians have a wealth of knowledge and are a superb resource for help in conducting a literature review.

Databases are changing rapidly with the advancement of computer systems. Most broad databases in horticulture, such as Agricola, are not good sources for horticultural therapy literature reviews. A database that does exist specifically for horticultural therapy is through the Department of Horticulture at Virginia Polytechnic Institute and State University. Another database is the Educational Resources Information Center (ERIC) national information system, supported by the U.S. Department of Education. Libraries usually charge a fee for the actual search, running time, and/or by the citation. Before attempting a search, ask for assistance to ensure the best possible search results.

Most of the above-listed databases can be accessed through electronic means via personal computer or through use of the internet. Gopher or internet searches can locate large amounts of sources.

Types of Research

The type of research should be determined at the beginning of the research process. There are two types of research:

- Experimentation includes setting up an experiment, running it, and collecting the data. The body of data is now entirely new and results from the researcher's project.
- After-the-fact research is looking at old information, formulating questions, and drawing new information out of what has already occurred.

A good example of after-the-fact research is Dr. Roger Ulrich's 1984 study looking at hospital records to determine what effect a patient's view through a hospital window had on their treatment and recovery. Another example in applied research is where a facility hires the researcher to analyze the effectiveness of a treatment regime looking at past patients' records.

Design Types

There are numerous design types (Table 16.4). Design types designate the random order of treatment and testing to take place during an experimentation process. A common design type is the pretest-posttest design where the experimental groups are tested before receiving the stimulus (testable treatment of the research) and following exposure to the stimulus (Cook and Campbell, 1979). There are design types to fit almost any situation, from unequal groups to interrupted time series designs. Consultation with a statistician can prove helpful.

TABLE 16.4. Common Research Design Types

Design Types*	Design Schematic	Explanation
Pretest-Posttest (Before-and-after test)	Control: 0_1 0_2 Treatment: 0_3 S 0_4	First, both groups are measured. The treatment group receives stimulus and then both groups are measured a second time.
Posttest Only (After only)	Control: 0_1 Treatment: S 0_2	After randomly selecting two groups, stimulus is applied to the treatment group and then both groups are measured.
Solomon Four-Group Design	Control$_1$: 0_1 0_2 Treatment$_1$: 0_3 S_1 0_4 Control$_2$: 0_5 Treatment$_2$: S_2 0_6	The Solomon four-group design combines the pre-test-posttest and the post-test-only designs. It allows the researcher to distinguish if any pretest bias is involved with the outcome results.
Time Series Designs	Treatment: 0_1 0_2 0_3 S 0_4 0_5 0_6	This type of design allows the researcher to measure gradual change before and after the stimulus was applied. Time series designs come in different types.

*These represent only a small number of the different design types. It is recommended to look at a good research design book, explore the alternatives, and choose the one best suited for the research project.

0_1 = observation number
S_1 = stimulus

Implementing the Research

It is always recommended that at least one researcher visit each site in the study. The researcher needs to be familiar with facility administration, the facility's physical site, the staff, and how human subjects are scheduled. Table 16.5 provides site considerations with basic questions to ask potential facilities.

TABLE 16.5. Site Considerations for Facility and Staff

Facility	SITE CONSIDERATIONS	Staff
• Appropriate physical setting and horticultural requirements to complete the research on site • Adequate space to accommodate treatment groups and control groups • Length of program (year round or seasonal in nature) • Financial support from facility • Administrative support and approval • Authorized portion of the staffs' work time allocated toward gathering data and scheduling subjects • Enough subjects to meet the study's needs • Special arrangements to accommodate scheduling needs		• Horticultural therapist or other trained professional responsible for overseeing on-site details • Staff properly trained or familiar with the type of research • Staff willing to accept and support research • Program dependent on full-time or part-time paid staff or volunteers for running programs • Amount of time staff wants to expend gathering data • Availability of staff to commit from start to finish (high staff turnover) • Researchers oversee the project full-time on site, part-time at site, or from a distant location

Funding the Research

The two primary sources for funding are private and public:

- Public funding is far more competitive and as a result harder to come by. These may include state set-aside grants, federal monies, or organizations like the National Institutes of Health (NIH).
- Private funding has less competition normally, but finding the sources of private funding can be more challenging.

Libraries have books that list philanthropists, companies, associations, civic clubs, and others willing to finance research projects. Researchers associated with a university may find staff who can assist and identify funding sources. Various consulting firms can also assist in procuring grants or other types of funding.

Some funding agencies, whether private or public, do not allow private individuals to apply for grants or contracts. They require applicants to be affiliated with an organization, such as a not-for-profit entity or a university. Most will ask applicants to have prior experience with managing funds. It is often wise to have someone co-author the funding proposal who has an established track record.

There are several sources within the horticultural field that fund proposals in the horticultural therapy realm:

- American Floral Endowment (AFE)
- Friends of Horticultural Therapy (FOHT)
- Horticultural Research Institute (HRI)
- Local and state horticultural societies
- Arboretums and botanical gardens
- Host agency

Raising money for research can be a full-time job. Most large organizations that rely on a percentage of outside funding hire full- or part-time staff members to raise money. Funding does not materialize overnight. Application deadlines of funding agencies can make it necessary to start the funding process several months to a year ahead of time. Funding is never a sure bet. A project that can not move forward without funding should make it a major priority.

Data Collection

There are different methods for gathering data, depending on the research question. Collection methods may include interviews, questionnaires, field observations, surveys, case studies, or experimental activities. At this point, the process should be well under way. The forms or procedures for collecting data should be complete. A statistician should have been consulted and an agreement drawn on how to analyze the data. This analysis will dictate how to make final preparations on data-collecting forms. It will also ensure that useful information toward answering the research question is being collected.

Record Keeping

Procedures and forms for gathering data should be laid out clearly before the actual collection begins. Sometimes a mock trial run will indicate whether the research procedures and forms are complete. A key management component is staff competence. Members of the research team should know the procedures and how to conduct themselves in a professional manner. When procedures run smoothly, staff and subjects will feel good about participating.

Human Subjects

The government has established a set of standards outlined in *The Belmont Report* (1978) on the appropriate management of human subjects in research. Universities have internal review boards that may supplement additional rules and regulations on conducting research. Some clinical facilities have their own research departments, and practitioners should understand relevant regulations before beginning the research process.

All subjects have the same rights regardless of their cognitive or physical status, or the stated rules for conducting research. Informed consent forms must be signed and kept on record for all individuals participating in the research. When an individual is not able to sign documentation, a legal caretaker must sign the consent form on the subject's behalf. Individuals may drop out of the study at any time for any reason. The researchers must not coerce their subjects.

Processing of human subject information should begin as soon as a research topic and strategy have been selected. Do not underestimate the length of time needed for this aspect of the research. Depending on the research population, governmental approval may be required

Each facility and university will provide forms and models to follow, such as consent forms, debriefing statements, and a statement on procedural applications. The study will dictate what data-gathering instrumentation will be needed. Researchers may use a standardized form or create their own. Researchers may be asked to provide information and justification regarding human subjects. Investigators should then adhere to guidelines and inquire about deadlines. A missed deadline can set the study behind schedule. Approval must be received from the facility or university before research can proceed.

Researchers and staff should strive to make the subjects' experiences positive. Be upbeat, constructive, and optimistic during the study. Tell subjects the benefits from participating and the overall good that the research outcome will produce. Thank them for participating.

Information Disclosure

There are differing opinions on how much information to disclose regarding the study's purpose. One group feels that as long as the researcher discloses the pertinent details, i.e., length, subject's duties, and participation criteria, that this is sufficient. These researchers believe that not disclosing the entire purpose of the research is allowable. Others feel that only 100 percent honesty is to be practiced for a viable and reliable study.

Researchers must be careful in wording the purpose when conveying information to the research subjects. A simple statement such as "the study's purpose is to explore the impact of plants on people" could influence the subject's outcome. If there is withholding of information, such as not disclosing the study's purpose for fear of influencing how the subject will react, then procedures must be outlined and provisions made in an exit (debriefing) statement to inform the individuals on why obscurity occurred during the study.

There is some confusion in research regarding the meaning and use of anonymity and confidentiality. The researchers should not know the subjects' true identity in anonymity. The data will be encoded and subjects are referred to by a code consisting of numbers and/or letters. In contrast with confidentiality, the researcher does know the identity of the subject(s). Both anonymity and confidentiality guarantee that individuals' identities are never made known in any written or published results of the research.

ETHICAL ISSUES IN RESEARCH

Responsibilities to Professional Code of Ethics

Everyone involved in the study should conduct themselves in a professional manner. There can be no tolerance for deception in conveying results to anyone. No one can alter or falsify any information related to the study. All results to be published and disseminated must be 100 percent correct. Copyright rules must be observed and citations stated.

Biases Involved in Researching Individuals

Two main types of bias may affect research:

- Research staff have personal biases or prejudices against the subject population. They may either consciously or subconsciously discriminate against certain or all research subjects.
- A professional bias occurs when the research staff, whether knowingly or not, coaches or encourages subjects to give favorable responses regarding experimental stimuli. This can easily take place when record-

ing verbal responses. It happens because the research staff wants the experiment to succeed and produce positive results.

Consistency in collecting data is of major importance, and in both instances of personal and professional bias, the data collected will not be true or valid.

DATA ANALYSIS

Data analysis encompasses two areas: actual analysis of data and interpretation of analyses. Analysis looks at the collected data through statistical applications, and interpretation occurs when the analyses are put into words for explanation. Both are critical steps in the research process.

STATISTICAL ANALYSIS

Analyzing data depends on the research design that encompasses statistics and data types. The majority of researchers use a computerized statistical package. The common software program for the social sciences is SPSS® (Statistical Package for the Social Sciences). It is available in a mainframe version or a Windows-based application. A statistician can be helpful in suggesting appropriate methods for analyzing data. An easy way to find a statistician is through universities.

Statistics come in two types: descriptive and inferential. Descriptive statistics use means, medians, standard deviations, confidence intervals, and interquartile range to classify and summarize numerical data sets. Inferential statistics use statistical tests that consist of procedures for generalizing results from the sample population (research subjects) to the broader population. The types of statistics used will depend on the data.

Data also come in two types: quantitative and qualitative. Analysis of quantitative data is commonly referred to as crunching the numbers; that is, inputting the collected data (processing) from interval (e.g., temperature in °F) and ratio scale (e.g., height) measures into a statistical formula and getting out numerical results. Quantitative data have true mathematical properties associated with their numeric values. There are many types of statistical analyses such as t-tests, ANOVA's (analysis of variance), correlations, and regression models. Find the one that best fits the situation.

Qualitative data use measures that do not contain true mathematical properties. Qualitative data is nominal (e.g., gender or religious affiliation) or ordinal (e.g., measures on a Likert scale) in nature.

Discussion of Results

Once the statistical analysis is complete, the next step is interpretation or discussion. The interpretation phase involves understanding the statistical output produced by the study and relating it back to the initial research questions. Statistical analyses will point out significant differences, but the researchers must still explain what those numbers mean. The interpretation should carry as much weight and importance as the analysis portion.

Recommendations

Once the analysis/interpretation phase is complete, researchers will make recommendations on application of the study. This can take the form of replication for confirming results, ways to improve the research design, and procedures or future implications of the results.

After the research is concluded, depending on the significance of outcome, the next step is disseminating the information to others and publishing the results. This can occur in both an oral presentation and a published journal article.

PUBLISHING THE RESULTS

Dissemination of the findings is one of the key reasons for conducting research. Professional journals offer one means for dissemination. Several points need to be considered when selecting a journal.

- Appropriateness of journal for the subject matter
- Reputation of the journal, refereed or not refereed
- Journal that gives the best exposure to the research findings

Table 16.6 outlines the components of a research article. Other descriptions and procedures for writing research articles can be found in any good research and methods book. Each journal will have specific guidelines for submitting articles. Narrow down and prioritize the choices. Once the possible source is contacted, information on the journal's manual style to follow for submissions will be sent. The most commonly used manual styles can be found in libraries. As with any work, it is always wise to have several peers or mentors review the work for grammatical and procedural errors before submitting the article.

Once the article is submitted and accepted for publication, the research process is complete. This has been a quick glance at the research process.

Many sources of information exist on how to conduct research. If this is the first attempt at research, find someone who is familiar with the procedures to help walk through the initial study.

TABLE 16.6. Components of a Research Article

• **Abstract**	Summarizes the entire paper and is usually written last.
• **Introduction**	Begins research article, may be labeled or not.
Problem Statement	Identifies problem, states purpose of study, and makes reference to why it might be of interest to the reader.
Literature Review	Provides conceptual framework for why study was completed, what prior work had been done in the research area before the present study, and how current study will add to the existing knowledge.
Research Hypothesis	States research hypothesis and the relationship among variables.
• **Methods**	Should be clear enough for anyone else to replicate the study just completed, includes all details.
Subjects	Describes subject characteristics and sampling techniques.
Instruments	Specifies instrumentation, scales, etc., that were used, along with different measurement levels.
Procedure	Discusses conditions under which data was collected.
• **Results**	States what the study found. Includes both descriptive and inferential statistics, whether in narrative or charts. Does not analyze results.
• **Discussion**	Considers study results, looks at support for the hypothesis, findings that are consistent with previous studies, strengths and weaknesses of study, and implementations for future research in this area.

GLOSSARY

Belmont Report: A government mandate on the handling of human subjects for research purposes

control groups: Subjects that have not received the experimental stimuli, allowing comparison of effects between control and experimental groups

gopher: A searchable electronic index

hypothesis: An unproven scientific conclusion drawn from known facts; a conjecture about one or more population parameters

Likert scale: A ranked-ordered (ordinal) scale that consists of a response range; examples are strongly agree, agree, neutral, disagree, strongly disagree

project timeline: The schedule that plots out the order of events necessary to complete the research project

qualitative data: Measurements on a nominal or ordinal scale consisting of unordered or ordered (ranked) discrete variables; examples are gender, race, religion, or Likert scale measurements

quantitative data: Measurements on interval or ratio scales consisting of continuous variables; examples are temperature, weight, and height

theory: A systematic explanation for a set of facts or laws

track record: A proven history of prior success in conducting research or securing grants

treatment groups: Subjects that have received the experimental stimuli in connection with the research project

BIBLIOGRAPHY

Babbie, E. 1992. *The practice of social research,* sixth edition. Belmont, CA: Wadsworth Publishing Co.

Belmont Report: Ethical principles and guidelines for the protection of human subjects of research. 1978. The National Commission for the Protection of Human Subjects of Biomedical and Behavioral Research. Washington, DC: Department of Health, Education, and Welfare, National Commission for the Protection of Human Subjects of Biomedical and Behavioral Research. For sale by the Supt. of Docs., U.S. Govt. Print. Office.

Cook, T. D. and D. T. Campbell. 1979. *Quasi-experimentation: Design and analysis issues for field settings.* Boston: Houghton Mifflin Company.

Diener, E. and R. Crandall. 1978. *Ethics in social and behavioral research.* Chicago: The University of Chicago Press.

Hinkle, D. E., W. Wiersman, and S. G. Jurs. 1994. *Applied statistics for the behavioral sciences,* third edition. Boston: Houghton Mifflin Company.

Likert, R. 1967. *The method of constructing an attitude scale. Readings in Attitude Theory and Measurement.* New York: John Wiley and Sons.

Selltiz, C., L. S. Wrightsman, and S. W. Cook. 1976. *Research methods in social relations,* third edition. New York: Holt, Rinehart and Winston.

Ulrich, R. S. 1984. View through a window may influence recovery from surgery. *Science,* 224, 420-421.

Wuebben, P. L., B. C. Straits, and G. I. Schulman. 1974. *The experiment as a social occasion.* Berkeley, CA: The Glendessary Press, Inc.

Chapter 17

Documentation, Program Evaluation, and Assessment

Vera Roth
Martha C. Straus

INTRODUCTION

As health care reform continues, it is important to recognize the value of documenting treatment services provided. Good documentation sends a clear message, both to the service payer and to the treatment team, that your modality is directly related to reducing the client's presenting problem(s) and symptoms. This chapter provides an overview of treatment planning. The learning objectives of this chapter are to

1. define the purpose and importance of documentation in the course of assessment, treatment, and for program evaluation;
2. review assessment and documentation standards of governing bodies;
3. define key principles of documentation including goals, objectives and provide examples; and
4. explain how horticultural therapy can be incorporated into the treatment planning process.

Documentation also provides information to the rest of the treatment team and, in many cases, insight into therapeutic issues. This information is relevant to the decision-making process regarding the client's treatment and to the professional integrity of the program. Documentation helps each clinician examine his or her treatment methods and assesses its effectiveness in meeting the client's needs and goals.

As the length of stay continues to decline, insurance reviewers are carefully screening and reviewing services and treatment practices. Everything is geared to the presenting symptoms and reducing the length of time patients are in therapy. It is vital that all services be documented in a timely fashion.

PURPOSE OF DOCUMENTATION

The significance of documentation is to report services provided and to let treatment team members and reimbursement parties know your observations, results, and recommendations. It also allows for a behavioral reference over time. This is particularly useful to show progress or a lack thereof. Documentation also serves as justification for continuation of services. One of the greatest benefits of good record keeping is that it shows the relevance of your modality to treatment. To facilitate documentation, many programs have developed forms that may include checklists or fill in the blanks to expedite this process.

When recording clinical information into the medical record, there are a few documentation guidelines that should be considered. All documentation should be done in ink, typically in blue or black, although some institutions have very clear policies for this. The information should be written in a neat and legible manner. Use only abbreviations that are approved by your facility and understood by other clinicians. Upon completion of the entry, the clinician should sign his or her name, credentials, and title.

GOVERNING BODIES

Most hospitals, nursing homes, or treatment centers seek accreditation. This recognition says to the public that this facility meets the minimum standards set by a particular governing body and that a certain standard of care is available. Facilities will seek accreditation under the governing body that represents their particular population and treatment focus. Minimum requirements for documentation and assessment are usually included in the regulations. In addition, facilities may have their own internal requirements. The two largest governing bodies for hospitals and rehabilitation facilities are the Joint Commission on Accreditation of Healthcare Organizations (JCAHO) and the Commission for Accreditation of Rehabilitation Facilities (CARF).

JCAHO

The Joint Commission on Accreditation of Healthcare Organizations is one of the largest of governing bodies. The purpose of JCAHO is to improve the quality of care provided to the public. This organization accredits healthcare facilities such as hospitals and rehabilitation programs and sets minimum requirements. JCAHO accreditation is required to obtain state licensure and federal reimbursement of Medicare and Medicaid. It also promotes a positive reputation to the public and other healthcare organizations and attracts other payers for financial reimbursement.

Each facility is surveyed every three years by a team of experts. Sometimes an institution is issued a "provisional" certification. This means certain points need to be corrected in a specific time or the facility could forfeit their accreditation. The standards are updated periodically, so it is important to review the most recent and relevant manual to your program to make sure that you are meeting the requirements. There is a separate manual for hospitals, and one for mental health, chemical dependency, and mental retardation/developmental disabilities programs.

The 1995 manual addresses changes in the focus of the JCAHO requirements from capability to performance standards. This means that the standards used today assist the healthcare facility to assess and improve its service performance. The quality assurance department would know what standards the facility is using. JCAHO no longer defines or sets documentation standards or time frames. Each facility sets its own polices and procedures for documentation and JCAHO reviews each program based on its own internal standards. These standards include type and frequency of documentation, information included in the initial assessment and the time frame in which this must be completed, discharge planning, procedures for medication, and other special treatments.

CARF

The Commission on Accreditation of Rehabilitation Facilities (CARF) accredits rehabilitation programs and promotes quality services. CARF requires three things in its documentation standards: An initial evaluation, progress notes, and a discharge summary. It is up to the facility to decide the time frames and frequency for each of these areas. Generally the initial evaluation must be done with in the first seventy-two hours. Progress notes can be written daily, weekly, biweekly, or monthly depending on the focus and requirements of the facility. Standards and internal forms are usually designed and set by each rehabilitation facility.

Medicare

Many nursing homes and long-term care facilities are under Medicare guidelines. Medicare is a federal program of health insurance for the disabled and people over sixty-five. Medicare is divided into Part A, which covers hospital stays, and is financed through Social Security, and Part B supplementary medical insurance covers tests, doctor's visits, therapy, medication, etc. Part B is optional and requires a participant to pay a monthly premium (Atchley, 1994).

Medicare documentation standards are primarily used by nursing homes. They use a documentation form called "MDS," which stands for Minimum Data Set–Version 2.0. This is used for nursing home residents as an assessment, care screening, and as a tracking form required by those programs receiving federal and state funds. This is completed initially upon admission as an assessment tool and then quarterly by the interdisciplinary treatment team. Each team member completes his or her section using the same form. Typically, the treatment team consists of nursing, social work, activities staff, and occupational and physical therapy. This form serves as a central functional sheet to identify problem areas and then formulate goals and objectives that become the treatment plan.

This MDS is an eight-page form divided into sections, which include identification information, demographic information, customary routine, cognitive patterns, communication/hearing patterns, vision patterns, mood and behavior patterns, psychosocial well-being, physical functioning, and structural problems. Continence in the last fourteen days, disease diagnosis, health conditions, oral/nutritional status, oral/dental status, skin condition, activity pursuit patterns, medications, special treatment and procedures, discharge potential and overall status, assessment information, and signatures are included in this comprehensive form.

Progress notes are written every ninety days to update and reassess the function of each patient. If the length of stay is anticipated to be short, documentation is every thirty days. A new treatment plan is written if there are significant changes that require any change in services. Treatment goals are written, which may include "maintenance goals" for a patient to continue at a certain level of performance.

Within the activities department, one area that is heavily explored with the patient is past and present leisure interests and activities. This assists the activities specialist with program planning and treatment planning for the patient. Nursing home residents do have the right to refuse activity therapies. Every effort is made to engage the patient in some type of daily activity schedule. Attendance and level of participation are documented in the quarterly notes.

Medicaid

Medicaid is a federal program managed by each state and provides health care to the disadvantaged. To be eligible one must have few assets and be in a low-income bracket. Eligibility may differ from state to state. Both Medicaid and Medicare fund only certain programs within health care facilities, and certain minimum requirements must be met.

DOCUMENTATION

Documentation relating to treatment usually consists of an initial assessment, from which goals and objectives are written. Some facilities also require a written plan explaining the steps that will be taken to reach the goal and time frames and the measurement criteria to know if the goal has been met. We will discuss each of these parts.

Assessment

One of the essential parts of documentation is an assessment. An assessment is defined as an organized collection of data and review of client-specific information required to determine treatment needs. The purpose of an assessment is to collect data to determine a baseline of functioning and recommend what kind of treatment is necessary to meet the patient's presenting problems. There are three steps in the assessment process. Initially, data is collected, analyzed, and then an initial treatment plan is prepared.

JCAHO requires that certain information be included in the assessment:

- A history of emotional, behavioral, and substance-abuse problems or treatment
- Current emotional and behavioral functioning
- Maladaptive or problem behaviors
- Functional evaluations of language, self-care, and visual-motor and cognitive functioning

When appropriate, a vocational and/or educational assessment is recommended. This information is all entered in the medical record.

When a person is admitted to a hospital, typically one of the first questions asked is the reason for the admission or the presenting problem(s). The specific details required are dependent on the hospital's own policy and the type of health care facility. It is important to get the patient's perspective, and if possible, the family may also be of assistance in this area.

The information gathered at the initial patient contact may indicate that a more thorough detailed assessment is needed. Exactly what type of further assessments are dependent on the diagnosis, the care requested, the type of health care facility, responses to previous treatment, and the patient's cooperation and consent.

The assessment should be completed by qualified personnel to adequately address the area of interest. This is usually determined by the healthcare organization. Typically it is done by a staff member with the proper training and experience, and who often has some type of certification or license relevant to the field of service. The assessment process is ongoing. After the initial assessment is completed, the client is periodically reassessed based on the time frame set by each facility. As the length of stay has declined, the assessment process has become more efficient. In psychiatric hospitals in the past, when there was the luxury of time, the assessment was more detailed, the interview was longer and more in depth, and it typically included a task analysis. Now the assessment is more consolidated.

A horticultural therapist may be a member of a multidisciplinary team and may be required to do a rehabilitation assessment. Key sections should include strengths and needs in the following areas:

- social/emotion
- physical (any limitations)
- leisure/recreation
- vocational (level of education, work experiences, skills identification)
- activities of daily living

Within each section there are particular areas to inquire about. For example, in the "social/emotional" section, cover the client's lifestyle, marital status, living arrangements, and personal habits. It is important to ask the client their perception of what has precipitated the need for treatment and how they are feeling.

In the "physical" section, note any health problems, such as any medical conditions, physical problems, or disabilities. Assess whether they appear overweight or underweight. Examine their mobility and their gait. There should also be reference to any strengths, for example, if they regularly exercise.

Assessment of the "leisure/recreation" area includes information about how they spend their free time in the off-work hours. This is typically evenings and weekends, but it is necessary to clarify this for a complete assessment. A notation of the types of activities, the frequency, and duration is informative.

The "vocational" section provides information about the amount and level of education and/or training, work status, work experiences, and identification of employment and transferrable skills. Any problems or difficulties with work should be noted.

The "activities of daily living" section should include information about grooming, hygiene, and self-care. Household management areas such as cooking, cleaning, laundry, and budgeting may also be addressed. Safety issues should also be explored.

Throughout all of the areas assessed there should be reference to strengths and weaknesses. Many institutions have a preset format, but the topic areas mentioned above are relevant to your assessment of the client.

Once an initial general assessment is completed, typically each clinician completes a more specific assessment relevant to their specialty area. Another type of assessment is a "situational assessment." This is often used in vocational evaluations when a client is set up in a work placement and his or her skill is monitored for speed, accuracy, self-monitoring, and independent problem solving. Because horticultural therapy programs include real-life work skills and activities that are required in real work situations, situational assessments may be done in the greenhouse environment.

A situational assessment is used to gather data about a client's work potential in a particular job setting. Having the client simulate work skills in the exact work environment he or she is considering is more useful than conducting paper-and-pencil aptitude testing. This is an important way for the client and the evaluator to explore a client's job potential. This is very similar to a "job tryout," but is typically a limited period of time.

A work sample is a standardized work task that is used in a particular job setting to predict the client's job potential. Work samples should be clearly written with instructions and diagrams if applicable. Sometimes the work sample needs to be demonstrated to the client first. The work samples usually have two types of norms or industrial standards by which the client is rated. The client's score is rated by the number of errors or accuracy and by work speed. Norms should be rated according to the competitive job market and job expectation. Refer to Table 17.1 for an example of a horticulture work sample and Figures 17.1 and 17.2 for market pak planting. Refer to Table 17.2 for a functional screen for rehabilitation services.

The horticulture assessment should be geared to the needs identified in the initial treatment plan and the focus on treatment. For example, a therapist in a psychiatric facility would concentrate more on the behavior and participation of a patient then on the range of motion of their right hand. Information may be gathered in the following areas:

- ability to follow directions
- concentration and attention span
- physical limitations
- safety relating to self and others
- organization on task
- leisure skill development
- affect, mood, and level of cooperation

TABLE 17.1. Horticultural Work Sample #1

Task: Repotting/Transplanting Seedlings

Objective: To Assess Work Skills for Greenhouse Production Worker or Nursery Assistant

Supplies Needed:

1. tray of seedlings adequate to plant "market paks" (801 plastic divider)
1. package of individual plant containers (eight paks per carry tray) (KORD 1020 Carry Tray, KORD 801 Divider)*

    moist potting soil (PRO-MIX)
1. planting tool—#2 lead pencil with eraser
1. sprinkling can containing one quart of water

*client may choose to sit or stand

Directions to Evaluator

1. Evaluator will give verbal and demonstrative directions to the client and allow client to observe diagrams.
2. Begin by pointing out supplies listed above.
3. Evaluator will demonstrate repotting and watering of seedlings in one "pak."
4. Record start and finish time.

Directions to Client

1. Put market paks into carry tray (Figure 17.1). Fill the empty planting containers full of potting soil. (Soil should be level with top of container. Do not pack down.)
2. Using the planting tool and with a twirling motion make holes in the described pattern in the soil* (Figure 17.2).
3. Gently separate the seedlings from the rows.
4. Place a seedling into each hole in the market pak, making sure the roots are near the bottom with the leaves and stem above the soil.
5. Gently push soil around the seedling to make it secure, using planting tool.
6. When all eight market paks are planted, water seedlings, entire tray, with sprinkling can until evenly distributing the entire container of water.
7. Return all supplies to storage area.
8. Call your evaluator when you are finished.

*Note: If client completes in less than the average time, check for neatness, space of seedlings from each other, and number of plants per section, and then encourage them to make adjustments.

Stress the Following with Client

1. Arrange each seedling an equal amount of space from the next.
2. Plant seven seedlings per container.

Guidelines for Scoring

A. Time frame—client should be able to complete one flat of seedlings in ten to fifteen minutes (fifty-six plants). Rate the client's performance satisfactory or unsatisfactory.
B. Observe how careful they are with handling seedlings.
C. Observe how they separate the seedlings from each other.
D. Observe dexterity with use of tools and plants.
E. Observe the depth in which the plant and roots are placed in the soil.

(Roth & Bruce, 1989)

FIGURE 17.1. Market Pak Planting

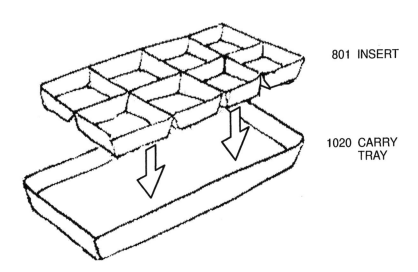

801 INSERT

1020 CARRY TRAY

FIGURE 17.2. Planting Pattern

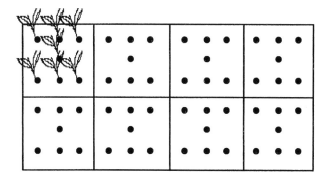

THE TREATMENT PLAN

Once the initial assessment is completed, the clinician identifies problem areas from the assessment and formulates behavioral goals to address them. This becomes part of the treatment plan. A treatment plan is a written action plan that spells out the services and the service providers based on the assessment data. It is a strategic plan to address the presenting problems and meet those needs through treatment goals and objectives, and criteria for terminating specific interventions. The treatment goals need to be measurable. Each goal should have objectives of how the problem area will be approached, the intervention, and what is the expected as the outcome. There should be review and documentation of the treatment, generally called progress notes. At this time, the interventions are documented with results and the treatment plan revised. If goals are accomplished then new goals are formulated or the treatment is discontinued if no longer indicated. This monitors treatment. It is based on reassessment, the needs for further care, and the achievement of identified goals.

Treatment Planning

Treatment planning is a process in which qualified staff examine the needs of the patient and identify a systematic approach to treating the presenting problems identified through the assessment process. It states specific objectives relating to the goals identified on the assessment form. The treatment plan also specifies the services, the frequency, and the

TABLE 17.2. Sheppard Pratt Health System Functional Screen for Rehabilitation Services (to be completed within twenty-four hours)

Addressograph

ACTIVITIES OF DAILY LIVING

	WFL	IMP
Self-care/grooming		
Home management		

1. Observe appearance.
2. Eating/showering/personal appearance? (Impaired if 1 out of 3)
3. Meals, shopping, chores, finances, child care (Impaired if 2 out of 5)

SENSORIMOTOR

	WFL	IMP
Energy level		
Fine/gross motor coordination		

1. More tired/sleeping a lot?
2. Observe tremors, gait, balance.
3. Problems with shaking, falling, bumping into things?

VOCATIONAL

	WFL	IMP
Managing work/school		
Productive activity		

1. Problems at work/school/day program?
2. Work/school affecting mental health?
3. How many days/week attend productive activity? (Impaired if <3)

COGNITIVE INTEGRATION

	WFL	IMP
Attention span		
Problem-solving skills		

1. Difficulty concentrating while at work, school, reading, watching TV?
2. Difficulty solving day-to-day problems? (examples: overslept or got lost)

LEISURE

	WFL	IMP
Identified leisure interests		
Structuring free time		

1. ID three interests (excluding TV, sleep, listening to music).
2. How do you structure free time?
3. When did you last engage in this? (Impaired if >1 month)

PSYCHOSOCIAL

	WFL	IMP
Interpersonal/social skills		
Coping skills		

1. Conflicts with family, friends, co-workers?
2. Engaged with family/friends in spare time?
3. How do you handle stress? (name two effective coping mechanisms)

Key: WFL = Within Functional Limits
IMP = Evidence of Impairment/Dysfunction

Comments: _____

_____

_____

Further Assessment Needed? ☐ Yes ☐ No _____

TABLE 17.2 (continued)

INITIAL RECOMMENDATIONS:

☐ Occupational Therapy   ☐ Movement Therapy
☐ Therapeutic Recreation   ☐ Vocational Services/Counseling
☐ Art Therapy   ☐ Other_____

Patient Participation in Goal Development:

Signature: _____ Date: _____

PARAMETERS/QUESTIONS FOR REHABILITATION SERVICES SCREENING FORM
ADULT DIVISION

ADLs

Self-Care/Grooming: (1) Observe patient's appearance: Is he/she disheveled or dirty? If so, this is evidence of impairment. (2) Have you had difficulty with regular eating, showering/bathing, or keeping up your personal appearance? (Impaired if yes to any of these)

Home Management: Have you had difficulty keeping up with meals, shopping, household chores, finances, or child care? (Impaired if yes to two out of five)

VOCATIONAL

Managing Work/School: (1) Do you have problems/concerns at work/school/day program caused by your illness or other factors? (2) Is work/school stress contributing to your mental health problems? (A "yes" answer to either question indicates impairment)

Engagement in Productive Activity: Are you engaged in any vocational activity such as a job, school, volunteer work, homemaking, or a day program? How many days/week? (Impaired if less than three days/week with no other responsibilities)

LEISURE

Identified Interests: Can you name at least three leisure interests you engage in (other than TV, sleeping, listening to music)? (Impaired if less than three)

Structuring Free Time: (1) How do you spend your free time? (Impaired if TV is the only thing, or if answers indicate drug/alcohol abuse or boredom) (2) When did you last engage in this (these) activity (activities)? (Impaired if over one month ago)

SENSORIMOTOR

Energy Level: (1) Have you felt more tired than usual? Are you sleeping more? (Impaired if "yes" to either)

Fine/Gross Motor Coordination: (1) Observe patient. Are there tremors, gait or balance problems, or other evidence of motor dysfunction? If yes, mark impaired. (2) Have you had problems lately with shaking, falling, or bumping into things? (Impaired if "yes")

COGNITIVE INTEGRATION

Attention Span: Have you had difficulty concentrating at work or home, or while reading or watching TV? (Impaired if "yes")

Problem Solving: Do you have difficulty solving day-to-day problems? (Impaired if "yes," or if ineffective reply is given to examples such as "What would you do if you overslept and were late for work/school/day program?" or "What would you do if you found yourself lost downtown [or in the woods]?")

PSYCHOSOCIAL

Interpersonal/Social Skills: (1) Have you had serious conflicts lately with family, friends, or co-workers? (Impaired if "yes") (2) Do you have friends and/or family that you do things with in your spare time? (Impaired if "no")

Coping skills: How do you handle stress? (WFL if can name at least two effective coping mechanisms/strategies. Impaired if answers indicate substance/alcohol abuse, destructive/self-destructive behavior, or withdrawal and excessive sleep, or dissociation.)

Goal Development: "Are there any areas we discussed that you would like to work on during your stay here?"

Referral for occupational therapy and/or therapeutic recreation if impairment exists in any category.

Referral to art and/or movement therapy if dissociative symptoms contribute to impairment.

Referral to vocational services if work/school problems exist, or if patient needs to find work, volunteer work, or vocational rehabilitation services.

settings required to meet the client's needs and goals. Typically, this plan reflects the facility's treatment philosophy and is completed by the various disciplines that make up the treatment team. The treatment plan provides guidelines and contains specific behavioral goals that the patient must achieve to attain, maintain, and/or reestablish emotional and/or physical well-being or adaptive behaviors. It may also indicate referrals or consultations for recommended services within or outside the healthcare facility. For example, updated or further psychological or neuropsychological testing may be recommended in a psychiatric facility.

The care, treatment, and goals are periodically revised based on the reassessment of the current clinical presentation, client's needs, and response to treatment. This is typically done when major clinical changes occur with the patient and at regular time intervals required by the institution. The treatment plan also spells out the specific criteria to be met for terminating or reducing treatment frequency.

Multidisciplinary treatment teams meet regularly to update patient needs.

Goals

Within a psychiatric hospital, the treatment plan has behavioral goals. In a rehabilitation facility the terminology is "functional goals." A goal is defined as "the purpose toward which an endeavor is directed," and "an end" is something worked toward and what you hope to accomplish, for example, a finish line in a race. These are written in a format in which they are clear and concise, observable, and measurable. The goals should be reasonable and attainable within the time frame of services. Sometimes goals are general goals of the program, and then there are client-specific goals addressing individual needs, problems, and symptomatology.

Goals may be short term, long term, and for maintenance. Short-term goals can be accomplished or addressed in a short period of time, perhaps in a session or two. Long-term goals are ones that may take a considerable period of time to achieve. This can be several days, a week, a month, or even a year. Maintenance goals are those behaviors or functions the patient has achieved, and it is recommended that they work to maintain or continue that level of performance.

Good questions to ask yourself when writing goals are the following: What behavior/function needs to be changed, maintained, or modified? How and when can it be observed and measured, and how many times must it occur before the objective is met?

Objectives

Objectives are written for each goal. These are the steps taken to meet the goal(s). An objective specifies the measurable changes to be perused. They should state the method/media/and approach. For each objective, a projected date of achievement is specified. For example, if an identified problem from the assessment and on the treatment plan is withdrawal and isolated behavior, the goal is to increase interaction. An objective could be for the patient to greet the leader or respond to a question three times per group for one week.

As client goals and objectives are determined it is important to include the client in the process if at all possible, that is, if the client is psychologically, physically, and educationally available to participate in the development and evaluation of the goals. See Table 17.3 for examples of goals and objectives relating to horticulture.

Progress Note/Report

A progress note/report is a written presentation of the observed behaviors. It should include a discussion of the actions toward the goals and

TABLE 17.3. B-1 Report of Therapeutic Activities and Group Sessions

Monday Date: _____

Initials	Signature	Title

Key

√ = Achieved
- = Not Achieved
P = Partially Achieved
E = Excused
R = Refused
N/A = Not Applicable

Initials	Treatment Activity	Attendance	Actively Participated	Follows Directions/Sequences Tasks	Cooperates with Others	Expresses Self Clearly	Identifies Goals/Interests Regarding Treatment	Concentrates (# Minutes)	Other:	Comments
	Planning Meeting									
	Movement Group									
	Discharge Planning									
	Horticulture									

objectives identified on the treatment plan. Modification of the goals or frequency of services can be documented at this time. It is targeted for a specific audience. In many cases, this is the treatment team, the facility's quality assurance reviewers, and insurance reviewers who are paying for the services. It is important to follow the guidelines and policies and procedures set up by the governing bodies of the institution for documentation standards and format.

Progress notes are always confidential and typically do not circulate outside of the institution, unless the patient has signed a consent form to release certain information to a particular person or program. They are considered a legal document.

A program may require a written progress note or may have a simple check sheet form with space for comments. Certain guidelines must be followed when writing a progress note. Primarily it must be a factual account of what happened, what was said, or what was done. Say only what is important, without judgment or opinions. Documentation must be in chronological order. Report any changes noticed, whether positive or negative. Any concerns or significant statements made by the patient must be documented. All notes must be dated and signed. For example, you notice that a patient looks more depressed. The note should read, "patient is more withdrawn, affect flatter, less energy, less interaction, etc." Whatever is happening that makes you think the patient is more depressed, not just that he seems more depressed, is what should be reflected in the note.

Any major changes, concerns, or statements seen or heard from a patient should be also reported to the charge nurse as well as documented. Notes may not always be read right away and concerns need to be immediately addressed.

Avoid global statements, such as "the patient participated well." What does that mean? Did he stay on task, follow directions, show initiative, work independently, interact with others, etc.?

In an effort to reduce paperwork, many programs have developed forms for progress documentation. See Table 17.3 for a daily note form developed by the Inpatient Psychotic Disorders Program at Sheppard Pratt Hospital. See Chapter 4 on physical disabilities for a computerized progress note form developed in horticultural therapy at the Rusk Institute.

TIME FRAMES

Goals and objectives are often required to have time frames associated with them. This is the usual and expected time in which the patient should be able to accomplish that goal. It is usually included at the end of the objective and written, "by 7/96," for example. Some facilities may limit the length of time a goal can be projected and carried. For example, a long-term care facility may project a goal for only three months and this goal may be continued two times. If the patient has not met this goal in this time frame, the goal needs to be discontinued and a new, more reachable goal written.

Program Evaluation

Health care needs are constantly changing so it is important to evaluate programs periodically to see if it is meeting the needs of the industry. This is particularly necessary to see if the program is efficient, up to date with the latest equipment, materials, trends, and information to meet the changing demands of the population(s) served.

WHY HORTICULTURAL THERAPY?

Clients and patients are placed in programs for a variety of reasons. Each "special population" chapter of this book includes goals, objectives, and reasons for using horticulture as a treatment modality. Most facilities require that each group have a set of protocols that outlines the focus of the group, the physical and cognitive level of functioning the patient needs, criteria for patient selection, methods, and special precautions. See Table 17.4. In the past, many psychiatric hospitals had a "menu" of groups, and the therapist who evaluated the patient would place him or her in a group that met the patient's level of functioning, needs, as well as interest. For example a severely depressed patient with poor concentration may be placed in a "low level," highly structured group in order to work with living plants to have contact with life again. A patient may have been placed in horticulture therapy to develop better leisure skills, to motivate them to participate, to increase their sense of responsibility, or to improve their work skills.

Today many hospitals have developed unit-based programs and patients have access to only those groups that are offered by that unit. Criteria for placement may not be as strict and activities may be more general in focus. Horticultural therapy group goals for unit-based programs may be more general, but each patient should have their individual goals for the day.

Patients may wonder why they are placed in horticulture therapy, and why are they working in a greenhouse when they are in the hospital. Because of the rapidly changing census and the reduced lengths of stays,

TABLE 17.4. The Sheppard and Enoch Pratt Hospital Rehabilitation Services Department, Horticulture Task Skills III Protocol

**THE SHEPPARD AND ENOCH PRATT HOSPITAL
REHABILITATION SERVICES DEPARTMENT**

HORTICULTURE TASK SKILLS III PROTOCOL

I. **FORMAT:**

    A. Time: Once a week for a fifty-minute session
    B. Size: six to eight patients
    C. Leader: OTR/L and Horticultural Therapist

II. **PURPOSE:**

Task Skills is designed for the patient functioning in the range of 4.0–5.9 of the Cognitive Disability Theory, as identified by the Allen Cognitive Level Test (ACL). Complex tasks involving problem-solving skills are therapeutically selected with the individual patient. The primary focus of the group is on the cognitive processes involved in their performance. Task skills also provide an opportunity to improve communication skills.

III. **GOALS:**

    A. Cognitive/Task Skills
      1. Improve decision-making skills in the areas of impulsivity and predictability.
      2. Improve ability to follow two or more directions through serial imitation.
      3. Improve ability in overt trial and error problem-solving skills.
    B. Interpersonal/Social
      1. Improve ability to cooperate with others regarding cleanup and sharing of tools and space.
      2. Improve ability to compromise and negotiate in a manner that is neither overly compliant nor aggressive.
      3. Improve ability to give and seek assistance.
    C. Emotional/Behavioral
      1. Improve ability to respect others and seek others' opinions and feelings.
      2. Improve ability to express satisfaction regarding the creative experience.
      3. Improve ability to increase self-esteem through identifying and completing a project.

IV. **CRITERIA FOR PATIENT SELECTION:**

    A. Patients are evaluated by an occupational therapist to be within the range of level 4.0–5.9 of the Allen Cognitive Level Test.
    B. Attention span is at least fifteen minutes.
    C. There is the ability to follow and act on a series of two-step directions.
    D. Parient has the ability to tolerate a highly stimulating environment in terms of noise, clutter, and group interaction.
    E. If changes in cognitive level are noted, patient should be retested.

TABLE 17.4 (continued)

Horticulture Task Skills III
Page 2

V. LEADERSHIP AND APPROACH:

A. Staff qualification to provide leadership:
  1. Qualifications specific to an occupational therapist:
     a. Knowledge of theory and application of Allen's Cognitive Disability Theory and Cognitive Level Test (ACL).
     b. Ability to structure and modify environments to enhance patient performance and tasks.
     c. Knowledge of craft media and ability to adapt them to patients' level of functioning.
  2. Approach:
     The role of the leader is to instruct, to set appropriate task and behavioral limits, to give accurate performance feedback, and to explain treatment goals. The leader also supports patients, encourages independence and social interaction, and assists patients in developing a sense of mastery and creativity. Patients will identify and discuss their cognitive and emotional functioning and its affect on their task skills.

B. Modalities:
     A variety of horticultural/craft media will be utilized. Age-appropriate task selection will be based on patient's cognitive functioning and familiarity with particular projects. Staff and interns will be developing task resources.

C. Methodology:
     Selection of horticultural projects will be predetermined by group leaders. As patients are able, they will be encouraged to participate in group interaction through cooperative patient teaching and assistance, joint cleanup, sharing of tools, and mutual discussion. The role of the leader is to instruct, to set appropriate task and behavioral limits, to give accurate performance feedback and to explain treatment goals. The leader also supports patients, encourages independence and social interaction, and assists patients in developing a sense of mastery and creativity.

VI. EVALUATION:

Patients will be evaluated regarding the frequency of their observed demonstration of the specific objectives. Documentation will be in progress notes. The group leader may use ACL to reassess the patient as necessary.

patients may only be seen once or twice in a group. The therapist must be ready to answer questions, to both patients and administration, relating specific goals and objectives that can be met by participation in horticulture therapy. These same goals may be met in another modality, so why horticulture therapy? This goes back to the beginning when we talk about working with something that is living, the reverse dependency, and the motivating factors of gardening. A patient must first attend and then participate. If this does not happen, the best goals in the world can not be met. Clients are often more willing to participate in a horticulture group then some other craft or work activity. It is something real, and often something from their past, where they can remember their parents, grandparents, or themselves gardening at home. Gardening is the number one outside leisure activity.

In summary, with the changing health care system, less money, and the reduced length of stay, good documentation is essential. Appropriate goals and objectives with good records of outcomes will help support your program and encourage effective treatment.

TREATMENT GOALS AND OBJECTIVES

The following are some examples of problems, goals, and objectives appropriate for horticultural therapy.

Problem: Inability to follow directions. This client needs to be able to follow four-step directions in order to be eligible for supportive employment.

HT Goal: Client will follow four-step directions consistently and independently.

Objective 1: Client will correctly report geraniums 80 percent of the time with staff support daily for one week.

Objective 2: Client will correctly repot geraniums 80 percent of the time, using a cue sheet, daily for one week.

Procedure: 1. Staff will demonstrate the four-step procedure

2. Using a cue card, staff will explain each step.

3. Client will work with staff reminders.

4. Continue step 3 until client has met first objective.

5. Staff will fade from client, only using cues to look at cue card when mistakes are made.

6. Continue step 5 until objective 2 is met.

Problem: Patient is depressed with poor self-esteem.

HT Goal: Patient will increase self-esteem by completing a daily project.

Objective 1: Patient will choose one small project per group and complete it daily.

Procedure: 1. Staff will help patient choose an appropriate project.

2. With staff support, patient will complete that project daily.

3. With staff support, patient will process feelings about being successful.

Problem: Patient has had a stroke and has poor use of her right hand.

HT Goal: Patient will increase right-hand range of motion.

Objective 1: Patient will disbud chrysanthemums daily for five minutes without a break.

Objective 2: Patient will disbud chrysanthemums daily for ten minutes without a break.

Procedure: 1. Staff will demonstrate correct method for disbudding and assist patient for ten minutes daily.

2. Staff will reduce the amount of assistance daily until patient is working on her own.

GLOSSARY

assessment: An organized collection of data to determine treatment needs

CARF: The Commission on Accreditation of Rehabilitation Facilities

goal: Something you work toward, a desired outcome

JCAHO: The Joint Commission on Accreditation of Healthcare Organizations

Medicare: A federal health insurance program for the disabled and people over age sixty-five

objective: Steps taken to work toward a goal

progress note: A written presentation of observed behaviors

situational assessment: Observation and data collection of a client in a natural or simulated work environment

work sample: A standardized work task used to test and predict a client's job potential.

BIBLIOGRAPHY

The American Heritage Dictionary of the English Language. (1980). Boston, MA: Houghton Mifflin Co.

Atchley, R.C. (1994). *Social Forces and Aging,* seventh edition. Belmont, CA: Wadsworth.

Joint Commission on Accreditation of Healthcare Organizations (1994). *1995 MHM Accreditation Manual for Mental Health, Chemical Dependency, and Mental Retardation/Developmental Disabilities Services Volume 1 & 2.* Oakbrook Terrace, IL. JCAHO.

Joint Commission on Accreditation of Healthcare Organizations. (1995). *1996 Comprehensive Accreditation Manual for Hospitals.* Oakbrook Terrace, Illinois.

Roth, Vera and Weaver, Bruce. (1989). *Horticulture Work Samples.* Baltimore, MD: Sheppard Pratt Health System. Unpublished.

Index

Page numbers followed by the letter "i" indicate illustrations; those followed by the letter "n" indicate footnotes; those followed by the letter "t" indicate tables.

Order Your Own Copy of
This Important Book for Your Personal Library!

HORTICULTURE AS THERAPY
Principles and Practice

_____ in hardbound at $79.95 (ISBN: 1-56022-859-8)

COST OF BOOKS _____	☐ **BILL ME LATER:** ($5 service charge will be added) (Bill-me option is good on US/Canada/Mexico orders only; not good to jobbers, wholesalers, or subscription agencies.)
OUTSIDE USA/CANADA/ MEXICO: ADD 20% _____	☐ Check here if billing address is different from shipping address and attach purchase order and billing address information.
POSTAGE & HANDLING _____ (US: $3.00 for first book & $1.25 for each additional book) Outside US: $4.75 for first book & $1.75 for each additional book)	
	Signature _____
SUBTOTAL _____	☐ **PAYMENT ENCLOSED: $**_____
IN CANADA: ADD 7% GST _____	☐ **PLEASE CHARGE TO MY CREDIT CARD.**
STATE TAX _____ (NY, OH & MN residents, please add appropriate local sales tax)	☐ Visa ☐ MasterCard ☐ AmEx ☐ Discover ☐ Diners Club Account # _____
FINAL TOTAL _____ (If paying in Canadian funds, convert using the current exchange rate. UNESCO coupons welcome.)	Exp. Date _____ Signature _____

Prices in US dollars and subject to change without notice.

NAME _____

INSTITUTION _____

ADDRESS _____

CITY _____

STATE/ZIP _____

COUNTRY _____ COUNTY (NY residents only) _____

TEL _____ FAX _____

E-MAIL_____

May we use your e-mail address for confirmations and other types of information? ☐ Yes ☐ No

Order From Your Local Bookstore or Directly From
The Haworth Press, Inc.
10 Alice Street, Binghamton, New York 13904-1580 • USA
TELEPHONE: 1-800-HAWORTH (1-800-429-6784) / Outside US/Canada: (607) 722-5857
FAX: 1-800-895-0582 / Outside US/Canada: (607) 772-6362
E-mail: getinfo@haworth.com
PLEASE PHOTOCOPY THIS FORM FOR YOUR PERSONAL USE.

BOF96

OVERSEAS DISTRIBUTORS OF HAWORTH PUBLICATIONS

AUSTRALIA
Edumedia
Level 1, 575 Pacific Highway
St. Leonards, Australia 2065
(mail only) PO Box 1201
Crows Nest, Australia 2065
Tel: (61) 2 9901–4217 / Fax: (61) 2 9906-8465

CANADA
Haworth/Canada
450 Tapscott Road, Unit 1
Scarborough, Ontario M1B 5W1
Canada
(Mail correspondence and orders only. No returns or telephone inquiries. Canadian currency accepted.)

DENMARK, FINLAND, ICELAND, NORWAY & SWEDEN
Knud Pilegaard
Knud Pilegaard Marketing
Mindevej 45
DK-2860 Soborg, Denmark
Tel: (45) 396 92100

ENGLAND & UNITED KINGDOM
Alan Goodworth
Roundhouse Publishing Group
62 Victoria Road
Oxford OX2 7QD, U.K.
Tel: 44–1865–521682 / Fax: 44–1865-559594
E-mail: 100637.3571@CompuServe.com

GERMANY, AUSTRIA & SWITZERLAND
Bernd Feldmann
Heinrich Roller Strasse 21
D–10405 Berlin, Germany
Tel: (49) 304–434–1621 / Fax: (49) 304–434–1623
E-mail: BFeldmann@t-online.de

JAPAN
Mrs. Masako Kitamura
MK International, Ltd.
1–50–7–203 Itabashi
Itabashi–ku
Tokyo 173, Japan

KOREA
Se–Yung Jun
Information & Culture Korea
Suite 1016, Life Combi Bldg.
61–4 Yoido–dong
Seoul, 150–010, Korea

MEXICO, CENTRAL AMERICA & THE CARIBBEAN
Mr. L.D. Clepper, Jr.
PMRA: Publishers Marketing & Research Association
P.O. Box 720489
Jackson Heights, NY 11372 USA
Tel/Fax: (718) 803–3465
E-mail: clepper@usa.pipeline.com

NEW ZEALAND
Brick Row Publishing Company, Ltd.
Attn: Ozwald Kraus
P.O. Box 100–057
Auckland 10, New Zealand
Tel/Fax: (64) 09–410–6993

PAKISTAN
Tahir M. Lodhi
Al-Rehman Bldg., 2nd Fl.
P.O. Box 2458
65–The Mall
Lahore 54000, Pakistan
Tel/Fax: (92) 42–724–5007

PEOPLE'S REPUBLIC OF CHINA & HONG KONG
Mr. Thomas V. Cassidy
Cassidy and Associates
470 West 24th Street
New York, NY 10011 USA
Tel: (212) 727–8943 / Fax: (212) 727–8539

PHILIPPINES, GUAM & PACIFIC TRUST TERRITORIES
I.J. Sagun Enterprises, Inc.
Tony P. Sagun
2 Topaz Rd. Greenheights Village
Ortigas Ave. Extension Tatay, Rizal
Republic of the Philippines
P.O. Box 4322 (Mailing Address)
CPO Manila 1099
Tel/Fax: (63) 2–658–8466

SOUTH AMERICA
Mr. Julio Emöd
PMRA: Publishers Marketing & Research Assoc.
Rua Joauim Tavora 629
São Paulo, SP 04015001 Brazil
Tel: (55) 11 571–1122 / Fax: (55) 11 575-6876

SOUTHEAST ASIA & THE SOUTH PACIFIC, SOUTH ASIA, AFRICA & THE MIDDLE EAST
The Haworth Press, Inc.
Margaret Tatich, Sales Manager
10 Alice Street
Binghamton, NY 13904–1580 USA
Tel: (607) 722–5857 ext. 321 / Fax: (607) 722–3487
E-mail: getinfo@haworth.com

RUSSIA & EASTERN EUROPE
International Publishing Associates
Michael Gladishev
International Publishing Associates
c/o Mazhdunarodnaya Kniga
Bolshaya Yakimanka 39
Moscow 117049 Russia
Fax: (095) 251–3338
E-mail: russbook@online. ru

LATVIA, LITHUANIA & ESTONIA
Andrea Hedgecock
c/o Iki Tareikalavimo
Kaunas 2042
Lithuania
Tel/Fax: (370) 777-0241 / E-mail: andrea@soften.ktu.lt

SINGAPORE, TAIWAN, INDONESIA, THAILAND & MALAYSIA
Steven Goh
APAC Publishers
35 Tannery Rd.
#10–06, Tannery Block
Singapore, 1334
Tel: (65) 747–8662 / Fax: (65) 747–8916
E-mail: sgohapac@signet.com.sg

7/97